HP-41

Reservoir Engineering Manual

HP-41 Reservoir Engineering Manual

D. Nathan Meehan and Eric L. Vogel

PennWell Books
PennWell Publishing Company
Tulsa, Oklahoma

PennWell Publishing Company
1421 South Sheridan Road/P.O. Box 1260
Tulsa, Oklahoma 74101

Library of Congress cataloging in publication data
Meehan, D. Nathan.
HP-41 reservoir engineering manual.
1. Oil reservoir engineering — Computer programs.
2. HP-41 (Calculating-machine) I. Vogel, Eric L.
II. Title.
TN871.M393 622'.3382 81-22704
ISBN 0-87814-186-3 AACR2

Printed in the United States of America

1 2 3 4 5 86 85 84 83 82

Table of Contents

Acknowledgments vi

Introduction vii

Notes on the Documentation ix

Section 1: Oil, Gas, and Water PVT Properties 1

1. OILPVT — Oil PVT Properties 2
2. GASPVT — Gas PVT Properties 7
3. DEW — Dew-Point Pressure from Gas Composition 14
4. GOR — Forecasting GOR Behavior for Gas-Condensate Reservoirs 18
5. H2OPVT — Water PVT Properties 23
6. GASPROD — Material-Balance Gas Production 27

Section 2: Decline Curve Analysis 30

7. DECLINE — Exponential, Harmonic, Hyperbolic, and Multiple-Well Decline Curve Analysis 31

Section 3: Volumetric Calculations and Reserves 55

8. OIP — Calculating Oil in Place and Reserves Volumetrically 56
9. GIP — Calculating Gas in Place and Reserves Volumetrically 61

Section 4: Material Balance 65

10. OILMBE — Calculating N or We from Material Balance 66
11. KG/KO — Gas-Oil Relative-Permeability Ratio 73
12. OILPRED — Predicting Solution-Gas-Drive Performance 81
13. QOVST — Rate-Time Forecast from Material Balance 91
14. INFCOEF — Calculating the Water Influx Coefficient 98
15. INFLUX — Predicting Water Influx 109
16. GASMBE — Gas Material Balance 113

Section 5: Natural Gas Engineering 121

17. BHPWHP — Bottom-Hole or Surface Pressures for Flowing or Static Gas Wells 122
18. GASDEL — Single-Well Gas Deliverability 129
19. STAB — Stabilized Flow Coefficient 135

Section 6: Pressure Transient Analysis 141

20. BLDUTIL — Utilities for Pressure Buildup Analysis 142
21. BUILDUP — Analyzing Pressure Buildups 151
22. DRAW — Multiple Rate Pressure Drawdown Analysis 160

Section 7: Waterflooding 170

23. FWVSW — Fractional Flow Calculations 171
24. 5SPOT — Five-Spot Waterflood Performance 179
25. SUMWF — Combined Performance for Stratified Waterfloods 195
26. INJ — Water Injectivity for Regular Patterns with Unit Mobility Ratio 200
27. CUTCUM — Forecasting Mature Waterfloods 205

Section 8: Well-Log Analysis 216

28. RW — Calculating RW from SP 217
29. XPLOT — Cross Plot Porosities 221
30. H2OSAT — Water Saturation Calculations 227
31. SLANT — True Stratigraphic and Vertical Thicknesses 233

Section 9: Economics 236

32. DISC — Discounted Cash Flow Analysis 237

Section 10: Bar Code 347

Acknowledgments

The authors express their appreciation to Champlin Petroleum Company and Hewlett-Packard Company, without whose support and assistance this book would not have been possible, and to Carole Vogel, whose endless hours of typing produced a manuscript of exceptional quality.

I dedicate this book to my father, Don Meehan, a ditch-jumping, dirt-road sport, and to the women in my life: Jan, Hannah, Sarah, and Rebekah.

D. NATHAN MEEHAN

I dedicate this book to my loving wife Carole, whose patience, endurance, support, and encouragement were my guiding light throughout this project.

ERIC L. VOGEL

Introduction

The purpose of this book is to present practical programs to solve some common reservoir engineering problems. We have concentrated on problems whose computations fall between the two extremes of the very complex (such as a 3-D compositional numerical simulation model) and the very trivial (such as X = a * b * c). Specifically, we selected problems that would benefit from the ability to calculate fluid properties (such as Z factors, viscosities, and formation volume factors), that lend themselves to interchangeable solutions, that are too complex for normal hand calculations, and that required flexibility in dealing with units.

We selected problems sensible for the range of capability provided by modern programmable calculators. These problems can be characterized as algorithm intensive rather than data intensive; that is, small amounts of data undergoing extensive numerical manipulation. Modern programmable calculators have extensive sets of functions for high-accuracy computations, relatively small amounts of memory for data storage, and small keyboards that limit the ease of data entry. On the other hand numerical simulation generally involves massive number crunching and large-scale computers designed to perform these tasks. Many programs for larger machines also involve large amounts of input and output with corresponding requirements for mass-storage and high-speed output capabilities.

Programmable calculators have made major strides toward solving problems that not too many years ago would have been the sacred domain of large time-sharing systems. The introduction of HP-IL, a compact, low-cost interface for battery-operable systems along with the first HP-IL peripherals, has opened a whole new world of input/output and data storage capabilities. Although we deliberately selected algorithm-intensive problems, there are now exciting possibilities for data-intensive calculator programs as well.

One group of algorithm-intensive problems that are readily tackled by programmable calculators are the correlations for petroleum fluid properties. Formerly, the petroleum engineer relied on charts and nomographs to provide estimates for oil, gas, and water PVT properties. With very few initial calculations, graphs are available to provide estimates of practically all of the desired properties. Many of these correlations are in the form of nomographs that nearly always can be expressed as a fairly simple set of equations. Many of the charts in common use today are curves or families of curves that were more or less hand-drawn through measured data points. It is sometimes surprising to see how few data points were used to generate these curves. Nonetheless, many of these correlations have become industry standards.

For the accuracy required by most reservoir engineering problems, many of the correlations are entirely adequate. A Craft and Hawkins example illustrates the large error in calculating oil in place from material balance when neglecting rock and water compressibilities (86.25 (10^6)BBL compared to 51.73 (10^6)BBL, 66.7% too high). Even a 50% error in estimating any one of the fluid properties results in a fairly small error in the estimate of oil in place. Some calculations require fairly accurate estimates of the fluid properties, or at least consistent estimates. Graphs of PVT properties are fairly difficult to read accurately and repeatably.

Calculator solutions of fluid properties are often faster, especially for repetitive calculations, and usually more accurate and consistent. The HP-41 Petroleum Fluids Pac goes a major step beyond previous programs by providing complete fluid property correlations for gas, oil, and water in one self-contained module. However, the real attractiveness of this software is the fact that all of the solutions contained in the Pac can be called as subroutines. Now programs can be written that calculate PVT properties on the fly, avoiding the hassle of either inputing the PVT properties at each point in the analysis or the need to write programs for the fluid properties themselves. Some of these programs are a great deal faster than other techniques. For example, the Pac subroutine CZ calculates the gas Z factor from a fairly complex correlation in 2–9 seconds because it is written in machine language. A solution of a less-accurate Z factor routine using normal program steps may take at least 30 seconds for a solution.

Other advantages of using the Pac subroutines are the I/O routines available. The Unit Management System built into these routines allows tremendous flexibility with units. Default English or SI units can be selected by the user, and different units than the defaults can be input or selected for output. Many of these user-input units are retained through the remainder of the program. The I/O routines make prompting and using a printer very easy, relieving the programmer of these tasks and providing consistent operation from program to program.

The features of the Pac are powerful enough that we chose to use it as the foundation on which to construct all of the programs in the book. Most of the programs make extensive use of the fluid property correlations, although some of the programs use only I/O subroutines and the Unit Management System. The user could alter some of these programs to operate without the Pac, but it is not recommended. We have taken advantage of all of the features of the Pac and its subroutines, and the user would have to be very familiar with these routines to make modifications and still expect the programs to work properly. If you understand the Pac that well, you probably will want to use it with these programs. However, you may want to modify the programs to use an alternative PVT correlation. In these cases the programs are structured so it would not be difficult to add different PVT correlation subroutines.

The quality of the programs in the book is quite high. Engineering programs for programmable calculators are often written by engineers whose main expertise is petroleum engineering. Many programs of this nature have been published in various trade journals and books. Frequently the result is software that does not represent the best programming practice, the most thorough knowledge of the machine, or sufficient generality to meet the needs of many different users. The authors comprise a team with the expertise of a practicing petroleum engineer who has published a number of programs and an expert in writing calculator programs who was responsible for writing the Petroleum Fluids Pac. This combination has led to programs that represent sound reservoir engineering practices, a standard of performance for flexible reservoir engineering solutions, and state-of-the-art calculator software for petroleum applications.

Much of the flexibility stems from the continuity among the different programs. For example, you can calculate PVT properties for an oil field using OILPVT and then calculate the volumetric oil in place and solution gas-drive recovery factor with the OIP program. Initial oil in place is estimated from performance using the material balance equation in OILMBE. The relative permeability ratio can be calculated from performance using KG/KO, or a correlation can be developed to represent KG/KO field data. Then, using that correlation, the performance of the reservoir can be predicted by OILPRED using material balance above the bubble point and the Tarner method below the bubble point. Finally, a rate-time forecast for the reservoir can be predicted with QOVST.

Many of these programs are quite long and will take some time to key in. Bar code for use with the HP 82153A Wand is in Section 10. If you have a Wand, we suggest you scan the programs in and then record the programs onto magnetic cards with the HP 82104A Card Reader or onto a cassette with the HP 82161A Digital Cassette Drive. For information about purchasing either magnetic cards or cassettes with the programs already recorded, contact either one of the authors:

D. Nathan Meehan
Champlin Petroleum Company
500 N Water, Suite 500 N
Corpus Christi, TX 78471

Eric L.Vogel
Hewlett-Packard Company
1000 NE Circle Blvd.
Corvallis, OR 97330

We feel strongly that petroleum reservoir engineers should thoroughly understand the programs and the calculations that they are performing when they use any type of software. This applies both to the programs in this book and to commercial or in-house software. For example, serious errors can result from blindly using an API water-drive recovery factor as a waterflood recovery. Slider makes a point in his book that the petroleum engineer should know programs well enough so that, given enough time, he could perform the calculations by hand. Many of these programs require iterative solutions, perform many PVT calculations at different pressures, and require keeping track of a fairly large number of variables. It is unlikely that most reservoir engineers would attempt to solve some of these programs by hand because of their complexity. However, this is not an excuse for not understanding the principles and theory prior to using these programs.

We have performed extensive functional testing and have cross-checked programs on a line-by-line basis to ensure the programs are error-free. This in itself was a massive task for 32 programs totaling over 24,000 bytes of program code (the equivalent of three Petroleum Fluids Pacs). Despite our best efforts, there may be some errors in these programs. Because of this possibility, and the possibility of misuse of the programs or misinterpretation of the results, we must make the following disclaimer:

The program material contained herein is supplied without representation or warranty of any kind. The authors and PennWell Books therefore assume no responsibility and shall have no liability, consequential or otherwise, of any kind arising from the use of this program material or any part thereof.

It is the authors' sincere hope and belief that this book will be a valuable addition to your professional library. Questions and comments about any aspect of this book are appreciated.

Notes on the Documentation

Read through the next section of notes on the documentation. This section will give you a much better understanding of how to get the most usefulness out of the programs in the book.

The programs in the book follow the same guidelines and use the same techniques that are established by the Petroleum Fluids Pac. Therefore, all of the programs assume the user is very familiar with the use of the Pac, particularly the section on the Petroleum Engineering Unit Management System. If you have not already done so, it is strongly recommended that you familiarize yourself with the Pac before you use these programs. Very little documentation of the Pac appears in this book; so to get the most benefit from these programs, you must build for yourself this fundamental understanding of the Pac.

The documentation for each program consists of several parts: overview, equations, nomenclature, yes/no questions, examples, user instructions, general information, registers, flags, and program listing. Variable names and units are represented in symbols corresponding to what the user sees in the display of his calculator.

Overview

The overview provides an introduction to the program — the theory involved, what the program does, what solution techniques are used, and references for further information. The overviews assume the reader is familiar with reservoir engineering principles and techniques. They are not intended to substitute for this understanding nor are they a tutorial on the subject of the program.

Equations

The equations section lists the equations used by the program to solve for the different variables. This section expands on the overview by providing additional details about the solution technique used. (Note that the programs may calculate a rearranged form of the same equation that is easier to compute with.)

The equations use the same symbols that appear in the display of the calculator when the program prompts for an input or calculates an output. These symbols are defined in the nomenclature section. (Any symbols used by the equations that are not displayed by the calculator are defined in the equations section.)

Some of the equations used by the programs are solved iteratively. All programs in the book that use iterative procedures to solve an equation will iterate until the calculated value is within 0.01% of the previously calculated value.

To avoid duplication with the Pac, equations for Pac correlations have not been reproduced in this book. Refer to the Pac manual for these equations.

Nomenclature

The nomenclature section defines the variables used for input and output by the program. To avoid duplication with the Pac, the nomenclature section only defines variables other than those defined by the Pac or that have a different meaning than those defined by the Pac. The variables defined by the Pac appear in Appendix C of the Pac manual and in the Quick Reference Card for the Pac. All Pac variables are used by the programs in the book except CP and CV (specific heats at constant pressure and constant volume), NHV, GHVD, and GHVW (net heating value and dry and wet gross heating values), and PSAT (saturation pressure of water).

The different columns of the nomenclature table are almost identical to Appendix C of the Pac manual. One additional column has been added — a column that tells whether a variable is an input to a program, an output from a program, or both. As in the Pac, the units are shown in the symbols that are used by the program (all uppercase, digits used for exponents instead of superscripts).

The variables for which units are saved are denoted by a footnote. In the Pac, only the units for the primary output variable were saved, and no units for input variables were saved. In the book some input and output units are saved for you.

Yes/No Questions

This section explains the meaning of any prompts that appear with the Y/N:Y or Y/N:N suffix. As in the Pac, these are questions that require a Y or N (yes or no) response followed by [R/S]. The last character of the prompt is Y if the prompt is currently true or N if the prompt is currently false. If the current state is acceptable, press [R/S] in response.

Examples

The examples show the keystrokes required to solve sample problems using the program. The only keystroke you may not recognize is [//], the symbol for the gold shift key. The examples assume the program has already been loaded into the calculator, and that the Petroleum Fluids Application Module is plugged into one of the ports of the HP-41. Because all of the programs use input and I/O subroutines from the Pac, they will work properly with or without a printer (the 41's 82143A or the HP-IL 82162A and 82905B Option 48X printers).

Unlike the examples in the Pac, all examples in the book assume a printer is present. Since many of the examples output tables of values, the assumption of a printer simplifies the keystrokes in the examples. For details on how to respond to outputs without the printer, see the Petroleum Engineering Unit Management System in the Pac manual.

Unless specified differently in the examples, all examples assume flags 09 and 10 are clear, the display format is FIX 4, and the display mode is flag 28 set and flag 29 set. Because of the rounding done by some programs, it is recommended that FIX (or SCI or ENG) 8 or 9 not be used when running these programs.

The HP 82143A and 82162A printers will print an input or output that is up to 24 characters wide. There are instances when the variable name, the equals sign, the value, and the units are longer than 24 characters, and leading characters will not appear on the line that is printed. To correct this situation, use a shorter display format (such as FIX 2), different units (such as MMCF/DAY instead of MCF/DAY), or clear flag 29 (removes the commas).

Certain programs use tones to help inform you of what is going on. Tones are used during iterative solutions and during some repetitive looping operations. All programs in the book using tones for these purposes will sound a TONE 5 at the start of each iteration or loop and a TONE 9 at the end of the process. Any programs that generate an error message will sound a TONE 3 when the program halts to display the message. Tones can be turned off by clearing flag 26 ([//] [CF] 26).

User Instructions

The user instructions presented in the book are a major departure from the user instruction forms used in the Pac. We felt that a flow chart, in conjunction with the examples, would better illustrate the various options available to the user while running the program.

Five symbols appear in flow charts in this book. The first is an oval that says START PROGRAM, END PROGRAM, START, or END. This is just a marker to indicate the beginning or end of the program, or the beginning or end of a section of the program.

The second symbol is a circle with a number in it. This refers to a section of the flow chart that appears on another page. The circle with the number in it will appear in the upper-left corner of the appropriate page.

The third symbol is a box with pointed ends. The pointed box is used for yes/no questions. When the yes/no question is posed by the program, you must respond with Y [R/S], N [R/S], or just [R/S].

The fourth symbol is a box. The box is used in four circumstances, and the action you must take in each instance is determined by what is inside the box.

If the first word in the box is INPUT followed by one or more variable names, the program will prompt you for each variable in the order they appear in the box. As you know from using the Pac, a prompt for an input variable is NAME=?, and you respond by keying in the number, units if desired, and pressing [R/S].

If the first word in the box is OUTPUT followed by one or more variable names, the program will output each variable in the order it appears in the box. If a printer is plugged into the 41, the variable name, its value, and its units will be printed. If not, the program will halt with the variable name and its value in the display and its units in the ALPHA register.

If the first word in the box is ERROR: followed by a message, the program will beep and halt with that message in the display. Leading out of the box, marked by the key [R/S], is the action that the program will take when you press [R/S] to restart the program.

If INPUT, OUTPUT, or ERROR do not appear in the box, the box represents a menu of choices available to you. Under each option will be a key such as [A], [E], or [R/S], indicating which key you should press to invoke that particular option. None of these options will be available if you have made any key reassignments to those keys.

In most cases, the symbol describing the option is for the information you know. For example, in the GASPVT user instructions, the choice is GASG or COMP. The one you select depends on what you know, the gas gravity or the composition. Most of the menus used by the programs follow this convention. There are a few that do not, but the meaning of their menu symbols will be fairly obvious.

The last symbol that appears in a flow chart is a diamond. This is a condition that the program uses to decide a course of action it should take. A diamond requires no response from you — it gives you information that explains what the program is doing.

For example, in the GASPVT user instructions, the diamond contains the question IS P < END P?. This shows you that while P is less than END P, the program will continue to output the table of P, Z, CG, BG, and UG. When P is greater than or equal to

END P, the program will stop outputing the table and will let you input BEG P, END P, and P INC for a new table.

General Information

The general information is just that — useful information about the program that you may need to know. The memory requirements describe how much memory is needed to run the program. The program length assumes you load the program and press [//] [GTO] [.] [.] to add a local END to the program and pack it. The hidden options are those options that must be invoked by setting or clearing a flag, i.e., options that are not controlled by a response to a yes/no question. The Pac subroutines called give you an idea of how dependent the program is on the Pac in case you want to rewrite it to work without the Pac. The subroutines' names are listed in the order they appear in the program.

Registers

This is a list of registers used by the program, what variable(s) are stored in the registers, and the units that those variables are stored in. To avoid duplication with the Pac, the register list does not include registers that have been defined by the Pac unless they are used for different variables than in the Pac. This means that if a register is used for the same purpose as in the Pac, it will not be mentioned either in the list of registers or in the list of unused registers.

In writing these programs, we made a concerted effort to use the same registers for the same variables. The programs always preserve registers 10–25, which are the standard fluid property variables defined by the Pac, and usually use registers beyond 25 for their own purposes. But even for variables not defined by the Pac, in most cases you will not have to reenter a value that you already keyed in.

For the programs you use regularly, it will become apparent which variables are preserved from program to program and which are not. One way to tell which are preserved is to compare the register lists for the programs of interest. A more operational approach is to press [←] when a prompt appears for a variable whose value you entered previously in the same program or in a different program. You will be pleasantly surprised at the number of times that variables have been saved for you, particularly within a section, where the programs are all related or where the output of one program is needed for use by another.

Reproduced below is the list of registers defined by the Pac and their contents.

Registers

00	Input, output, scratch
01	English units, scratch
02	English units, scratch
03	Known output units
04	Known output units
05	Input, output, scratch
06	Scratch
07	Scratch
08	Unused
09	Unused
10	Tc, Tc* (R)
11	Pc, Pc* (PSI)
12	OIL G (API)
13	RS, RSI, RSb (SCF/BBL)
14	PBP (PSI)
15	GAS G
16	T (F)
17	P (PSI)
18	%POR
19	%NACL
20	%SO
21	%SW
22	STD T (F)
23	STD P (PSI)
24	SEP T (F)
25	SEP P (PSI)
26	%N2
27	%CO2
28	%H2S
29	%METH
30	%ETH
31	%PROP
32	%IBUT
33	%N-BUT
34	%IPEN
35	%N-PEN
36	%N-HEX
37	%N-HEP
38	%N-OCT
39	%N-NON
40	%N-DEC
41	%O2
42	%H2
43	%He
44	%H2O

Flags

The flags section describes which flags are used by the program and what the set and clear conditions indicate. There are certain flags used by all programs that are not mentioned in the flag list. These are:

08 Set: First pass through the program.
Clear: Any other pass through the program.

09 Set: Use SI default units.
Clear: Use English default units.

10 Set: Halt and prompt for units on output.
Clear: Do not halt on output.

The programs also use flags 12 (double-wide), 21 (printer enable), 22 (numeric input), 23 (ALPHA input), 25 (error ignore), 27 (USER mode), 29 (digit grouping), 36-39 (number of display digits), 40-41 (display format), and 55 (printer existence).

Program Listing

The program listing is what you will use if you key in the program yourself. The program listings are shown with XROMs, as if the programs were keyed in with the Petroleum Fluids Application Module plugged into a port of the HP-41C. For further information on XROMs, refer to Section 11 of the *HP-41C/41CV Operating Manual: A Guide for the Experienced User.*

Modifying Programs

If you modify programs to work independently of the Pac, be aware that every call to a Pac subroutine must be replaced by something of your own. Calls to calculation subroutines must be replaced or omitted. If replaced, your own calculation subroutines must be placed in program memory; if omitted, you must input the fluid properties yourself. Any use of the unit management functions [CON] and [INCON] must be replaced by appropriate conversion factors, being careful about overflowing the stack.

If you remove calls to input or I/O subroutines, replace them with simplified versions of the subroutines. This will help preserve the modularity of the program and avoid duplication of code. A pattern you may wish to follow is using subroutines called TITLE, IN, and OUT that appeared in Meehan's article (*OGJ*, 26 May 1980).

The program steps you use to replace calls to input or I/O subroutines will always take more program space than the calls to routines in the Pac. But if you are stripping out the calls to those routines, the sections of the program that deal with units (for example, to save the units of a particular variable) can be removed as well. This will help offset some of the program space required by your own subroutines. Also, if you remove units that are saved by the program, those data registers can be used for other variables. This will reduce the minimum size needed by the program and thus the total memory requirements to run the program.

If you want to replace a Pac calculation subroutine with one of your own, either remove the Petroleum Fluids Application Module before you key in the subroutine call or load your subroutine into program memory before keying in the call. If you do not, the subroutine call will appear as an XROM and your subroutine in program memory will be ignored.

Section 1
Oil, Gas, and Water PVT Properties

1. OILPVT — Oil PVT Properties

Calculates dissolved gas-oil ratio, coefficient of isothermal compressibility, formation volume factor, viscosity, and two-phase formation volume factor above and below the bubble point.

2. GASPVT — Gas PVT Properties

Calculates Z factor, coefficient of isothermal compressibility, formation volume factor, and viscosity from gas gravity or composition.

3. DEW — Dew-Point Pressure from Gas Composition

Calculates dew-point pressure for retrograde condensate reservoirs from reservoir temperature and gas composition.

4. GOR — Forecasting GOR Behavior for Gas-Condensate Reservoirs

Estimates gas-oil ratio performance as a function of pressure for retrograde condensate reservoirs.

5. H2OPVT — Water PVT Properties

Calculates coefficient of isothermal compressibility, formation volume factor, viscosity, and gas-water ratio, taking into account salinity and gas saturation.

6. GASPROD — Material-Balance Gas Production

Estimates gas equivalent of produced condensate and water content of natural gas.

1. OILPVT — Oil PVT Properties

OILPVT uses the correlations from the HP-41 Petroleum Fluids Pac (the Pac) to generate a table of crude oil fluid properties based on data routinely acquired in the field. The calculated properties are:

1. Dissolved gas-oil ratio (RSb)
   Calculated only below the bubble point; above the bubble point it equals RSI.
2. Coefficient of isothermal compressibility
   COb below the bubble point; CO above the bubble point.
3. Oil formation volume factor
   BOb below the bubble point; BO above the bubble point; BOBP at the bubble point.
4. Oil viscosity
   UOb below the bubble point; UO above the bubble point; UOBP at the bubble point. UOd is the dead oil viscosity (RS = 0).
5. Two-phase formation volume factor (BT)
   Calculated only below the bubble point; at or above the bubble point it equals BO.

Equations

See the oil isothermal compressibility, oil formation volume factor, oil viscosity, and two-phase formation volume factor programs in the Pac.

Nomenclature

Symbol	*Variable Name*	*Input or Output*	*English Units*	*SI Units*
BEG P*	Beginning pressure of table	I	PSI	KPA
END P*	Ending pressure of table	I	PSI	KPA
P*	Pressure	O	PSI	KPA
P INC*	Pressure increment of table	I	PSI	KPA

*The units for these variables are saved by the program.

Yes/No Questions

SKIP? Yes: Skip input of PVT data.
No: Allow input of PVT data.

Example

The Northwest Franklin oil field in East Texas has the following fluid data:

Gas gravity = 0.762, no diluents
Gas gravity measured at a separator temperature of 95 F and pressure of 125 PSI
Crude gravity = 31.7 API
Reservoir temperature = 155 F
Initial dissolved gas-oil ratio = 510 SCF/BBL
Initial pressure = 2,850 PSI

Generate a table of crude oil properties from 100 PSI up to the initial pressure with 250 PSI increments. Calculate the dissolved gas-oil ratio, oil compressibility, formation volume factor, oil viscosity, and two-phase formation volume factor.

To calculate the pseudocritical properties, execute the TcPc program in the Pac. Since the gas is in equilibrium with crude oil, answer Y to the COND? Y/N prompt. GAS G = 0.762; %N2, %CO2, and %H2S are all zero. The calculated Tc and Pc values are 396.9440 F and 660.1595 PSI, respectively. The values for STD T and STD P for East Texas are 60 F and 14.65 PSI, respectively. (See Table 4, Standard Pressures by Location, in the Pac.) Figures 1-1 through 1-5 present the calculated results graphically. Note the discontinuity in the calculated value of CO. This is due partly to the differences in correlations used above and below the bubble point. Recall that the compressibility of oil has a $\partial RS/\partial P$ term in it. Since RS versus pressure has a cusp at the bubble point, the $\partial RS/\partial P$ term undergoes a discontinuity there (at least in the correlations used). Also notice that the PVT property names change when the pressure exceeds the bubble point.

Keystrokes (SIZE> =029)	**Display**	**Comments**
[XEQ] [ALPHA] TcPc [ALPHA]	*COND? Y/N:*	Last character is Y or N
Y [R/S]	*GAS G=?*	
.762 [R/S]	*%N2=?*	
0 [R/S]	*%CO2=?*	
0 [R/S]	*%H2S=?*	
0 [R/S]	*GAS G=?*	Tc and Pc printed
[XEQ] [ALPHA] OILPVT [ALPHA]	*SKIP? Y/N:*	Last character is Y or N
N [R/S]	*Tc=?*	Tc already calculated
[R/S]	*Pc=?*	Pc already calculated
[R/S]	*STD T=?*	
60 [R/S]	*STD P=?*	
14.65 [R/S]	*SEP T=?*	
95 [R/S]	*SEP P=?*	
125 [R/S]	*OIL G=?*	
31.7 [R/S]	*GAS G=?*	GAS G input previously

Keystrokes (SIZE> =029)	Display	Comments
[R/S]	*MW=?*	
[R/S]	*T=?*	GAS GS printed
155 [R/S]	*RSI=?*	
510 [R/S]	*BEG P=?*	PBP, BOBP, UOd, and UOBP printed
100 [R/S]	*END P=?*	
2850 [R/S]	*P INC=?*	
250 [R/S]	*BEG P=?*	Table of properties printed

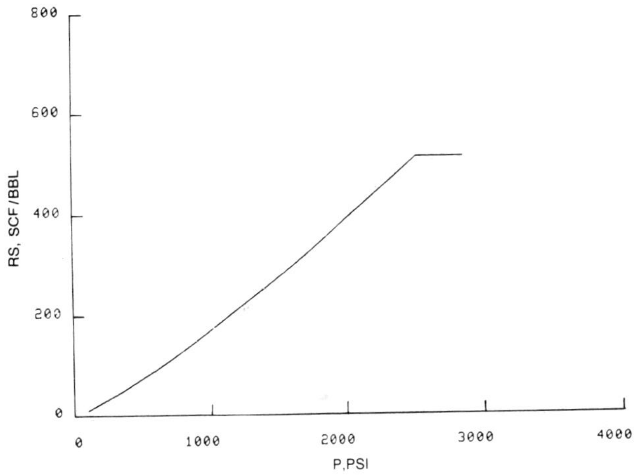

Figure 1–1

```
Tc Pc
COND: YES
GAS G=0.7620
%N2=0.0000
%CO2=0.0000
%H2S=0.0000

Tc=396.9440 R
Pc=660.1595 PSI

OIL PVT
SKIP: NO
STD T=60.0000 F
STD P=14.6500 PSI
SEP T=95.0000 F
SEP P=125.0000 PSI
OIL G=31.7000 API
GAS GS=0.7671
T=155.0000 F
RSI=510.0000 SCF/BBL
PBP=2,514.4215 PSI

BOBP=1.2840
UOd=3.9870 CP
UOBP=0.9112 CP

BEG P=100.0000 PSI
END P=2,850.0000 PSI
P INC=250.0000 PSI

P=100.0000 PSI
RSb=11.0981 SCF/BBL
COb=0.0038 1/PSI
BOb=1.0484
UOb=3.6550 CP
BTb=16.2365

P=350.0000 PSI
RSb=49.0975 SCF/BBL
COb=0.0012 1/PSI
BOb=1.0664
UOb=2.8614 CP
BTb=4.9393

P=600.0000 PSI
RSb=93.0927 SCF/BBL
COb=0.0007 1/PSI
BOb=1.0871
UOb=2.3074 CP
BTb=3.0615

P=850.0000 PSI
RSb=140.7571 SCF/BBL
COb=0.0005 1/PSI
BOb=1.1097
UOb=1.9223 CP
BTb=2.3033

P=1,100.0000 PSI
RSb=191.1540 SCF/BBL
COb=0.0004 1/PSI
BOb=1.1335
UOb=1.6445 CP
BTb=1.9054

P=1,350.0000 PSI
RSb=243.7566 SCF/BBL
COb=0.0003 1/PSI
BOb=1.1583
UOb=1.4367 CP
BTb=1.6691

P=1,600.0000 PSI
RSb=298.2226 SCF/BBL
COb=0.0002 1/PSI
BOb=1.1840
UOb=1.2761 CP
BTb=1.5192

P=1,850.0000 PSI
RSb=354.3097 SCF/BBL
COb=0.0002 1/PSI
BOb=1.2105
UOb=1.1487 CP
BTb=1.4202

P=2,100.0000 PSI
RSb=411.8362 SCF/BBL
COb=0.0001 1/PSI
BOb=1.2377
UOb=1.0453 CP
BTb=1.3531

P=2,350.0000 PSI
RSb=470.6605 SCF/BBL
COb=0.0001 1/PSI
BOb=1.2655
UOb=0.9598 CP
BTb=1.3067

P=2,600.0000 PSI
CO=1.2606E-5 1/PSI
BO=1.2826
UO=0.9184 CP

P=2,850.0000 PSI
CO=1.1500E-5 1/PSI
BO=1.2791
UO=0.9407 CP
```

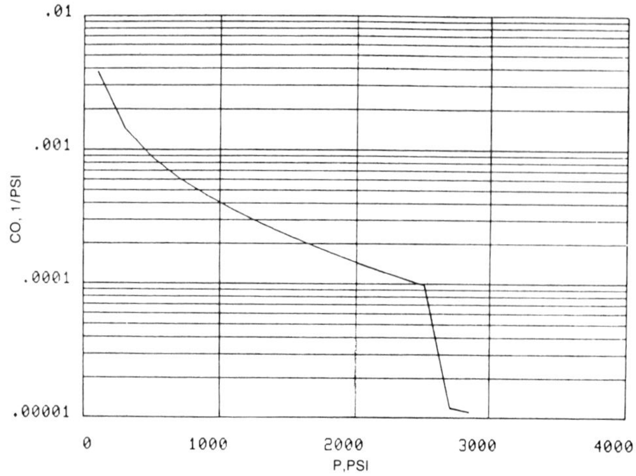

Figure 1–2

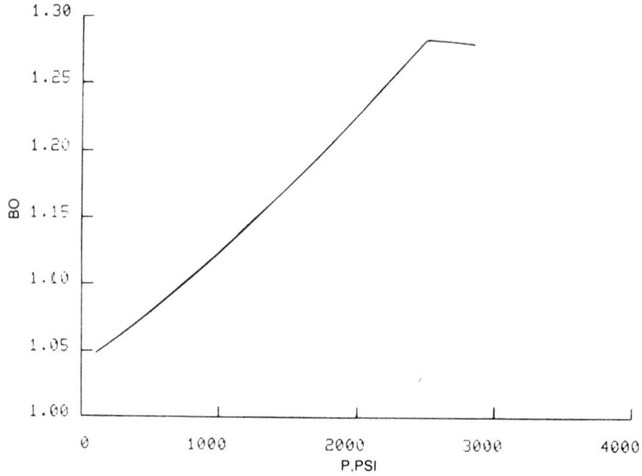

Figure 1–3

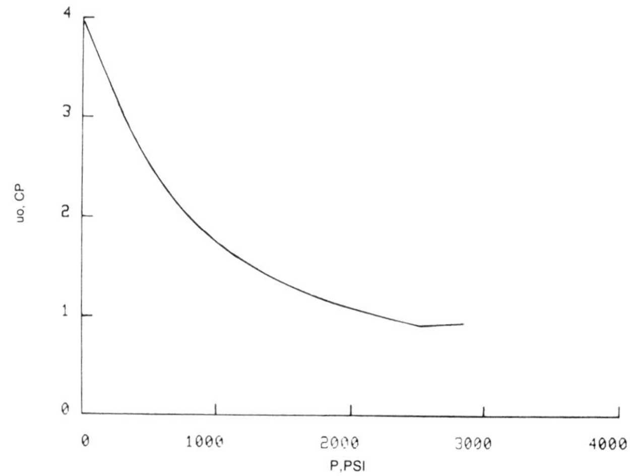

Figure 1–4

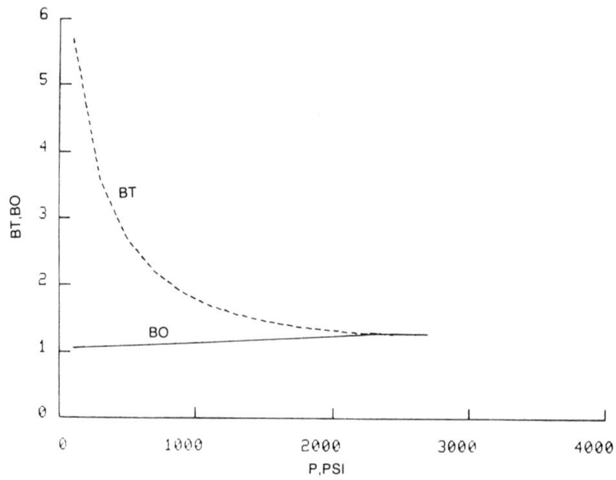

Figure 1–5

General Information

Memory Requirements

Program length:	351 bytes (2 cards)
Minimum size:	028
Minimum hardware:	41C + 1 memory module

Hidden Options

None

Pac Subroutines Called

TITLE, Y/N?, W7, CBOb, OUT, CUOd, CUOb, INK, OUTK, CRSb, X8, CON, CCOb, CBTb, CCO, CBO, CUO, OUTU

Registers

03	Pressure units
04	Pressure units
06	UOd (CP)
07	UOBP (CP)
08	END P (PSI)
09	P INC (PSI)
17	BEG P, P (PSI)
26	BOBP
27	RSb (SCF/BBL)

Registers 18–21 unused

Flags

02	Set:	Skip input of PVT data.
	Clear:	Allow input of PVT data.

User Instructions

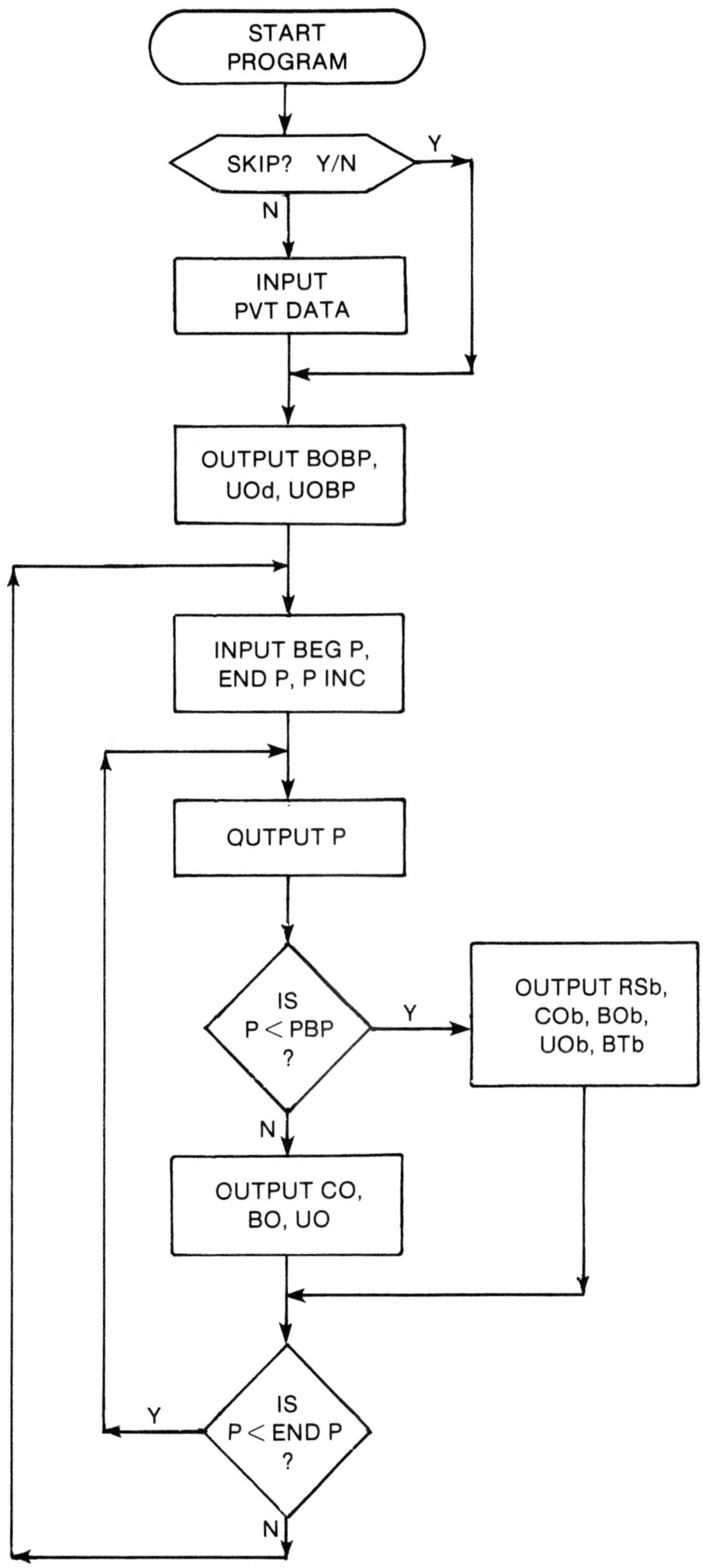

Program Listing

```
 01♦LBL "OILPVT"
"OIL PVT" 28
XROM "TITLE" FC?C 25
PROMPT "KPA" ASTO 03
CLA ASTO 04 "SKIP" 2
XROM "Y/N?" FC? 02
XROM "W7" XROM "CBOb"
STO 26 ADV "BOBP"
XROM "OUT" XROM "CUOd"
STO 06 "UOd" XEQ 05
RCL 13 X<>Y
XROM "CUOb" STO 07
"UOBP" XEQ 05 ADV

 32♦LBL 15
"PSI" ASTO 01 CLA
ASTO 02 RCL 04 RCL 03
16 STO 00 "BEG P"
XROM "INK" RDN STO 03
X<>Y STO 04 X<>Y
CF 08 7 STO 00
"END P" XROM "INK" RDN
STO 03 X<>Y STO 04
X<>Y R↑ "P INC"
XROM "INK" RDN STO 03
RDN STO 04 ADV

 66♦LBL 16
"PSI" ASTO 01 CLA
ASTO 02 RCL 04 RCL 03
RCL 17 "P" XROM "OUTK"
RDN STO 03 X<>Y
STO 04 R↑ RCL 14 X<Y?
GTO 00 RDN XROM "CRSb"
STO 27 XROM "X8"
RCL 16 "F-R" CON
RCL 10 / RCL 17
RCL 11 / RCL 17
XROM "CCOb" "COb"
XEQ 04 RCL 17
XROM "CBOb" "BOb"
XROM "OUT" RCL 27
RCL 06 XROM "CUOb"
"UOb" XEQ 05 RCL 17
XROM "CBTb" "BTb"
XROM "OUT" GTO 01

114♦LBL 00
XROM "CCO" "CO" XEQ 04
RCL 26 XROM "CBO" "BO"
XROM "OUT" RCL 07
XROM "CUO" "UO" XEQ 05

126♦LBL 01
ADV RCL 08 RND RCL 17
RCL 09 + RND X<Y?
GTO 02 X<>Y RCL 17
RND X=Y? GTO 03
RCL 08 RND

143♦LBL 02
LASTX STO 17 GTO 16

147♦LBL 03
RCL 09 ST+ 17 GTO 15

151♦LBL 04
ASTO T "1/PSI" ASTO 01
CLA ASTO 02 ASTO Z
"1/KPA" GTO 06

160♦LBL 05
ASTO T "CP" ASTO 01
CLA ASTO 02 ASTO Z
"PA*S"

168♦LBL 06
ASTO Y CLA ARCL T
XROM "OUTU" END
```

2. GASPVT — Gas PVT Properties

GASPVT uses the Pac correlations to generate a table of gas properties as a function of pressure. The pseudocritical properties can be input, calculated from gas gravity, or calculated from gas composition. GASPVT calculates the gas deviation factor (Z), coefficient of isothermal compressibility (CG), gas formation volume factor (BG), and gas viscosity (UG). The Pac program PROP can be used to calculate gas gravity, net and gross heating values, and specific heats and specific heat ratios.

Equations

See the Z Factor, Gas Isothermal Compressibility, Gas Formation Volume Factor, Gas Viscosity, Pseudocritical Temperature and Pressure from Gas Gravity, and Gas Properties from Composition programs in the Pac.

Nomenclature

Symbol	*Variable Name*	*Input or Output*	*English Units*	*SI Units*
BEG P*	Beginning pressure of table	I	PSI	KPA
END P*	Ending pressure of table	I	PSI	KPA
P*	Pressure	O	PSI	KPA
P INC*	Pressure increment of table	I	PSI	KPA

*The units for these variables are saved by the program.

Yes/No Questions

CLEAR? Yes: Clear constituent registers.
No: Leave constituent registers unchanged.

SKIP? Yes: Skip input of PVT data.
No: Allow input of PVT data.

Example 1: Using Gas Composition

The Plum Nearly (Smackover) field is a single-well sour-gas reservoir in East Mississippi. Table 2–1 gives the composition taken from a gas analysis from the only test of this well. Initial reservoir conditions are as follows:

Temperature = 255 F
Pressure = 13,600 PSI

Generate a table of gas properties including Z factor, coefficient of isothermal compressibility, gas formation volume factor, and gas viscosity. Recall that the standard pressure in Mississippi is 15.025 PSI. Note that the pseudocritical properties, the Wichert-Aziz correction and the corrected pseudocritical properties are also calculated.

Figures 2–1 through 2–4 depict the calculated values for Z factor, CG, BG, and UG. Note that the initial Z factor is over 1.7 for this hot, sour-gas well.

Table 2–1

Component	*Mole %*
Nitrogen	0.51
Carbon dioxide	6.25
Hydrogen sulfide	3.40
Methane	74.73
Ethane	4.85
Propane	3.26
I-Butane	1.11
N-Butane	0.54
I-Pentane	0.71
N-Pentane	0.32
Hexane	0.68
Heptane +	3.64
Total	100.00

Keystrokes (SIZE> =045)	**Display**	**Comments**
[XEQ] [ALPHA] GASPVT [ALPHA]	*SKIP? Y/N:*	Last character is Y or N
N [R/S]	*GASG COMP*	
[E]	*CLEAR? Y/N:*	Gas composition known. Last character is Y or N
Y [R/S]	*%N2=?*	
.51 [R/S]	*%CO2=?*	
6.25 [R/S]	*%H2S=?*	
3.4 [R/S]	*%METH=?*	
74.73 [R/S]	*%ETH=?*	
4.85 [R/S]	*%PROP=?*	
3.26 [R/S]	*%I-BUT=?*	
1.11 [R/S]	*%N-BUT=?*	
.54 [R/S]	*%I-PEN=?*	
.71 [R/S]	*%N-PEN=?*	
.32 [R/S]	*%N-HEX=?*	
.68 [R/S]	*%N-HEP=?*	
3.64 [R/S]	*%N-OCT=?*	

Keystrokes (SIZE> =045)	Display	Comments
[//] [E]	*STD T=?*	%TOT, GAS G, Tc, Pc, CWA, Tc*, and Pc* printed
60 [R/S]	*STD P=?*	
15.025 [R/S]	*T=?*	
255 [R/S]	*BEG P=?*	
500 [R/S]	*END P=?*	
13600 [R/S]	*P INC=?*	
500 [R/S]	*BEG P=?*	Table of properties printed

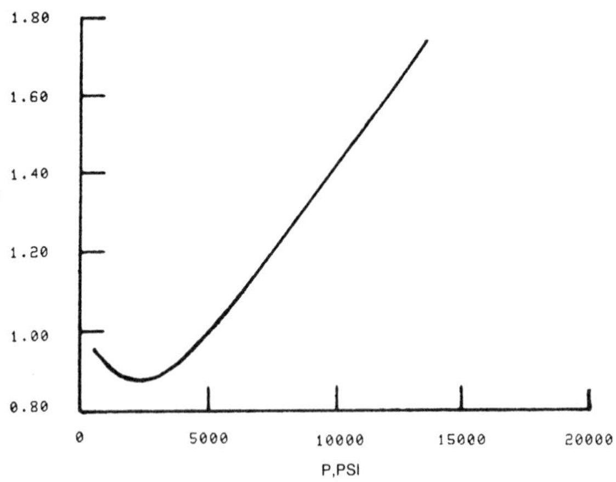

Figure 2–1

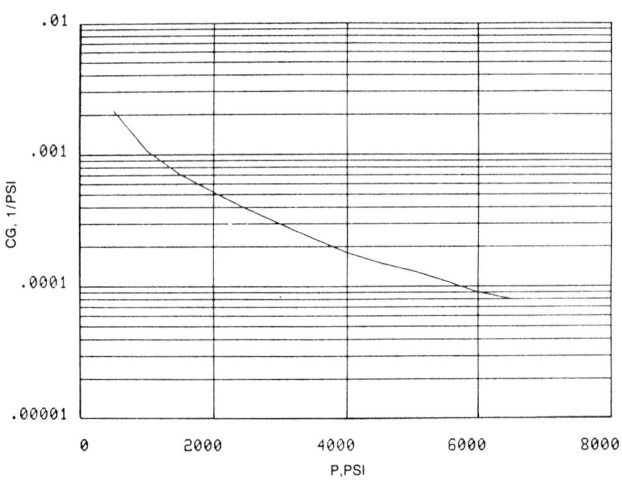

Figure 2–2

```
GAS PVT

SKIP: NO
CLEAR: YES
%N2=0.5100
%CO2=6.2500
%H2S=3.4000
%METH=74.7300
%ETH=4.8500
%PROP=3.2600
%IBUT=1.1100
%N-BUT=0.5400
%IPEN=0.7100
%N-PEN=0.3200
%N-HEX=0.6800
%N-HEP=3.6400

%TOT=100.0000
GAS G=0.8588
Tc=425.4512 R
Pc=698.6290 PSI
CWA=14.5490 F
Tc*=410.9022 R
Pc*=673.9813 PSI
STD T=60.0000 F
STD P=15.0250 PSI
T=255.0000 F
BEG P=500.0000 PSI
END P=13,600.0000 PSI
P INC=500.0000 PSI

P=500.0000 PSI
Z=0.9568
CG=0.0021 1/PSI
BG=0.0395 FT3/SCF
UG=0.0140 CP

P=1,000.0000 PSI
Z=0.9204
CG=0.0011 1/PSI
BG=0.0190 FT3/SCF
UG=0.0149 CP

P=1,500.0000 PSI
Z=0.8936
CG=0.0007 1/PSI
BG=0.0123 FT3/SCF
UG=0.0162 CP

P=2,000.0000 PSI
Z=0.8790
CG=0.0005 1/PSI
BG=0.0091 FT3/SCF
UG=0.0178 CP

P=2,500.0000 PSI
Z=0.8769
CG=0.0004 1/PSI
BG=0.0072 FT3/SCF
UG=0.0196 CP

P=3,000.0000 PSI
Z=0.8864
CG=0.0003 1/PSI
BG=0.0061 FT3/SCF
UG=0.0217 CP

P=3,500.0000 PSI
Z=0.9050
CG=0.0002 1/PSI
BG=0.0053 FT3/SCF
UG=0.0238 CP

P=4,000.0000 PSI
Z=0.9306
CG=0.0002 1/PSI
BG=0.0048 FT3/SCF
UG=0.0260 CP

P=4,500.0000 PSI
Z=0.9613
CG=0.0002 1/PSI
BG=0.0044 FT3/SCF
UG=0.0282 CP

P=5,000.0000 PSI
Z=0.9957
CG=0.0001 1/PSI
BG=0.0041 FT3/SCF
UG=0.0304 CP

P=5,500.0000 PSI
Z=1.0329
CG=0.0001 1/PSI
BG=0.0039 FT3/SCF
UG=0.0326 CP

P=6,000.0000 PSI
Z=1.0722
CG=0.0001 1/PSI
BG=0.0037 FT3/SCF
UG=0.0346 CP

P=6,500.0000 PSI
Z=1.1129
CG=0.0001 1/PSI
```

```
BG=0.0035 FT3/SCF
UG=0.0367 CP

P=7,000.0000 PSI
Z=1.1547
CG=0.0001 1/PSI
BG=0.0034 FT3/SCF
UG=0.0386 CP

P=7,500.0000 PSI
Z=1.1974
CG=0.0001 1/PSI
BG=0.0033 FT3/SCF
UG=0.0405 CP

P=8,000.0000 PSI
Z=1.2406
CG=0.0001 1/PSI
BG=0.0032 FT3/SCF
UG=0.0424 CP

P=8,500.0000 PSI
Z=1.2843
CG=4.9386E-5 1/PSI
BG=0.0031 FT3/SCF
UG=0.0442 CP

P=9,000.0000 PSI
Z=1.3282
CG=4.4739E-5 1/PSI
BG=0.0030 FT3/SCF
UG=0.0459 CP

P=9,500.0000 PSI
Z=1.3724
CG=4.0793E-5 1/PSI
BG=0.0030 FT3/SCF
UG=0.0476 CP

P=10,000.0000 PSI
Z=1.4167
CG=3.7412E-5 1/PSI
BG=0.0029 FT3/SCF
UG=0.0493 CP

P=10,500.0000 PSI
Z=1.4610
CG=3.4492E-5 1/PSI
BG=0.0029 FT3/SCF
UG=0.0509 CP

P=11,000.0000 PSI
Z=1.5054
CG=3.1951E-5 1/PSI
BG=0.0028 FT3/SCF
UG=0.0525 CP

P=11,500.0000 PSI
Z=1.5498
CG=2.9724E-5 1/PSI
BG=0.0028 FT3/SCF
UG=0.0541 CP

P=12,000.0000 PSI
Z=1.5941
CG=2.7760E-5 1/PSI
BG=0.0027 FT3/SCF
UG=0.0556 CP

P=12,500.0000 PSI
Z=1.6383
CG=2.6018E-5 1/PSI
BG=0.0027 FT3/SCF
UG=0.0570 CP

P=13,000.0000 PSI
Z=1.6825
CG=2.4464E-5 1/PSI
BG=0.0027 FT3/SCF
UG=0.0585 CP

P=13,500.0000 PSI
Z=1.7266
CG=2.3071E-5 1/PSI
BG=0.0026 FT3/SCF
UG=0.0599 CP

P=13,600.0000 PSI
Z=1.7354
CG=2.2810E-5 1/PSI
BG=0.0026 FT3/SCF
UG=0.0602 CP
```

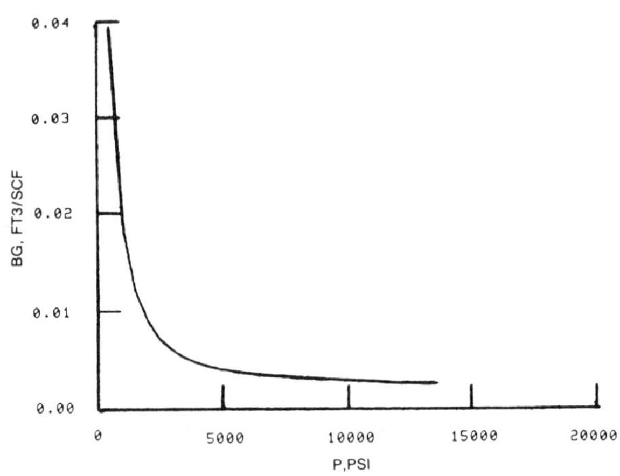

Figure 2–3

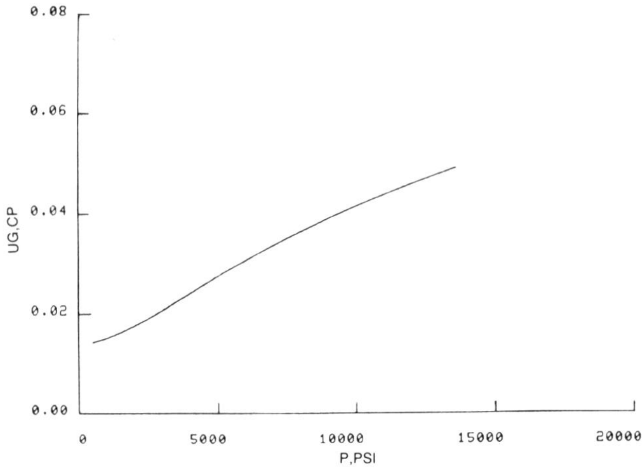

Figure 2–4

Example 2: Using Gas Gravity

A Vicksburg gas reservoir has been discovered in South Texas. Only the gas gravity (0.872) is known from a jug test. Reservoir temperature is 180 F, initial pressure is 6200 PSI, and no diluents are expected. The results are presented graphically in Figures 2–5 through 2–8.

Keystrokes	Display	Comments
[XEQ] [ALPHA] GASPVT [ALPHA]	*SKIP? Y/N:N*	
[R/S]	*GASG COMP*	
[A]	*GAS G=?*	Gas gravity known
.872 [R/S]	*%N2=?*	
0 [R/S]	*%CO2=?*	
0 [R/S]	*%H2S=?*	
0 [R/S]	*STD T=?*	Tc and Pc printed
[R/S]	*STD P=?*	
14.65 [R/S]	*T=?*	
180 [R/S]	*BEG P=?*	
500 [R/S]	*END P=?*	
6200 [R/S]	*P INC=?*	
500 [R/S]	*BEG P=?*	Table of properties printed

```
GAS PVT

GAS G=0.8720
%N2=0.0000
%CO2=0.0000
%H2S=0.0000
Tc=441.8952 R
Pc=661.5656 PSI
STD P=14.6500 PSI
T=180.0000 F
BEG P=500.0000 PSI
END P=6,200.0000 PSI
P INC=500.0000 PSI

P=500.0000 PSI
Z=0.9158
CG=0.0022 1/PSI
BG=0.0330 FT3/SCF
UG=0.0127 CP

P=1,000.0000 PSI
Z=0.8367
CG=0.0012 1/PSI
BG=0.0151 FT3/SCF
UG=0.0140 CP

P=1,500.0000 PSI
Z=0.7746
CG=0.0008 1/PSI
BG=0.0093 FT3/SCF
UG=0.0161 CP

P=2,000.0000 PSI
Z=0.7432
CG=0.0005 1/PSI
BG=0.0067 FT3/SCF
UG=0.0189 CP

P=2,500.0000 PSI
Z=0.7437
CG=0.0004 1/PSI
BG=0.0054 FT3/SCF
UG=0.0221 CP

P=3,000.0000 PSI
Z=0.7667
CG=0.0003 1/PSI
BG=0.0046 FT3/SCF
UG=0.0253 CP

P=3,500.0000 PSI
Z=0.8035
CG=0.0002 1/PSI
BG=0.0041 FT3/SCF
UG=0.0284 CP

P=4,000.0000 PSI
Z=0.8485
CG=0.0001 1/PSI
BG=0.0038 FT3/SCF
UG=0.0312 CP

P=4,500.0000 PSI
Z=0.8984
CG=0.0001 1/PSI
BG=0.0036 FT3/SCF
UG=0.0338 CP

P=5,000.0000 PSI
Z=0.9513
```

CG=0.0001 1/PSI
BG=0.0034 FT3/SCF
UG=0.0362 CP

P=5,500.0000 PSI
Z=1.0060
CG=0.0001 1/PSI
BG=0.0033 FT3/SCF
UG=0.0385 CP

P=6,000.0000 PSI
Z=1.0617
CG=0.0001 1/PSI
BG=0.0032 FT3/SCF
UG=0.0405 CP

P=6,200.0000 PSI
Z=1.0842
CG=0.0001 1/PSI
BG=0.0032 FT3/SCF
UG=0.0413 CP

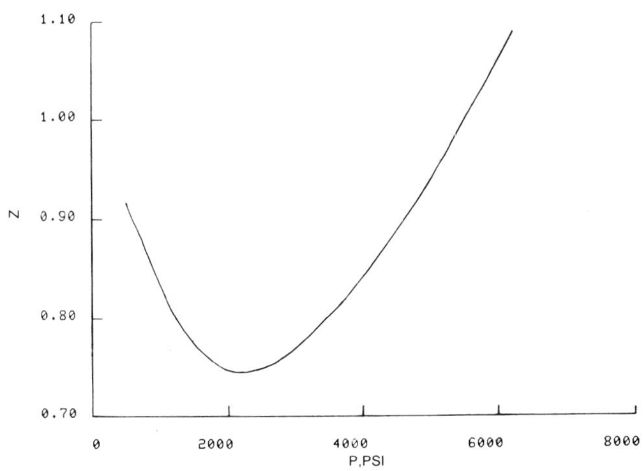

Figure 2–5

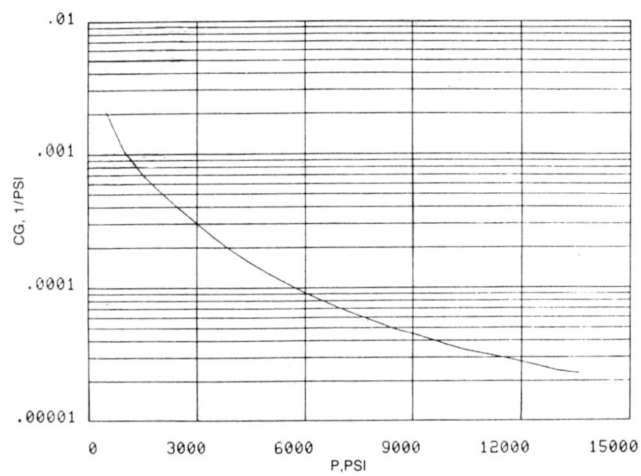

Figure 2–6

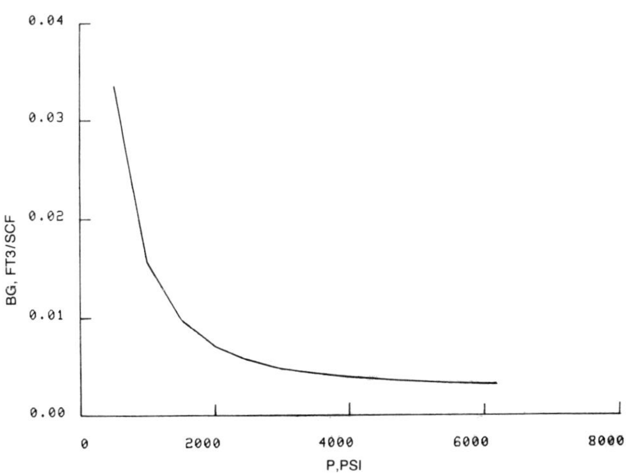

Figure 2–7

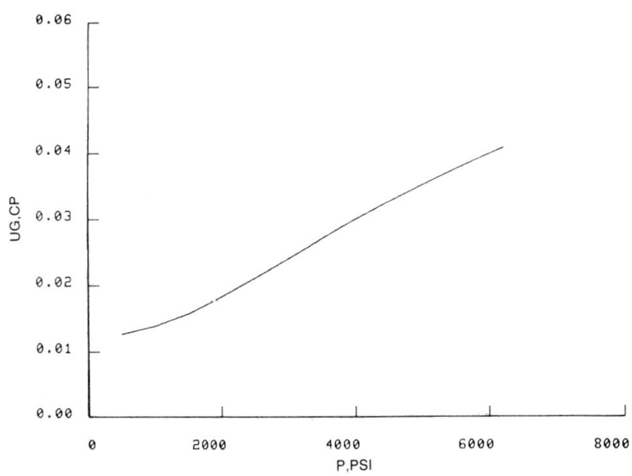

Figure 2–8

User Instructions

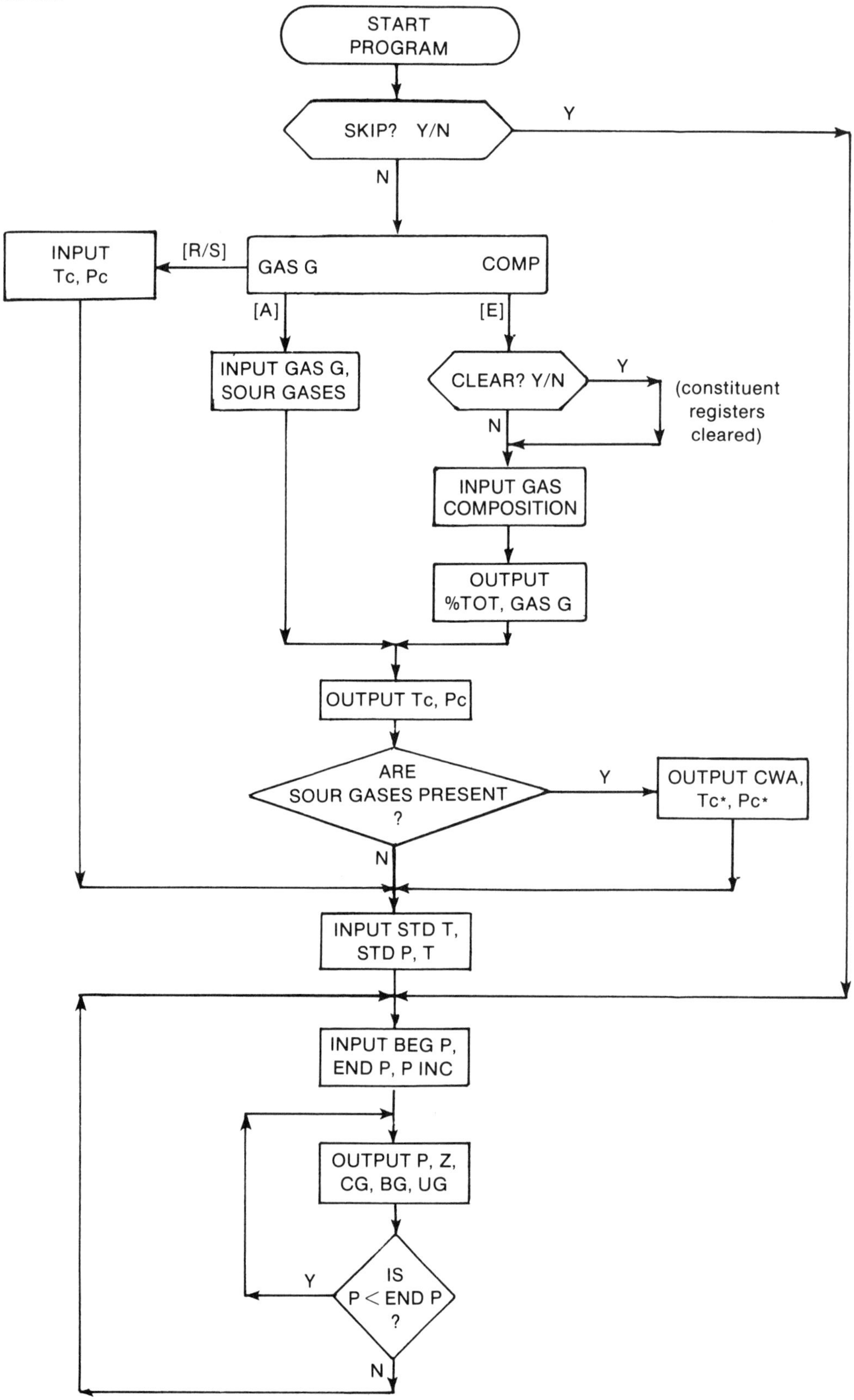

General Information

Memory Requirements

Program length:	388 bytes (2 cards)
Minimum size:	045
Minimum hardware:	41C + 1 memory module

Hidden Options

None

Pac Subroutines Called

TITLE, Y/N?, ITcPc, COMP, OUT, CGASG, CTPC, GASG, SOUR, CTcPc, X0, STDTP, T, INK, OUTK, CZ, CCG, OUTU, CBG, CUG, CON

Registers

03	Pressure units
04	Pressure units
06	TR
07	PR
08	END P (PSI)
09	P INC (PSI)
17	BEG P, P (PSI)

Registers 12–14, 18–21, 24, and 25 unused

Flags

02	Set:	Skip input of PVT data.
	Clear:	Allow input of PVT data.
07	Set:	Clear constituent registers.
	Clear:	Leave constituent registers unchanged.

Program Listing

```
 01♦LBL "GASPVT"
"GAS PVT" 45
XROM "TITLE" FC?C 25
PROMPT CF 05 SF 27
"KPA" ASTO 03 CLA
ASTO 04 "SKIP" 2
XROM "Y/N?" FS? 02
GTO 15

 18♦LBL 00
"GASG  COMP" PROMPT
XROM "ITcPc" GTO 03

 23♦LBL D
 24♦LBL E
"CLEAR" 7 XROM "Y/N?"
26.044 FC? 07 GTO 02
0

 32♦LBL 01
STO IND Y ISG Y GTO 01

 36♦LBL 02
"X9" XROM "COMP"

 39♦LBL "X9"
ADV "%TOT" XROM "OUT"
XROM "CGASG" STO 15
"GAS G" XROM "OUT"
XROM "CTPC" XROM "X0"
GTO 03

 50♦LBL A
 51♦LBL B
XROM "GASG" XROM "SOUR"
XROM "CTcPc" XROM "X0"

 56♦LBL 03
XROM "STDTP" XROM "T"

 59♦LBL 15
RCL 16 "F-R" CON
RCL 10 / STO 06 "PSI"
ASTO 01 CLA ASTO 02
RCL 04 RCL 03 16
STO 00 "BEG P"
XROM "INK" RDN STO 03
X<>Y STO 04 X<>Y
CF 08 7 STO 00
"END P" XROM "INK" RDN
STO 03 X<>Y STO 04
X<>Y R↑ "P INC"
XROM "INK" RDN STO 03
RDN STO 04 ADV

 99♦LBL 16
"PSI" ASTO 01 CLA
ASTO 02 RCL 04 RCL 03
RCL 17 "P" XROM "OUTK"
RDN STO 03 RDN STO 04
RCL 06 RCL 17 RCL 11
/ STO 07 CZ "Z"
XROM "OUT" RCL 06
RCL 07 XROM "CCG"
"1/PSI" ASTO 01 CLA
ASTO 02 ASTO Z "1/KPA"
ASTO Y "CG"
XROM "OUTU" RCL 06
RCL 07 XROM "CBG"
"FT3/SCF" ASTO 01 ASHF
ASTO 02 "M3/SCM"
ASTO Y CLA ASTO Z
"BG" XROM "OUTU"
RCL 06 RCL 07
XROM "CUG" "CP"
ASTO 01 CLA ASTO 02
ASTO Z "PA*S" ASTO Y
"UG" XROM "OUTU" ADV
RCL 08 RND RCL 17
RCL 09 + RND X<Y?
GTO 04 X<>Y RCL 17
RND X=Y? GTO 05
RCL 08 RND

174♦LBL 04
LASTX STO 17 GTO 16

178♦LBL 05
RCL 09 ST+ 17 GTO 15
END
```

3. DEW — Dew-Point Pressure from Gas Composition

Certain reservoirs have initial pressure and temperature conditions described as retrograde-condensate or dew-point reservoirs. Their temperature is greater than the critical point but less than the cricondentherm. The initial pressure also exceeds the cricondenbar. Initially, the fluid is a single phase and is frequently referred to as a gas because the temperature exceeds the critical point. As the reservoir pressure declines, the properties of the produced fluid remain constant until the reservoir pressure reaches the dew point (see Figure 3–1). At this pressure, liquid begins to condense out of the reservoir fluid (thus the dew point). The first liquid condensation has a relatively greater concentration of heavier molecules; the produced fluid begins to have decreasing liquid content, increasing the producing gas-oil ratio.

Nemeth and Kennedy generated a correlation for the dew-point pressure as a function of composition and temperature. Their correlation was based on 579 dew-point pressures determined from 480 different condensate systems. DEW solves their correlation and can be used when the expensive laboratory measurements are unavailable.

Although the program will allow inputs of esoteric constituents (%O2, %H2, %He, and %H2O), the correlation ignores these and only considers the sour gases and hydrocarbons. If the gas analysis includes octane, nonane, and decane, the program will calculate the molecular weight and density of the heptanes-plus fraction from these constituents.

Equations

$$\begin{aligned} Pd = \exp\{&A[0.2\ \%N2 + \%CO2 + \%H2S \\ &+ 0.4\ \%METH + \%ETH + 2(\%PROP \\ &+ \%I\text{-}BUT + \%N\text{-}BUT) + \%I\text{-}PEN \\ &+ \%N\text{-}PEN + \%N\text{-}HEX] + B\ DENC7+ \\ &+ C[\%METH/\%C7{+}+ 0.2)] + DT' \\ &+ EL + FL^2 + GL^3 + HM + IM^2 \\ &+ JM^3 + K\} \end{aligned}$$

$$\begin{aligned} A &= -2.0623054\ (10^{-2}) \\ B &= 6.6259728 \\ C &= -4.4670559\ (10^{-3}) \\ D &= 1.0448346\ (10^{-4}) \\ E &= 3.2673714\ (10^{-2}) \\ F &= -3.6453277\ (10^{-3}) \\ G &= 7.4299951\ (10^{-5}) \\ H &= -0.11381195 \\ I &= 6.2476497\ (10^{-4}) \\ J &= -1.0716866\ (10^{-6}) \end{aligned}$$

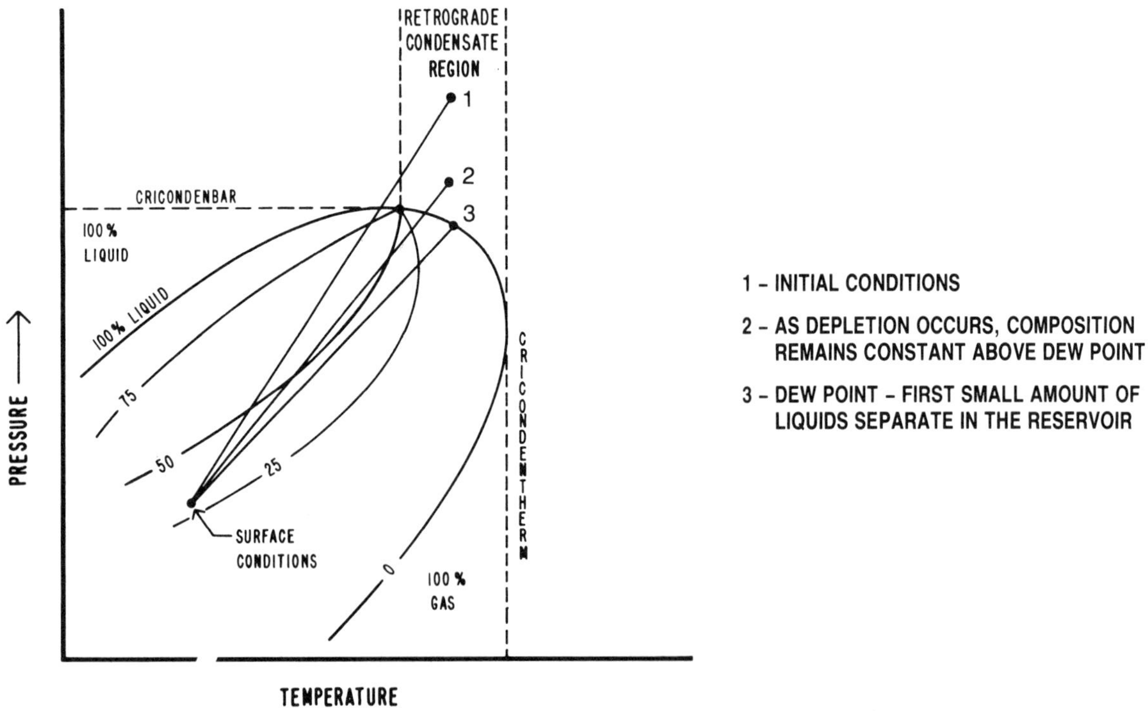

Figure 3–1

K = 10.746622
L = (C7+) (MWC7+)
M = MWC7+/(DENC7+ + 0.0001)
DENC7+ = (0.6882 %N-HEP
 + 0.7068 %N-OCT
 + 0.7217 %N-NON
 + 0.7342 %N-DEC)/%C7+
MWC7+ = (100.2 %N-HEP + 114.2 %N-OCT
 + 128.3 %N-NON
 + 142.3 %N-DEC)/%C7+
%C7+ = %N-HEP + %N-OCT + %N-NON
 + %N-DEC
C7+ = %C7+/100
T′ = T in R

Nomenclature

Symbol	*Variable Name*	*Input or Output*	*English Units*	*SI Units*
DENC7+	Density of heptanes plus	I,O	G/CM3	G/CM3
MWC7+	Molecular weight of heptanes plus	I,O	–	–
Pd	Dew-point pressure	O	PSI	KPA

Yes/No Questions

CLEAR? Yes: Clear constituent registers.
No: Leave constituent registers unchanged.

Example 1

The gas reservoir in Example 1 of GASPVT will be used to illustrate the DEW program. If you just ran that example, answer N to the CLEAR? Y/N prompt, [R/S] past the temperature prompt, and press [//] [E] at any of the subsequent composition prompts to skip past them. The molecular weight and density of the C7+ fraction are calculated by the program based on the input mole percentages. Since the composition obtained only gave a heptanes-plus percentage for the heavy ends, the calculated values are those of normal heptane. The calculated dew-point pressure is 1,956.6 PSI.

Keystrokes (SIZE> =045)	Display	Comments
[XEQ] [ALPHA] DEW [ALPHA]	*CLEAR? Y/N:*	Last character is Y or N
Y [R/S]	*T=?*	
255 [R/S]	*%N2=?*	
.51 [R/S]	*%CO2=?*	
6.25 [R/S]	*%H2S=?*	
3.4 [R/S]	*%METH=?*	
74.73 [R/S]	*%ETH=?*	
4.85 [R/S]	*%PROP=?*	
3.26 [R/S]	*%I-BUT=?*	
1.11 [R/S]	*%N-BUT=?*	
.54 [R/S]	*%I-PEN=?*	
.71 [R/S]	*%N-PEN=?*	
.32 [R/S]	*%N-HEX=?*	
.68 [R/S]	*%N-HEP=?*	
3.64 [R/S]	*%N-OCT=?*	
[//] [E]	*MWC7+=?*	%TOT and MWC7+ printed
[R/S]	*DENC7+=?*	DENC7+ printed
[R/S]	*1,956.5797*	Pd printed and program halts

```
DEW POINT

CLEAR: YES
T=255.0000 F
%N2=0.5100
%CO2=6.2500
%H2S=3.4000
%METH=74.7300
%ETH=4.8500
%PROP=3.2600
%IBUT=1.1100
%N-BUT=0.5400
%IPEN=0.7100
%N-PEN=0.3200
%N-HEX=0.6800
%N-HEP=3.6400

%TOT=100.0000
MWC7+=100.2000
DENC7+=0.6882 G/CM3

Pd=1,956.5797 PSI
```

Example 2

If the molecular weight and density of the heptanes-plus fraction are known, they can be input. Repeat Example 1 with MWC7+ = 114.5 and DENC7+ = 0.6942 G/CM3. The calculated dew-point pressure for this case is 2,148.8 PSI. Note that this value is nearly 10% greater than the value calculated without knowing the properties of the heptanes-plus fraction. However, the calculated dew points are so far below the initial reservoir pressure of 13,600 PSI that essentially constant reservoir composition can be assumed until the vast majority of the recoverable reserves have been obtained.

Keystrokes	Display	Comments
[XEQ] [ALPHA] DEW [ALPHA]	*CLEAR? Y/N:Y*	
N [R/S]	*T=?*	
[R/S]	*%N2=?*	
[//] [E]	*MWC7+=?*	%TOT and MWC7+ printed
114.5 [R/S]	*DENC7+=?*	DENC7+ printed
.6942 [R/S]	*2,148.7698*	Pd printed and program halts

```
DEW POINT

CLEAR: NO

%TOT=100.0000
MWC7+=100.2000
MWC7+=114.5000
DENC7+=0.6882 G/CM3
DENC7+=0.6942 G/CM3

Pd=2,148.7698 PSI
```

User Instructions

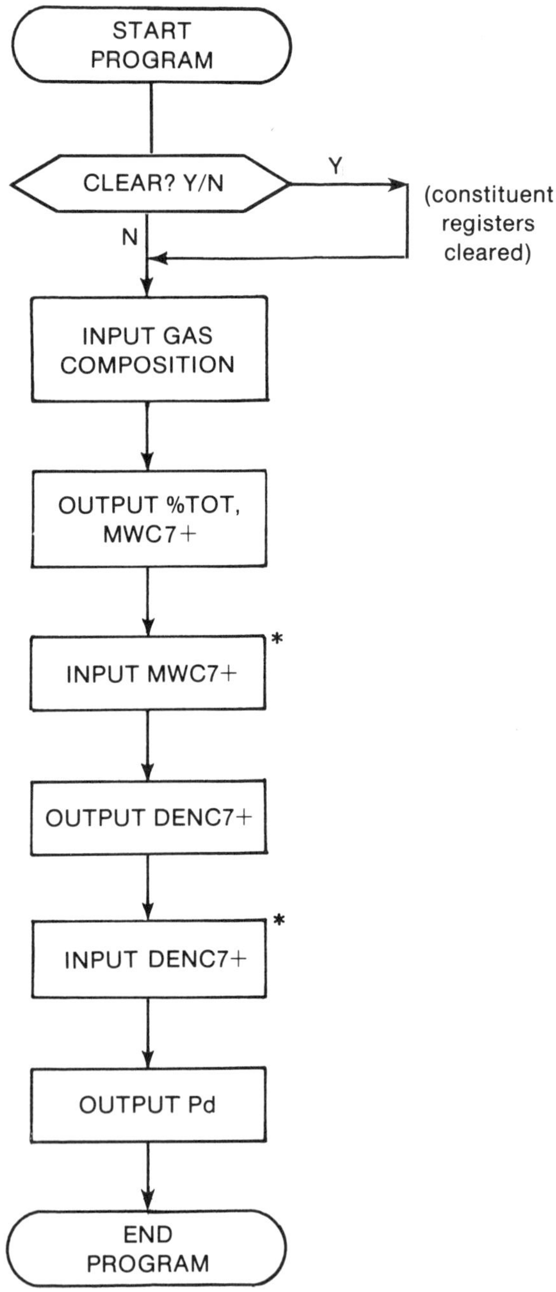

* If MWC7+ and DENC7+ are known from the compositional analysis, they should be input.

General Information

Memory Requirements

Program length:	466 bytes (3 cards)
Minimum size:	045
Minimum hardware:	41C and 1 memory module

Hidden Options

None

Pac Subroutines Called

TITLE, Y/N?, COMP, OUT, W3, IN, OUTU, INU, CON

Registers

06	Scratch
07	MWC7+
08	%C7+
09	DENC7+ (G/CM3)
14	Pd (PSI)

Registers 03, 04, 10–13, and 15–25 unused

Flags

07 Set: Clear constituent registers.
Clear: Leave constituent registers unchanged.

Program Listing

```
 01♦LBL "DEW"
45  "DEW POINT"
XROM "TITLE"  FC?C 25
PROMPT  SF 27  "CLEAR"
7  XROM "Y/N?"  26.044
FC? 07  GTO 01  0

 15♦LBL 00
STO IND Y  ISG Y  GTO 00

 19♦LBL 01
XROM "T"  "G7"
XROM "COMP"

 23♦LBL "G7"
ADV  "%TOT"  XROM "OUT"
36  STO 00  CLST  100.2
XROM "W3"  114.2
XROM "W3"  128.3
XROM "W3"  142.3
XROM "W3"  RCL 40
RCL 39  +  RCL 38  +
RCL 37  +  STO 08  X>0?
/  STO 07  "MWC7+"
XROM "OUT"  6  STO 00
"MWC7+"  XROM "IN"  36
STO 00  CLST  .6882
XROM "W3"  .7068
XROM "W3"  .7217
XROM "W3"  .7342
XROM "W3"  RCL 08  X>0?
/  STO 09  "G/CM3"
ASTO 01  ASTO Y  CLA
ASTO 02  ASTO Z
"DENC7+"  XROM "OUTU"
RCL 02  RCL 01  8
STO 00  "DENC7+"
XROM "INU"  6.6259728
RCL 09  *  RCL 29
RCL 08  .2  +  /
4.4670559 E-3  *  -
RCL 16  "F-R"  CON
1.0448346 E-4  *  +
RCL 08  1 E2  /  RCL 07
*  STO 06  3.2673714 E-2
*  +  RCL 06  X↑2
3.6453277 E-3  *
RCL 06  3  Y↑X
7.4299951 E-5  *  +
RCL 07  RCL 09  1 E-4  +
/  STO 06  .11381195  *
-  RCL 06  X↑2
6.2476497 E-4  *  +
RCL 06  3  Y↑X
1.0716866 E-6  *  -
10.746622  +  RCL 30
RCL 27  +  RCL 28  +
RCL 36  +  RCL 31
RCL 32  +  RCL 33  +
ST+ X  +  RCL 34  +
RCL 35  +  RCL 29  .4  *
+  RCL 26  5  /  +
2.0623054 E-2  *  -  E↑X
STO 14  ADV  "PSI"
ASTO 01  CLA  ASTO 02
ASTO Z  "KPA"  ASTO Y
"Pd"  XROM "OUTU"  END
```

4. GOR — Forecasting GOR Behavior for Gas-Condensate Reservoirs

This program illustrates an empirical technique for estimating retrograde GOR behavior in volumetric gas-condensate reservoirs. When more sophisticated compositional studies are unavailable, this method should provide a reasonably accurate forecast. The necessary data normally are acquired during well testing.

Either separator gas gravity or wellhead stream gas gravity may be input by the user. After this program has been executed, a table of predicted gas-oil ratios as a function of pressure may be developed, as well as the gas-oil ratio at any pressure. This program may be used in conjunction with the gas deliverability program if condensate production is to be included in a deliverability forecast.

The program requires the dew point of the gas as an input. Joiner and Long suggested using the initial reservoir pressure if the dew-point pressure is unknown. However, if the gas composition is known, the DEW program can be used to estimate the gas dew point pressure. Similarly, if gas composition is available, the values for %C4+ and %C5+ may be input by the user.

It should be noted that below 30% of Pd, a constant GOR is used. Occasionally, the calculated value of R50% will be greater than Rd. This can occur for high GOR wells (> 50 MCF/BBL) and indicates that no retrograde behavior can be expected. In these cases, a constant GOR should be used for prediction.

Equations

$$R50\% = \exp[31.49 - 1.085(10^{-4})Pd - 92.03\ \%C4+/T + 110.8\ \%C5+/T + 0.0215\ T + 6.833\ \text{WELL G} - 26.98/Rd - 6.632\ \ln T]$$

$$\%C4+ = 6.547 + 25.52\ \text{WELL G} + 30.38/Rd + 0.02633\ Rd - 30.3\ \text{OIL G}' - 0.00417\ T$$

$$\%C5+ = -8.53 + 7.83\ \text{WELL G} + 56.26/Rd + 0.0109\ Rd + 0.7286\ \text{OIL G} - 0.00424\ T$$

For $0.3\ Pd \le P < Pd$, $\log R = 2(1 - P/Pd)\log(R50\%/Rd) + \log Rd$

For $P < 0.3\ Pd$, R is evaluated at $P = 0.3\ Pd$

For $P \ge Pd$, $R = Rd$

OIL G′ = OIL G in SPGR

Nomenclature

Symbol	*Variable Name*	*Input or Output*	*English Units*	*SI Units*
BEG P*	Beginning pressure of table	I	PSI	KPA
END P*	Ending pressure of table	I	PSI	KPA
P*	Pressure	I,O	PSI	KPA
P INC*	Pressure increment of table	I	PSI	KPA
Pd*	Dew-point pressure	I	PSI	KPA
R*	Total surface GOR at P	O	MCF/BBL	SCM/M3
Rd*	Total surface GOR at Pd	I,O	MCF/BBL	SCM/M3
R50%*	Total surface GOR at 50% Pd	O	MCF/BBL	SCM/M3
WELL G	Total well stream gas gravity at Pd	I,O	–	–
%C4+	Mole percent butanes-plus components in dew-point fluid	I,O	–	–
%C5+	Mole percent pentanes-plus components in dew-point fluid	I,O	–	–

*The units for these variables are saved by the program.

Yes/No Questions

None

Example 1

Forecast the gas-oil ratio performance for the Vicksburg gas reservoir (Example 2 of GASPVT). The input data required are:

Initial GOR = 16,250 SCF/BBL
Estimated dew-point pressure = 5,000 PSI
Reservoir temperature = 180 F
Condensate gravity = 52.2 API
Gas gravity (separator) = 0.872

The estimated dew-point is from a similar field in the area. Joiner and Long suggest using the initial reservoir pressure if the dew point is unknown. The wellstream gas gravity calculated was 1.0423 with the %C4+ fraction equal to 11.36% and the %C5+ fraction equal to 6.31%. Note that the input units (SCF/BBL) were retained for output because we chose to use SCF/BBL instead of the default MCF/BBL. Also note that the forecast halted once P exceeded Pd, since R = Rd at pressures greater than the dew point.

Keystrokes (SIZE>=029)	Display	Comments
[XEQ] [ALPHA] GOR [ALPHA]	*Rd=?*	
16250 [ALPHA] SCF/BBL [R/S]	*Pd=?*	
5000 [R/S]	*T=?*	
180 [R/S]	*OIL G=?*	
52.2 [R/S]	*GASG WELLG*	
[A]	*GAS G=?*	Gas gravity known
.872 [R/S]	*%C4+=?*	WELL G, %C4+ printed
[R/S]	*%C5+=?*	%C5+ printed
[R/S]	*P TABLE*	R50% printed
[E]	*BEG P=?*	Table option
500 [R/S]	*END P=?*	
6200 [R/S]	*P INC=?*	
500 [R/S]	*P TABLE*	Table printed

```
GOR FORECAST

Rd=16,250.0000 SCF/BBL
Pd=5,000.0000 PSI
T=180.0000 F
OIL G=52.2000 API
GAS G=0.8720
WELL G=1.0423

%C4+=11.3552
%C5+=6.3110
R50%=50,349.6823 SCF/BBL

BEG P=500.0000 PSI
END P=6,200.0000 PSI
P INC=500.0000 PSI

P=500.0000 PSI
R=79,150.5766 SCF/BBL

P=1,000.0000 PSI
R=79,150.5766 SCF/BBL

P=1,500.0000 PSI
R=79,150.5766 SCF/BBL

P=2,000.0000 PSI
R=63,128.4911 SCF/BBL

P=2,500.0000 PSI
R=50,349.6823 SCF/BBL

P=3,000.0000 PSI
R=40,157.6288 SCF/BBL

P=3,500.0000 PSI
R=32,028.7056 SCF/BBL

P=4,000.0000 PSI
R=25,545.2827 SCF/BBL

P=4,500.0000 PSI
R=20,374.2692 SCF/BBL

P=5,000.0000 PSI
Rd=16,250.0000 SCF/BBL
```

Example 2

Repeat the previous example with the following known values from a compositional analysis. Also calculate the gas-oil ratio at the current pressure of 4,250 PSI. When you restart the program, default units for R (MCF/BBL) will be assumed again. Set flag 10 ([//] [SF] 10) to allow you to change the units of the first R-value that is output (R50%).

Wellstream gas gravity = 0.965
%C4+ = 10.22
%C5+ = 5.91
Dew-point pressure = 4952 PSI

Figure 4–1 compares the two predicted gas-oil ratio forecasts.

Keystrokes	Display	Comments
[//] [SF] 10 [XEQ] [ALPHA] GOR [ALPHA]	*Rd=?*	
[R/S]	*Pd=?*	
4952 [R/S]	*T=?*	
[R/S]	*OIL G=?*	
[R/S]	*GASG WELLG*	
[E]	*WELL G=?*	Well gravity known
.965 [R/S]	*%C4+=?*	%C4+ printed
10.22 [R/S]	*%C5+=?*	%C5+ printed
5.91 [R/S]	*R50%, MCF/BBL?*	
SCF/BBL [R/S]	*P TABLE*	R50% printed
[//] [CF] 10 [A]	*P=?*	Single pressure known
4250 [R/S]	*P TABLE*	R printed
[E]	*BEG P=?*	Table option
500 [R/S]	*END P=?*	
[R/S]	*P INC=?*	
[R/S]	*P TABLE*	Table printed

```
GOR FORECAST

Pd=4,952.0000 PSI
WELL G=0.9650

%C4+=9.3812
%C4+=10.2200
%C5+=5.7053
%C5+=5.9100
R50%=41,647.0239 SCF/BBL

P=4,250.0000 PSI
R=21,219.6104 SCF/BBL

BEG P=500.0000 PSI

P=500.0000 PSI
R=60,684.2815 SCF/BBL

P=1,000.0000 PSI
R=60,684.2815 SCF/BBL

P=1,500.0000 PSI
R=60,353.0338 SCF/BBL

P=2,000.0000 PSI
R=49,906.9024 SCF/BBL

P=2,500.0000 PSI
R=41,268.8269 SCF/BBL

P=3,000.0000 PSI
R=34,125.8622 SCF/BBL

P=3,500.0000 PSI
R=28,219.2289 SCF/BBL

P=4,000.0000 PSI
R=23,334.9380 SCF/BBL

P=4,500.0000 PSI
R=19,296.0387 SCF/BBL

P=4,952.0000 PSI
Rd=16,250.0000 SCF/BBL
```

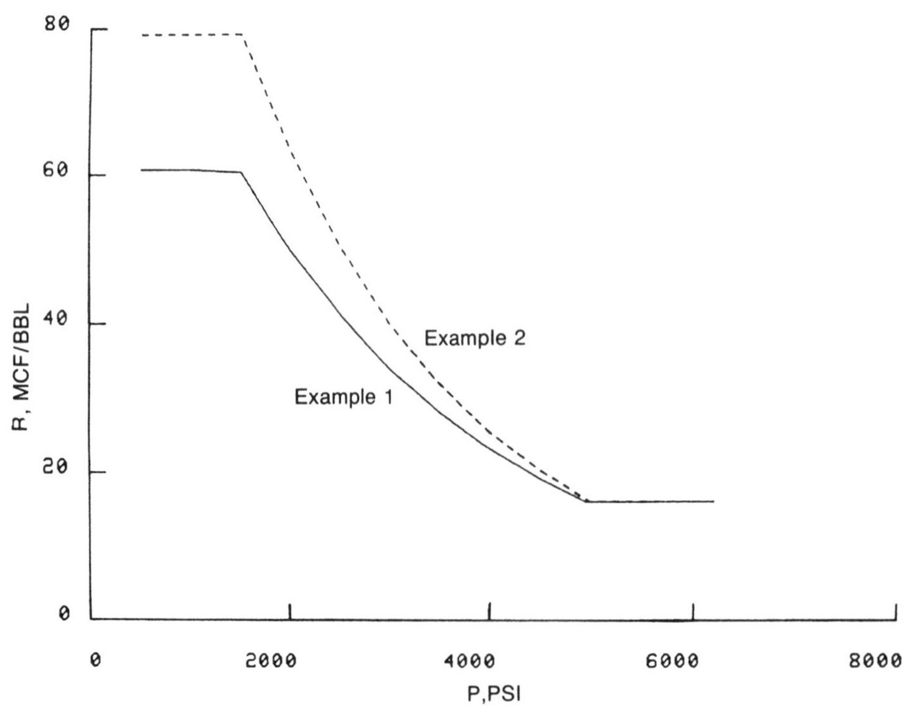

Figure 4–1

User Instructions

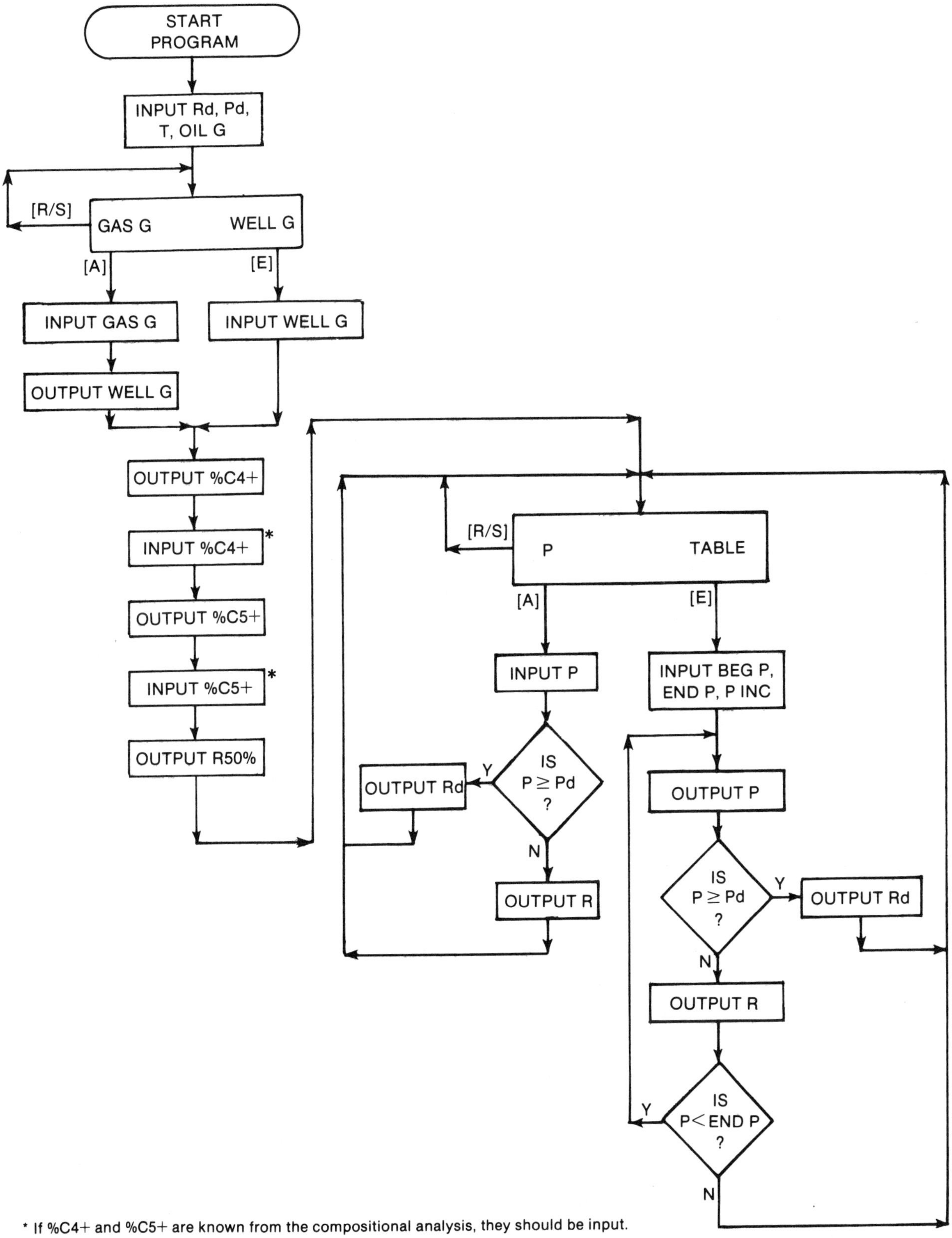

* If %C4+ and %C5+ are known from the compositional analysis, they should be input.

General Information

Memory Requirements

Program length: 672 bytes (3 cards)
Minimum size: 029
Minimum hardware: 41C + 1 memory module

Hidden Options

None

Pac Subroutines Called

TITLE, T, OILG, IN, GASG, CON, OUT, INK, OUTK

Registers

03	Pressure units
04	Pressure units
06	Gas-oil ratio units
07	Gas-oil ratio units
08	END P (PSI)
09	P INC (PSI)
13	Rd (MCF/BBL)
14	Pd (PSI)
17	BEG P, P (PSI)
26	%C4+, R50% (MCF/BBL)
27	%C5+, R (MCF/BBL)
28	WELL G

Registers 10, 11, and 18–25 unused

Flags

None

Program Listing

```
 01♦LBL "GOR"
"GOR FORECAST" 29
XROM "TITLE" FC?C 25
PROMPT SF 27 "KPA"
ASTO 03 "SCM/M3"
ASTO 06 CLA ASTO 04
ASTO 07 12 "Rd"
XEQ 11 13 "Pd" XEQ 08
CF 08 XROM "T"
XROM "OILG"

 24♦LBL 14
"GASG WELLG" PROMPT
GTO 14

 28♦LBL D
 29♦LBL E
27 STO 00 "WELL G"
XROM "IN" GTO 00

 35♦LBL A
 36♦LBL B
XROM "GASG" 4.6 RCL 13
/ RCL 12 "API-SPGR"
CON * + -.006 RCL 12
* 9.75 + RCL 12 *
251 + RCL 13 / 1 E3
/ 1 + / STO 28
"WELL G" XROM "OUT"

 65♦LBL 00
ADV 25.52 * RCL 13
.02633 * + 30.38
RCL 13 / + RCL 12
"API-SPGR" CON 30.3 *
- RCL 16 .00417 * -
6.547 + STO 26 "%C4+"
XROM "OUT" 25 STO 00
"%C4+" XROM "IN"
RCL 28 7.83 * 56.26
RCL 13 / + RCL 13
.0109 * + RCL 12
.07286 * + RCL 16
.00424 * - 8.53 -
STO 27 "%C5+"
XROM "OUT" 26 STO 00
"%C5+" XROM "IN" 110.8
* RCL 26 92.03 * -
RCL 16 / LASTX .0215
* + RCL 28 6.833 *
+ 26.98 RCL 13 / -
RCL 16 LN 6.632 * -
RCL 14 1085 E-7 * -
31.49 + E↑X STO 26
"R50%" XEQ 12 ADV

160♦LBL 01
" P     TABLE" PROMPT
GTO 01

164♦LBL A
16 "P" XEQ 08 XEQ 06
XEQ 05 GTO 01

171♦LBL D
172♦LBL E
16 "BEG P" XEQ 08 7
"END P" XEQ 08 8
"P INC" XEQ 08 ADV
RCL 17

184♦LBL 02
XEQ 06 STO 17 "P"
XEQ 09 XEQ 05 RCL 17
RND RCL 14 RND X=Y?
GTO 01 RCL 08 RND
RCL 17 RCL 09 + RND
X<Y? GTO 03 RCL 17
RND X=Y? GTO 04
RCL 08 RND

210♦LBL 03
LASTX STO 17 GTO 02

214♦LBL 04
RCL 09 ST+ 17 GTO 01

218♦LBL 05
RCL 13 RCL 27 "R"
X=Y? "Rd" XEQ 12 ADV
RTN

227♦LBL 06
RCL 14 / ENTER↑ RND
1 X<=Y? GTO 07 .3 *
X<=Y? RCL Z CHS 1 +
2 * RCL 26 RCL 13 /
LOG * RCL 13 LOG +
10↑X STO 27 RCL 17
RTN

256♦LBL 07
RCL 13 STO 27 RCL 14
RTN

261♦LBL 08
STO 00 XEQ 10
XROM "INK" RDN STO 03
X<>Y STO 04 R↑ RTN

271♦LBL 09
"P" XEQ 10 XROM "OUTK"
RDN STO 03 X<>Y
STO 04 R↑ RTN

281♦LBL 10
ASTO T "PSI" ASTO 01
CLA ASTO 02 CLA
ARCL T RCL 04 RCL 03
RCL Z RTN

293♦LBL 11
STO 00 XEQ 13
XROM "INK" RDN STO 06
X<>Y STO 07 R↑ RTN

303♦LBL 12
XEQ 13 XROM "OUTK" RDN
STO 06 X<>Y STO 07 R↑
RTN

312♦LBL 13
ASTO T "MCF/BBL"
ASTO 01 ASHF ASTO 02
CLA ARCL T RCL 07
RCL 06 RCL Z END
```

5. H2OPVT — Water PVT Properties

H2OPVT uses the Pac correlations to generate a table of water properties as a function of pressure. The program calculates the coefficient of isothermal compressibility (CW), water formation volume factor (BW), water viscosity (UW), and solution gas-water ratio (RSW). These properties can be calculated with or without gas in solution. It can generally be assumed that oil-field waters (such as connate water) that have been in contact with hydrocarbons for many years are gas saturated. Note that the viscosity correlation used does not reflect the effects of dissolved gas in the water. The effect of dissolved gas is (qualitatively) to decrease water viscosity. For this reason, the calculated water viscosity is denoted UW MAX when dissolved gas is present.

Equations

See the water isothermal compressibility, water formation volume factor, water viscosity, and gas-water ratio programs in the Pac.

Nomenclature

Symbol	*Variable Name*	*Input or Output*	*English Units*	*SI Units*
BEG P*	Beginning pressure of table	I	PSI	KPA
END P*	Ending pressure of table	I	PSI	KPA
P*	Pressure	O	PSI	KPA
P INC*	Pressure increment of table	I	PSI	KPA
UW MAX	Maximum UW if water is gas-saturated (RSW>0)	O	CP	PA*S

*The units for these variables are saved by the program.

Yes/No Questions

RSW>0? Yes: Gas-saturated water or brine.
No: Gas-free water or brine.

SKIP? Yes: Skip input of PVT data.
No: Allow input of PVT data.

Example

Calculate the aquifer properties for the South Diddley (Arbuckle) Field in Kansas. The water is gas-saturated in the reservoir.

Water salinity = 58,000 PPM
Reservoir temperature = 105 F
Initial reservoir pressure = 950 PSI
Abandonment pressure = 400 PSI

Notice that if gas-saturated water or brine is specified, UW MAX is output. Also, if gas-free water or brine is selected, RSW is not output since it equals zero. Figures 5–1 through 5–4 present the calculated values for CW, BW, UW, and RSW. The dotted lines show the calculated values assuming the water is not gas-saturated to illustrate the sensitivity of the calculated values to dissolved gas.

Keystrokes (SIZE>=020)	**Display**	**Comments**
[XEQ] [ALPHA] H2OPVT [ALPHA]	*SKIP? Y/N:*	Last character is Y or N
N [R/S]	*RSW>0? Y/N:*	Last character is Y or N
Y [R/S]	*%NACL=?*	
[R/S]	*PPM=?*	
58000 [R/S]	*T=?*	
105 [R/S]	*BEG P=?*	
400 [R/S]	*END P=?*	
950 [R/S]	*P INC=?*	
200 [R/S]	*BEG P=?*	Table of gas-saturated property values printed
[XEQ] [ALPHA] H2OPVT [ALPHA]	*SKIP? Y/N:N*	
[R/S]	*RSW>0? Y/N:Y*	
N [R/S]	*%NACL=?*	
[R/S]	*PPM=?*	
[R/S]	*T=?*	
[R/S]	*BEG P=?*	
400 [R/S]	*END P=?*	
[R/S]	*P INC=?*	
[R/S]	*BEG P=?*	Table of gas-free property values printed

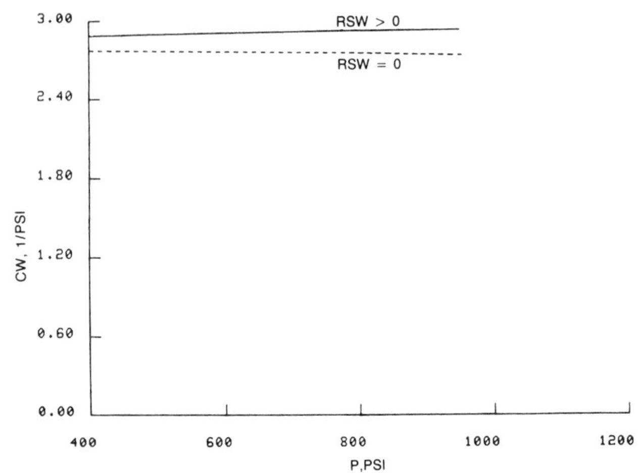

Figure 5–1

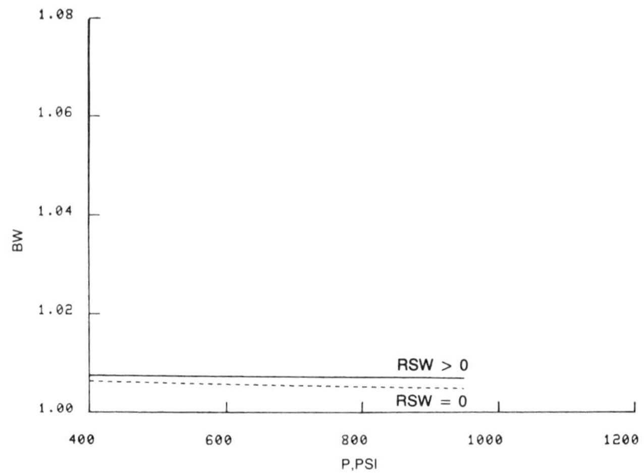

Figure 5–2

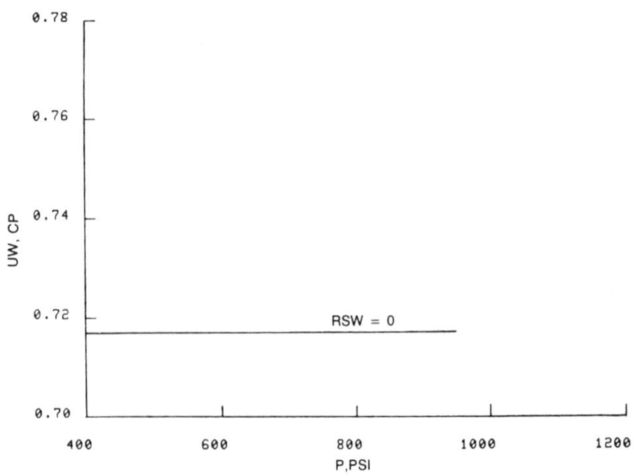

Figure 5–3

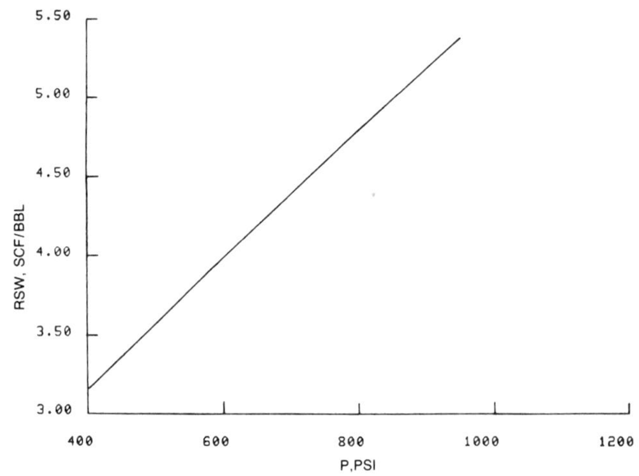

Figure 5–4

```
H2O PVT

SKIP: NO
RSW>0: YES
PPM=58,000.0000
T=105.0000 F
BEG P=400.0000 PSI
END P=950.0000 PSI
P INC=200.0000 PSI

P=400.0000 PSI
CW=2.8858E-6 1/PSI
BW=1.0077
UW MAX=0.7170 CP
RSW=3.1542 SCF/BBL

P=600.0000 PSI
CW=2.8991E-6 1/PSI
BW=1.0075
UW MAX=0.7171 CP
RSW=3.9874 SCF/BBL

P=800.0000 PSI
CW=2.9110E-6 1/PSI
BW=1.0073
UW MAX=0.7172 CP
RSW=4.7897 SCF/BBL

P=950.0000 PSI
CW=2.9189E-6 1/PSI
BW=1.0071
UW MAX=0.7173 CP
RSW=5.3711 SCF/BBL

H2O PVT

RSW>0: NO
BEG P=400.0000 PSI

P=400.0000 PSI
CW=2.7695E-6 1/PSI
BW=1.0065
UW=0.7170 CP

P=600.0000 PSI
CW=2.7530E-6 1/PSI
BW=1.0059
UW=0.7171 CP

P=800.0000 PSI
CW=2.7365E-6 1/PSI
BW=1.0054
UW=0.7172 CP

P=950.0000 PSI
CW=2.7242E-6 1/PSI
BW=1.0050
UW=0.7173 CP
```

User Instructions

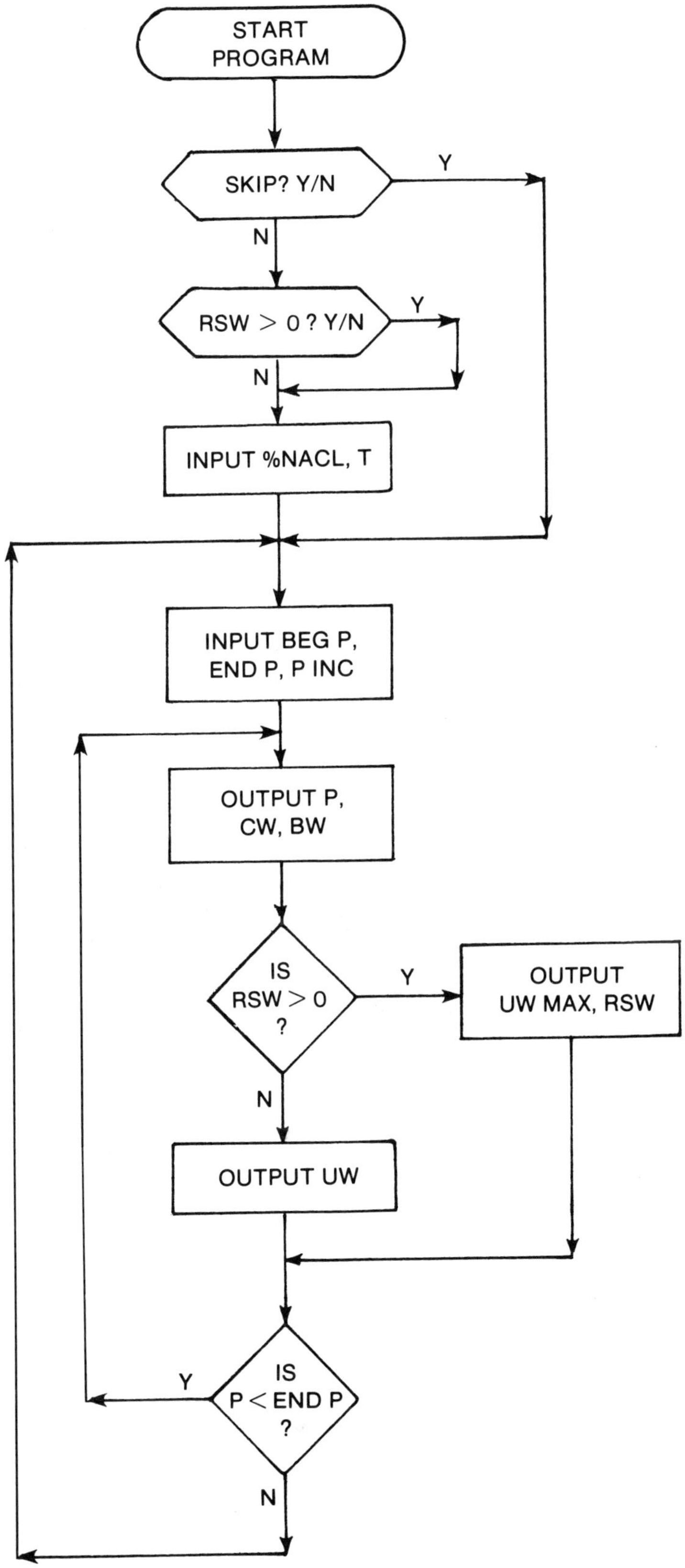

General Information

Memory Requirements

Program length: 285 bytes (2 cards)
Minimum size: 020
Minimum hardware: 41C

Hidden Options

None

Pac Subroutines Called

TITLE, Y/N?, %NACL, T, INK, OUTK, CCW, OUTU, CBW, OUT, CUW, CRSW

Registers

03	Pressure units
04	Pressure units
08	END P (PSI)
09	P INC (PSI)
17	BEG P, P (PSI)

Registers 06, 07, 10–15, and 18 unused

Flags

02	Set:	Skip input of PVT data.
	Clear:	Allow input of PVT data.
06	Set:	Gas-saturated water or brine.
	Clear:	Gas-free water or brine.

Program Listing

```
 01♦LBL "H2OPVT"
"H2O PVT" 20
XROM "TITLE" FC?C 25
PROMPT "KPA" ASTO 03
CLA ASTO 04 "SKIP" 2
XROM "Y/N?" FS? 02
GTO 15 "RSW>0" 6
XROM "Y/N?"
XROM "%NACL" XROM "T"

 21♦LBL 15
"PSI" ASTO 01 CLA
ASTO 02 RCL 04 RCL 03
16 STO 00 "BEG P"
XROM "INK" RDN STO 03
X<>Y STO 04 X<>Y
CF 08 7 STO 00
"END P" XROM "INK" RDN
STO 03 X<>Y STO 04
X<>Y R↑ "P INC"
XROM "INK" RDN STO 03
RDN STO 04 ADV

 55♦LBL 16
"PSI" ASTO 01 CLA
ASTO 02 RCL 04 RCL 03
RCL 17 "P" XROM "OUTK"
RDN STO 03 RDN STO 04
XROM "CCW" "1/PSI"
ASTO 01 CLA ASTO 02
ASTO Z "1/KPA" ASTO Y

"CW" XROM "OUTU"
XROM "CBW" "BW"
XROM "OUT" XROM "CUW"

"CP" ASTO 01 CLA
ASTO 02 ASTO Z "PA*S"
ASTO Y "UW" FS? 06

"⊢ MAX" XROM "OUTU"
FC? 06 GTO 00
XROM "CRSW" "SCF/BBL"
ASTO 01 ASHF ASTO 02
"M3/SCM" ASTO Y CLA
ASTO Z "RSW"
XROM "OUTU"

107♦LBL 00
ADV RCL 08 RND RCL 17
RCL 09 + RND X<Y?
GTO 01 X<>Y RCL 17
RND X=Y? GTO 02
RCL 08 RND

124♦LBL 01
LASTX STO 17 GTO 16

128♦LBL 02
RCL 09 ST+ 17 GTO 15
END
```

6. GASPROD — Material-Balance Gas Production

Higher pressure dry gas in the reservoir will generally drop out some of the heavier components upon production as condensate. Gas material-balance equation values for GP should include the gas equivalent of produced condensate. GASPROD calculates an approximate gas equivalent based on condensate API gravity.

Since the gas in the reservoir is saturated with water at reservoir conditions, and since essentially all of this water is removed prior to gas sales, the material-balance gas production should include the water vapor that was a part of the original gas in place. However, the water produced as a result of condensation from the gas should not be included as WP, the water production from the reservoir. GASPROD also calculates the water content of the gas.

Also, any vent gas from condensate storage should be included in material balance production. So, in general,

GP = dry gas sold
+ gas equivalent of produced condensate
+ original water vapor in produced gas
+ vent gas from condensate storage

Note that the GASMBE program automatically adjusts GP and WP to reflect the gas equivalent of the condensate and the water content of the gas, respectively. It is not necessary to use GASPROD to adjust GP and WP prior to running GASMBE.

Equations

GE (gas equivalent of condensate in MCF/BBL)
= 2.99965 (1.03 − COND G')

GVCOND = GE CONDP

COND G' = COND G in SPGR

WGR (water-gas ratio in LBM/MMCF)
= $A(T/100)^{B}$ S.C.*
A = 3:4 + 60630/P
$B = 3.2147 + 2.657(10^{-4})P - 2.27(10^{-8})P^2$

S.C. (salinity correction) =
$1 - 4.893(10^{-3})\%NACL$
$- 1.757(10^{-4})\%NACL^2$

H2OGAS' = WGR' GP
H2OGAS' = H2OGAS in LBM
WGR' = WGR in LBM/MCF

*Unpublished correlation D.N. Meehan, Champlin Petroleum Company

Nomenclature

Symbol	*Variable Name*	*Input or Output*	*English Units*	*SI Units*
COND G	Condensate gravity	I	API	KG/M3
CONDP*	Condensate production	I	BBL	M3
GP*	Gas production	I	MCF	SCM
GV COND*	Gas volume of condensate	O	MCF	SCM
H2O GAS*	Water content of gas	O	BBL	M3

*The units for these variables are saved by the program.

Yes/No Questions

None

Example

A well produces 2100 MCF/DAY, 35 BBL/DAY of 56.2 API condensate, and 10 BBL/DAY of water. What is the gas equivalent of the stock-tank condensate? How much of the water is from dissolved gas in the reservoir? Additional data required are as follows:

Reservoir pressure = 1,820 PSI
Reservoir temperature = 165 F
Water salinity (est.) = 2.4 %NACL

Keystrokes (SIZE> =027)	**Display**	**Comments**
[XEQ] [ALPHA] GASPROD [ALPHA]	*COND GP*	
[A]	*COND G=?*	Condensate production known
56.2 [R/S]	*CONDP=?*	
35 [R/S]	*COND G=?*	GVCOND printed
[XEQ] [ALPHA] GASPROD [ALPHA]	*COND GP*	
[E]	*%NACL=?*	Gas production known
2.4 [R/S]	*T=?*	
165 [R/S]	*GP=?*	
2100 [R/S]	*P=?*	
1820 [R/S]	*GP=?*	H2OGAS printed

```
GAS PROD

COND G=56.2000 API
CONDP=35.0000 BBL

GVCOND=28.9910 MCF

%NACL=2.4000
T=165.0000 F
GP=2,100.0000 MCF
P=1,820.0000 PSI

H2OGAS=1.3340 BBL
```

General Information

Memory Requirements

Program length:	384 bytes (2 cards)
Minimum size:	027
Minimum hardware:	41C + 1 memory module

Hidden Options

None

Pac Subroutines Called

TITLE, %NACL, T, P, INU, INK, OUTK, CON

Registers

03	Condensate and water production units
04	Condensate and water production units
06	Gas production units
07	Gas production units
08	COND G (API)
09	GP (MCF)
26	CONDP (BBL)

Registers 10–15, 18, and 20–25 unused

Flags

None

User Instructions

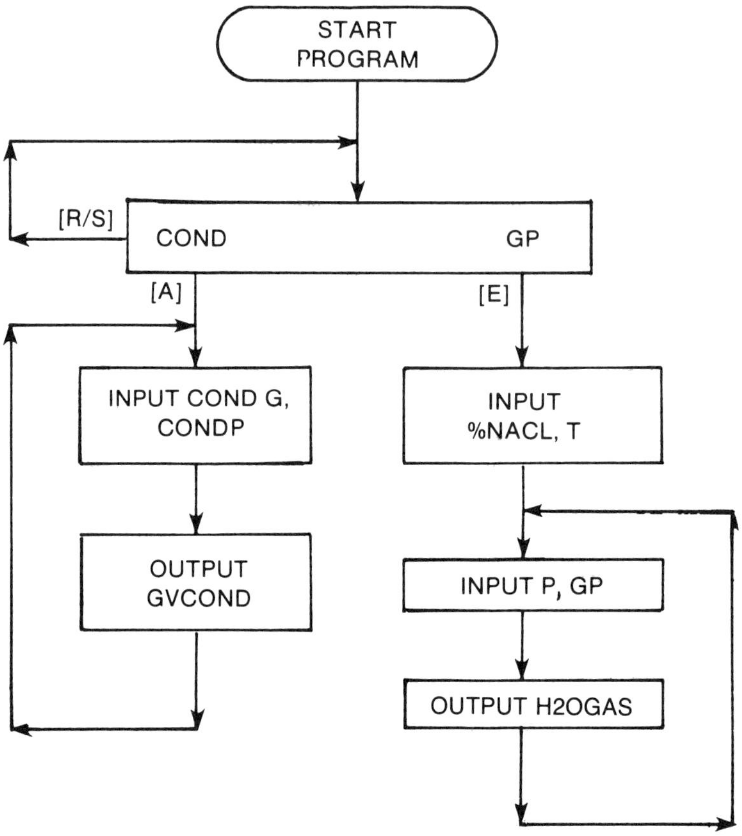

Program Listing

```
 01♦LBL "GASPROD"
"GAS PROD" 27
XROM "TITLE" FC?C 25
PROMPT SF 27 "M3"
ASTO 03 "SCM" ASTO 06
CLA ASTO 04 ASTO 07

 15♦LBL 00
"COND      GP" PROMPT
GTO 00

 19♦LBL E
XROM "%NACL" XROM "T"

 22♦LBL 01
8 "GP" XEQ 02
XROM "P" XEQ 08 FS? 08
ADV "H2OGAS" XEQ 06
CF 08 ADV GTO 01

 35♦LBL A
7 STO 00 "API"
ASTO 01 CLA ASTO 02
ASTO Z "KG/M3" ASTO Y
"COND G" XROM "INU" 25
"CONDP" XEQ 05 XEQ 10
FS? 08 ADV "GVCOND"
XEQ 03 CF 08 ADV
GTO A

 58♦LBL 02
STO 00 XEQ 04
XROM "INK" RDN STO 06
X<>Y STO 07 R↑ RTN

 68♦LBL 03
XEQ 04 XROM "OUTK" RDN
STO 06 X<>Y STO 07 R↑
RTN

 77♦LBL 04
ASTO T "MCF" ASTO 01
CLA ASTO 02 ARCL T
RCL 07 RCL 06 RCL Z
RTN

 88♦LBL 05
STO 00 XEQ 07
XROM "INK" RDN STO 03
X<>Y STO 04 R↑ RTN

 98♦LBL 06
XEQ 07 XROM "OUTK" RDN
STO 03 X<>Y STO 04 R↑
RTN

107♦LBL 07
ASTO T "BBL" ASTO 01
CLA ASTO 02 ARCL T
RCL 04 RCL 03 RCL Z
RTN

118♦LBL 08
RCL 16 1 % 2657 E-7
RCL 17 227 E-10 * -
RCL 17 * 3.2147 +
Y↑X 60630 RCL 17 /
3.4 + * RCL 09 *
1 E3 / "LBM/BBL-SPGR"
CON RCL 19 X≠0?
GTO 09 X<>Y RTN

149♦LBL 09
1757 E-7 * 4893 E-6 +
RCL 19 * 1 X<>Y - *
RTN

161♦LBL 10
1.03 RCL 08 "API-SPGR"
CON - 2.99965 *
RCL 26 * END
```

Section 2

Decline Curve Analysis

7. DECLINE — Exponential, Harmonic, Hyperbolic, and Multiple Well Decline Curve Analysis

Calculates exponential, harmonic, and hyperbolic decline curves for a wide variety of conditions and provides for single and multiple well forecasting.

7. DECLINE — Exponential, Harmonic, Hyperbolic, and Multiple Well Decline Curve Analysis

The extrapolation of plots of producing rate versus time or cumulative production on various scales is used widely to forecast future production and reserves. The extrapolation of producing rate plotted as a function of time on semilogarithmic paper is perhaps the most common method for predicting oil reserves after production has commenced. Other methods (e.g., material balance, volumetric calculations, water influx calculations) must be manipulated further to produce a rate versus time forecast. These methods may be more sound theoretically but by no means necessarily give better results. In fact, if the production predictions from these methods do not fit reasonably well with the decline curve, the error probably lies in the other methods.

The use of decline curves does have a better theoretical foundation than this may imply (see Fetkovich, Gentry). However, a few notes are in order. Oil material balances and rate-time forecasts almost always result in hyperbolic decline forecasts. Gas deliverability predictions are more frequently exponential than those for oil. When hyperbolic declines indicate a value of N (the hyperbolic exponent) greater than 1, consider the following as possible explanations.

a) *Fractured reservoir* — It is particularly common for tight, naturally fissured reservoirs to indicate values of N greater than 1.0. This is due to the transition in producing mechanism from the expansion of the oil in the fractures to the contribution of the matrix. These reservoirs are characterized by particularly steep declines that (on semilog paper) eventually exhibit extreme curvature and much shallower declines. Suggestions for analyzing these reservoirs include extrapolating the latter portion of the curves and the use of fissured-matrix, constant-pressure-type curves. These curves can also be developed from normalized dimensionless field data when a number of wells have produced for a relatively long time. These remarks also apply to certain other reservoirs when multiple producing mechanisms and/or layers are involved.

b) *Bad selection of data points* — Particularly when points are selected close together, the range of valid hyperbolic data may be surprisingly small when fitting three rate-time points or two rate-time points and the cumulative production. It is a particularly good idea to use smoothed data and perform sensitivities to the points selected, especially when data points are erratic and have not established a clear trend.

c) *Waterfloods* — The production past the peak producing rate of a waterflood will often appear hyperbolic with N greater than 1.0. It is better to forecast waterflood performance using another method.

The restriction of $0<N<1$ for hyperbolic decline curves may seem arbitrary to someone familiar with Arps' original work. Indeed, it is not difficult to calculate results for $N>1$. Nevertheless, this usually is not a good idea for reserve forecasting. The use of $N>1$ may fit the historical data well but will generally result in too optimistic a treatment for forecasts. This program handles hyperbolic, harmonic ($N = 1$), and exponential ($N = 0$) decline curves.

Another point to consider is the definition of decline rate. The decline equations use the *nominal decline rate* (D) for calculations. Two different relations exist for relating the nominal decline rate to an annual decline rate. The optional method is that used by Frick and in such commercial evaluation software as POGO*. In this method, the annual decline rate (%AI) is the decline slope that intersects the decline curve one year later. So for any curve $0<N<1$ with a value of %AI = 20 at T = 0 and initial rate Q = 100, the rate at time T = 1 year will be 80.

The default method is to interpret the annual decline rate to be equal to the instantaneous slope of the decline curve. This method is used by commercial evaluation programs such as OGRE**. In the case of exponential declines, these two methods are identical. However, a hyperbolic or harmonic decline curve will yield different results. When calculating decline slope from a curve or production, the calculated value of D will be the same regardless of the method selected.

This program always defaults to the second method, but either approach is valid. If you intend to use values calculated by DECLINE in commercial software packages, be sure to use the definition of decline rate appropriate for that package.

DECLINE presents a variety of methods to solve for the decline rate and/or hyperbolic exponent, forecast to a future rate, time, or cumulative production and generates forecasts with user-specified time steps. Multiple well forecasts are also provided for the sum of series of wells that start production at different times. Although a simple exponential decline program can be written for almost any programmable calculator and fairly elaborate programs are possible on the HP-67/97 and the TI-59, this

*A product of PSI Energy Software, Calgary

**A product of David P. Cook & Assoc., Dallas

HP-41 program using the Pac I/O routines is the most versatile and complete such program available. The addition of units for gas or oil result in a very versatile program. The modularized nature of the program makes it easy to understand and modify.

Equations

Default %AI definition (all decline types):

$$\%AI = 100(1 - e^{-D})$$

Exponential:

Optional %AI definition:

$$\%AI = 100(1 - e^{-D})$$

$$QO = \frac{QOI}{e^{DT}}$$

$$QG = \frac{QGI}{e^{DT}}$$

$$NP = \frac{365(QOI - QO)}{D}$$

$$GP = \frac{365(QGI - QG)}{D}$$

where D is the nominal decline rate

Harmonic:

Optional %AI definition:

$$\%AI = \frac{100D}{1 + D}$$

$$QO = \frac{QOI}{1 + DT}$$

$$QG = \frac{QGI}{1 + DT}$$

$$NP = \frac{365\ QOI}{D} \ln \left| \frac{QOI}{QO} \right|$$

$$GP = \frac{365\ QGI}{D} \ln \left| \frac{QGI}{QG} \right|$$

Hyperbolic:

Optional %AI definition:

$$\%AI = 100 \left[1 - \frac{1}{(1 + ND)^{1/N}} \right]$$

$$QO = \frac{QOI}{(1 + NDT)^{1/N}} \qquad QG = \frac{QGI}{(1 + NDT)^{1/N}}$$

$$NP = \frac{365\ QOI^N\ (QOI^{1-N} - QO^{1-N})}{D(1 - N)}$$

$$GP = \frac{365\ QGI^N\ (QGI^{1-N} - QG^{1-N})}{D(1 - N)}$$

$$D_{new} = D\ \frac{NEW\ QO}{QOI} \qquad D_{new} = D\ \frac{NEW\ QG}{QGI}$$

DECLINE uses different equations to solve for each unknown value, depending on which values are known. There are a total of eight equations to solve for %AI, QO or QG, NP or GP, T, and D, all of which can be derived from the %AI, QO or QG, and NP or GP equations given for each decline type. To solve for the hyperbolic exponent N, there are two cases (shown for oil only):

Hyperbolic case 1*:

$$f(N) = \frac{(QOI/QO1)^N - 1}{N\,T1} - \frac{(QOI/QO2)^N - 1}{N\,T2}$$

Hyperbolic case 2*:

$$f(N) = \frac{1 - (QOI/QO)^{N-1}}{(QOI/QO)^N - 1}\ \frac{N}{1 - N} - \frac{NP}{QOI\ T}$$

DECLINE attempts to minimize f(N). For the first three iterations, the half-interval method is used and the new N is predicted as:

$$N_{new} = (N_{left} + N_{right})/2$$

where N_{left} and N_{right} are the left and right boundaries of f(N). Initially, $N_{left} = 0$ (exponential decline) and $N_{right} = 1$ (harmonic decline).

Subsequent iterations use the regula falsi method, and the new N is predicted as:

$$N_{new} = \frac{N_{left}\ f(N_{right}) - N_{right}\ f(N_{left})}{f(N_{right}) - f(N_{left})}$$

Multiple Well:

At time T, $PROD_T = X\ NP\ NWELL$, $PROD_T = X\ GP\ NWELL$

*Hyperbolic case 1 has three rate-time points known. Hyperbolic case 2 has two rate-time points and the cumulative production known.

Total annual production PROD $= \sum_{w} PROD_{T}$,

where w is the number of wells that produced at time T.

$$PROD = \sum_{T=1}^{25} PROD$$

Nomenclature

Symbol	Variable Name	Input or Output	English Units	SI Units
BEG T*	Time after time zero at which production begins (default = 0 YR)	I	YR	YR
DGP*	Delta GP-change in cumulative gas production	O	MCF	SCM
DNP*	Delta NP-change in cumulative oil production	O	BBL	M3
DTI*	Initial delta T-size of first time step for single well forecast (default = 1 YR)	I	YR	YR
DTS*	Subsequent delta T-size of subsequent time steps for single well forecast (default = 1 YR)	I	YR	YR
GP*	Cumulative gas production	I,O	MCF	SCM
GPI*	Initial cumulative gas production for single well forecast (default = 0 MCF)	I,O	MCF	SCM
N	Hyperbolic decline exponent	I,O	-	-
NP*	Cumulative oil production	I,O	BBL	M3
NPI*	Initial cumulative oil production for single well forecast (default = 0 BBL)	I,O	BBL	M3
NEW QG*	New gas producing rate	I	MCF/ DAY	SCM/ DAY
NEW QO*	New oil producing rate	I	BBL/ DAY	M3/ DAY
NWELL	Number of wells that begin production at T (default = 1)	I	-	-
PROD*	Total production of all wells at T	O	BBL or MCF	M3 or SCM
QG*	Gas producing rate	I,O	MCF/ DAY	SCM/ DAY
QGI*	Initial gas producing rate	I	MCF/ DAY	SCM/ DAY
QG1*	Gas producing rate at T1	I	MCF/ DAY	SCM/ DAY
QG2*	Gas producing rate at T2	I	MCF/ DAY	SCM/ DAY
QO*	Oil producing rate	I,O	BBL/ DAY	M3/ DAY
QOI*	Initial oil producing rate	I	BBL/ DAY	M3/ DAY
QO1*	Oil producing rate at T1	I	BBL/ DAY	M3/ DAY
QO2*	Oil producing rate at T2	I	BBL/ DAY	M3/ DAY
T*	Time	I,O	YR	YR
TF*	Well life (default = T)	I	YR	YR
T1*	Time at first rate-time point	I	YR	YR
T2*	Time at second rate-time point	I	YR	YR
X	Multiplicative factor (default = 1)	I	-	-
XDGP*	X times DGP	O	MCF	SCM
XDNP*	X times DNP	O	BBL	M3
XPROD*	X times PROD	O	BBL or MCF	M3 or SCM
XQG*	X times QG	O	MCF/ DAY	SCM/ DAY
XQO*	X times QO	O	BBL/ DAY	M3/ DAY
%AI	Percent annual decline rate	I,O	-	-
ΣPROD*	Total production	O	BBL or MCF	M3 or SCM
ΣWELL	Total number of wells	O	-	-
ΣXPROD*	X times ΣPROD	O	BBL or MCF	M3 or SCM

*The units for these variables are saved by the program.

Yes/No Questions

CLEAR?	Yes: Clear production registers. No: Leave production registers unchanged.
OIL?	Yes: Oil reservoir No: Gas reservoir

Examples

Because of the wide variety of example problems, these examples will illustrate only a portion of the possible applications. It is recommended that you work through these examples carefully to get a feel for the different options available in DECLINE. You should also consult the user instructions to see all the options and how they relate to each other. Except where noted, the default method for calculating %AI from D is used.

Example 1

Figure 7–1 is a graph of the oil production for Lease A, which has cumulative production of 18,900 BBL. Forecast the production to an economic limit of 2 BBL/DAY, calculating annual gross and net production. The working interest is 100% with a net revenue interest of 80%. The initial rate of the harmonic decline portion is 115 BBL/DAY. The producing rate one year later is 25 BBL/DAY. Since the well has produced for six months in 1978, use an

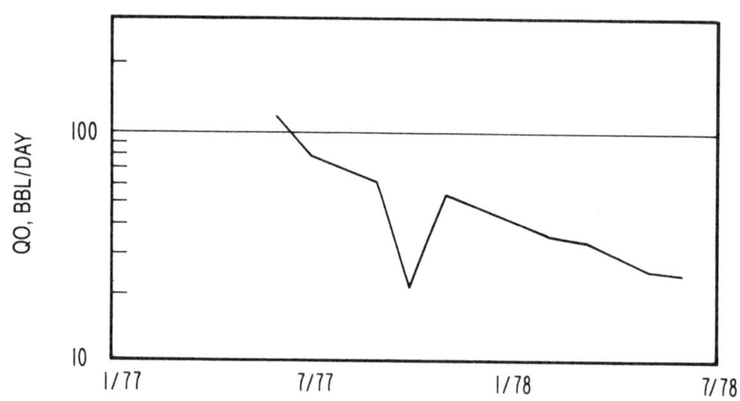

Figure 7–1

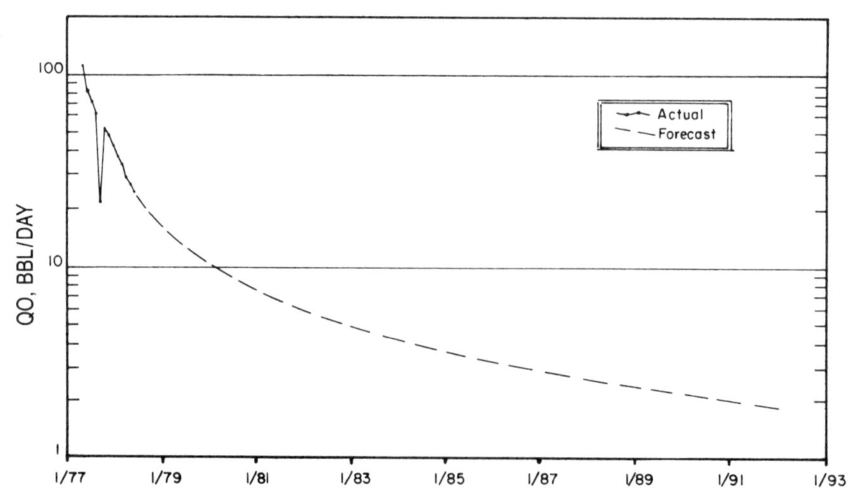

Figure 7–2 Harmonic decline forecast

initial time step size of 6 months with annual time steps thereafter. Figure 7-2 presents the historical and predicted production results.

Keystrokes (SIZE> =049)	Display	Comments
[XEQ] [ALPHA] DECLINE [ALPHA]	*OIL? Y/N*	Last character is Y or N
Y [R/S]	*EXP HAR HYP*	Select decline type
[C]	*QOI=?*	Harmonic decline option
115 [R/S]	*AI Q*	
[C]	*QO=?*	Q known
25 [R/S]	*T NP*	
[D]	*T=?*	T known
1 [R/S]	*QI AI Q T NP*	%AI and NP printed
[C]	*QO=?*	Forecast to future rate
2 [R/S]	*QI AI Q T NP*	T and NP printed
[//] [A]	*NEW QO=?*	Determine annual decline rate at QO = 25 BBL/DAY
25 [R/S]	*QI AI Q T NP*	%AI printed
[A]	*QOI=?*	
25 [R/S]	*AI Q*	
[B]	*%AI=?*	
54.2788 [R/S]	*Q T NP*	
[C]	*QO=?*	
[R/S]	*QI AI Q T NP*	Remaining T and NP printed
[H]	*DTI=?*	
6 [ALPHA] MO [R/S]	*DTS=?*	
1 [ALPHA] YR [R/S]	*TF=?*	Final time defaults to T
[R/S]	*NPI=?*	
18900 [R/S]	*X=?*	Multiplicative factor is net interest
.8 [R/S]		Forecast printed

```
DECLINE

OIL: YES

HARMONIC

QOI=115.0000 BBL/DAY
QO=25.0000 BBL/DAY

T=1.0000 YR
%AI=97.2676
NP=17,793.3926 BBL

QO=2.0000 BBL/DAY
T=15.6944 YR
NP=47,242.6870 BBL

NEW QO=25.0000 BBL/DAY
%AI=54.2788

QOI=25.0000 BBL/DAY
%AI=54.2788
T=14.6945 YR
NP=29,449.3168 BBL


SINGLE WELL

DTI=6.0000 MO
DTS=1.0000 YR
NPI=18,900.0000 BBL
X=0.8000

QOI=25.0000 BBL/DAY
NPI=18,900.0000 BBL

T=0.5000 YR
QO=17.9688 BBL/DAY
XQO=14.3750 BBL/DAY
NP=22,750.5268 BBL
DNP=22,750.5268 BBL
XDNP=18,200.4214 BBL

T=1.5000 YR
QO=11.5000 BBL/DAY
XQO=9.2000 BBL/DAY
NP=27,954.1121 BBL
DNP=5,203.5853 BBL
XDNP=4,162.8682 BBL
```

```
T=2.5000 YR
QO=8.4559 BBL/DAY
XQO=6.7647 BBL/DAY
NP=31,539.2999 BBL
DNP=3,585.1878 BBL
XDNP=2,868.1503 BBL

T=3.5000 YR
QO=6.6861 BBL/DAY
XQO=5.3488 BBL/DAY
NP=34,277.4658 BBL
DNP=2,738.1659 BBL
XDNP=2,190.5327 BBL

T=4.5000 YR
QO=5.5288 BBL/DAY
XQO=4.4231 BBL/DAY
NP=36,493.3227 BBL
DNP=2,215.8569 BBL
XDNP=1,772.6855 BBL

T=5.5000 YR
QO=4.7131 BBL/DAY
XQO=3.7705 BBL/DAY
NP=38,354.5669 BBL
DNP=1,861.2443 BBL
XDNP=1,488.9954 BBL

T=6.5000 YR
QO=4.1071 BBL/DAY
XQO=3.2857 BBL/DAY
NP=39,959.1950 BBL
DNP=1,604.6280 BBL
XDNP=1,283.7024 BBL

T=7.5000 YR
QO=3.6392 BBL/DAY
XQO=2.9114 BBL/DAY
NP=41,369.4697 BBL
DNP=1,410.2748 BBL
XDNP=1,128.2198 BBL

T=8.5000 YR
QO=3.2670 BBL/DAY
XQO=2.6136 BBL/DAY
NP=42,627.4259 BBL
DNP=1,257.9562 BBL
XDNP=1,006.3649 BBL

T=9.5000 YR
QO=2.9639 BBL/DAY
XQO=2.3711 BBL/DAY
NP=43,762.7824 BBL
DNP=1,135.3565 BBL
XDNP=908.2852 BBL

T=10.5000 YR
QO=2.7123 BBL/DAY
XQO=2.1698 BBL/DAY
NP=44,797.3282 BBL
DNP=1,034.5459 BBL
XDNP=827.6367 BBL

T=11.5000 YR
QO=2.5000 BBL/DAY
XQO=2.0000 BBL/DAY
NP=45,747.5150 BBL
DNP=950.1868 BBL
XDNP=760.1494 BBL

T=12.5000 YR
QO=2.3185 BBL/DAY
XQO=1.8548 BBL/DAY
NP=46,626.0691 BBL
DNP=878.5541 BBL
XDNP=702.8433 BBL

T=13.5000 YR
QO=2.1617 BBL/DAY
XQO=1.7293 BBL/DAY
NP=47,443.0380 BBL
DNP=816.9689 BBL
XDNP=653.5751 BBL

T=14.5000 YR
QO=2.0246 BBL/DAY
XQO=1.6197 BBL/DAY
NP=48,206.4930 BBL
DNP=763.4550 BBL
XDNP=610.7640 BBL

T=14.6945 YR
QO=2.0000 BBL/DAY
XQO=1.6000 BBL/DAY
NP=48,349.3168 BBL
DNP=142.8238 BBL
XDNP=114.2590 BBL
```

Example 2

This exponential decline forecast illustrates the variety of methods available in DECLINE. The initial oil production rate is 90 BBL/DAY with a 20% annual decline rate. What will the production rate and cumulative production be in 10 years? How long will it take to reach the economic limit of 2 BBL/DAY, and what will be the cumulative production at that time? When will the well have produced 100,000 BBL?

Keystrokes	Display	Comments
[/-/] [E]	*EXP HAR HYP*	Change decline type
[A]	*QOI=?*	Exponential decline option
90 [R/S]	*AI Q*	
[B]	*%AI=?*	AI known
20 [R/S]	*Q T NP*	
[D]	*T=?*	
10 [R/S]	*QI AI Q T NP*	QO and NP printed
[C]	*QO=?*	Forecast to future rate
2 [R/S]	*QI AI Q T NP*	T and NP printed
[E]	*NP=?*	Forecast to future cumulative production
100000 [R/S]	*QI AI Q T NP*	QO and T printed

```
EXPONENTIAL

QOI=90.0000 BBL/DAY
%AI=20.0000
T=10.0000 YR
QO=9.6637 BBL/DAY
NP=131,407.5981 BBL

QO=2.0000 BBL/DAY
T=17.0593 YR
NP=143,943.2142 BBL

NP=100,000.0000 BBL
QO=28.8648 BBL/DAY
T=5.0962 YR
```

Example 3

A certain oil well has calculated volumetric reserves of 60,000 M3 (cubic meters). The initial producing rate is 800 M3/MO. The economic limit should be approximately 30 M3/MO. Assuming an exponential decline, forecast the gross production for this well with annual time steps. (Run this example without using the SI mode allowed by the Unit Management System of the Pac.)

Keystrokes	Display	Comments
[XEQ] [ALPHA] DECLINE [ALPHA]	*OIL? Y/N:Y*	
[R/S]	*EXP HAR HYP*	
[A]	*QOI=?*	
800 [ALPHA] M3/MO [R/S]	*AI Q*	
[C]	*QO=?*	
30 [R/S]	*T NP*	
[E]	*NP=?*	
60000 [ALPHA] M3 [R/S]	*QI AI Q T NP*	%AI and T printed
[H]	*DTI=?*	Defaults to 1 YR
[R/S]	*DTS=?*	Defaults to 1 YR
[R/S]	*TF=?*	Defaults to T
[R/S]	*NPI=?*	Defaults to zero
[R/S]	*X=?*	Defaults to one
[R/S]	*QI AI Q T NP*	Forecast printed

```
DECLINE

EXPONENTIAL

QOI=800.0000 M3/MO
QO=30.0000 M3/MO
NP=60,000.0000 M3
%AI=14.2728
T=21.3209 YR

SINGLE WELL

QOI=800.0000 M3/MO
```

```
T=1.0000 YR
QO=685.8176 M3/MO
NP=8,897.3286 M3
DNP=8,897.3286 M3

T=2.0000 YR
QO=587.9323 M3/MO
NP=16,524.7594 M3
DNP=7,627.4308 M3

T=3.0000 YR
QO=504.0179 M3/MO
NP=23,063.5425 M3
DNP=6,538.7831 M3

T=4.0000 YR
QO=432.0804 M3/MO
NP=28,669.0582 M3
DNP=5,605.5157 M3

T=5.0000 YR
QO=370.4105 M3/MO
NP=33,474.5100 M3
DNP=4,805.4518 M3

T=6.0000 YR
QO=317.5425 M3/MO
NP=37,594.0894 M3
DNP=4,119.5794 M3

T=7.0000 YR
QO=272.2203 M3/MO
NP=41,125.6896 M3
DNP=3,531.6002 M3

T=8.0000 YR
QO=233.3669 M3/MO
NP=44,153.2316 M3
DNP=3,027.5420 M3

T=9.0000 YR
QO=200.0589 M3/MO
NP=46,748.6586 M3
DNP=2,595.4270 M3

T=10.0000 YR
QO=171.5049 M3/MO
NP=48,973.6457 M3
DNP=2,224.9870 M3

T=11.0000 YR
QO=147.0263 M3/MO
NP=50,881.0647 M3
DNP=1,907.4191 M3

T=12.0000 YR
QO=126.0416 M3/MO
NP=52,516.2418 M3
DNP=1,635.1770 M3

T=13.0000 YR
QO=108.0519 M3/MO
NP=53,918.0333 M3
DNP=1,401.7915 M3

T=14.0000 YR
QO=92.6299 M3/MO
NP=55,119.7499 M3
DNP=1,201.7166 M3

T=15.0000 YR
QO=79.4090 M3/MO
NP=56,149.9480 M3
DNP=1,030.1981 M3

T=16.0000 YR
QO=68.0751 M3/MO
NP=57,033.1079 M3
DNP=883.1600 M3

T=17.0000 YR
QO=58.3589 M3/MO
NP=57,790.2163 M3
DNP=757.1083 M3

T=18.0000 YR
QO=50.0294 M3/MO
NP=58,439.2641 M3
DNP=649.0478 M3

T=19.0000 YR
QO=42.8888 M3/MO
NP=58,995.6746 M3
DNP=556.4105 M3

T=20.0000 YR
QO=36.7674 M3/MO
NP=59,472.6698 M3
DNP=476.9952 M3

T=21.0000 YR
QO=31.5197 M3/MO
NP=59,881.5843 M3
DNP=408.9146 M3

T=21.3209 YR
QO=30.0000 M3/MO
NP=60,000.0000 M3
DNP=118.4157 M3
```

Example 4

Lease B's historical production is plotted in Figure 7–3. Cumulative production from June 1979 to July 1979 is 171,400 BBL, although the hyperbolic decline portion begins around January 1978. The initial rate is 330 BBL/DAY for the hyperbolic decline portion, declining to 240 BBL/DAY. What is the hyperbolic exponent N? What are the reserves to an economic limit of 2 BBL/DAY? What is the decline slope %AI at the initial point (QOI = 330 BBL/DAY) and currently (QO = 135 BBL/DAY)? What is the calculated value of QO at T = 1 YR? The remaining reserves are forecast in Figure 7–4.

Keystrokes	Display	Comments
[XEQ] [ALPHA] DECLINE [ALPHA]	*OIL? Y/N:Y*	
[R/S]	*EXP HAR HYP*	
[E]	*QOI=?*	Hyperbolic decline option
330 [R/S]	*N Q*	
[C]	*QO1=?*	
240 [R/S]	*T1=?*	
6 [ALPHA] MO [R/S]	*N Q NP*	
[C]	*QO2=?*	
135 [R/S]	*T2=?*	
1.5 [ALPHA] YR [R/S]	*QI N Q T NP*	N and %AI printed
[C]	*QO=?*	
2 [R/S]	*QI N Q T NP*	T and NP printed
[//] [A]	*NEW QO=?*	
135 [R/S]	*QI N Q T NP*	%AI printed
[D]	*T=?*	
1 [R/S]	*QI N Q T NP*	QO and NP printed

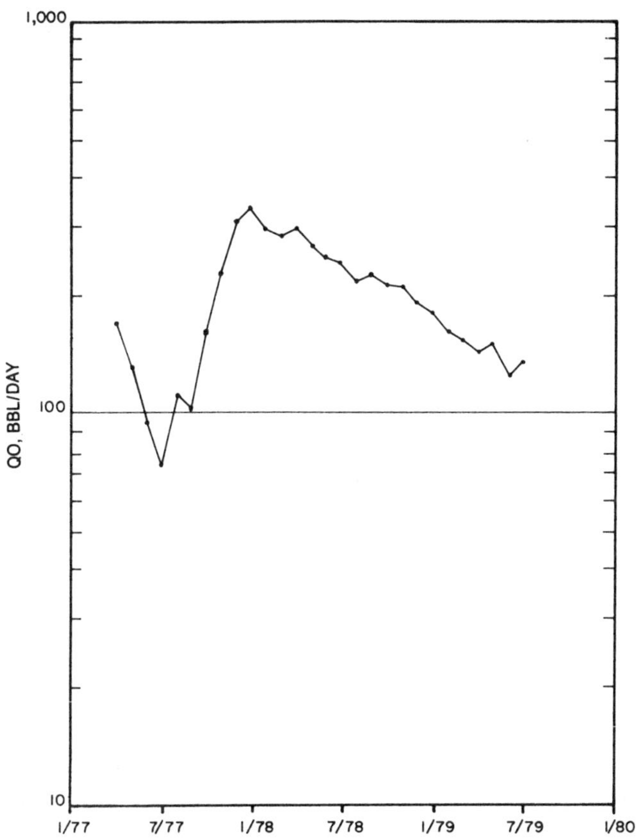

Figure 7–3 Single well production

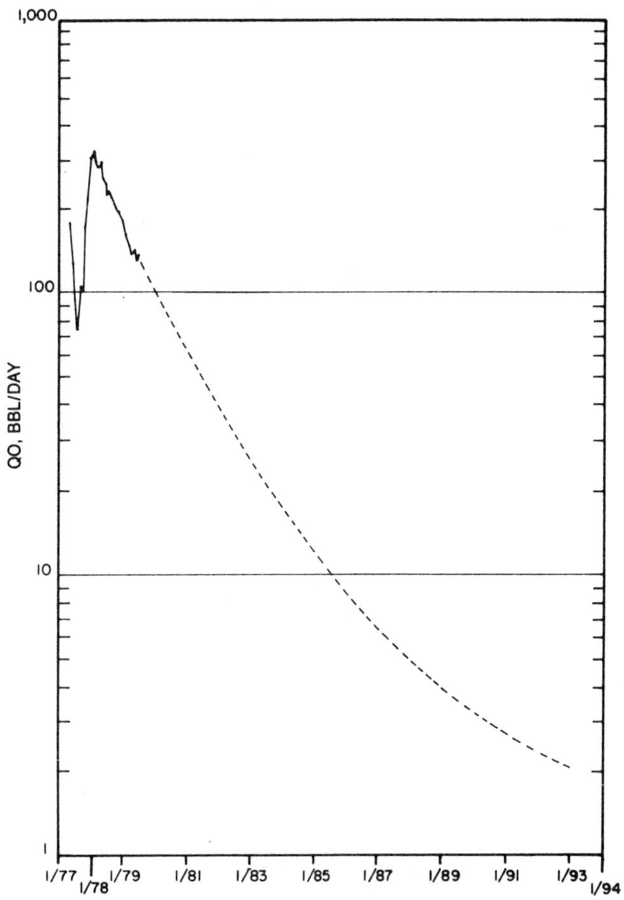

Figure 7–4 Production forecast

```
DECLINE

HYPERBOLIC

QOI=330.0000 BBL/DAY
QO1=240.0000 BBL/DAY
T1=6.0000 MO
QO2=135.0000 BBL/DAY
T2=1.5000 YR
N=0.2263
%AI=48.3364

QO=2.0000 BBL/DAY
T=14.5559 YR
NP=231,188.7966 BBL

NEW QO=135.0000 BBL/DAY
%AI=41.6952

T=1.0000 YR
QO=178.3234 BBL/DAY
NP=89,309.7955 BBL
```

Example 5

Solve Example 4 using the optional definition of %AI.

Keystrokes	Display	Comments
[XEQ] [ALPHA] DECLINE [ALPHA]	*OIL? Y/N:Y*	
[R/S]	*EXP HAR HYP*	
[/] [SF] 00 [E]	*QOI=?*	Select optional %AI definition
[R/S]	*N Q*	
[C]	*QO1=?*	
240 [R/S]	*T1=?*	
6 [ALPHA] MO [R/S]	*N Q NP*	
[C]	*QO2=?*	
135 [R/S]	*T2=?*	

Keystrokes	Display	Comments
1.5 [ALPHA] YR [R/S]	*QI N Q T NP*	N and %AI printed
[C]	*QO=?*	
2 [R/S]	*QI N Q T NP*	T and NP printed
[/] [A]	*NEW QO=?*	
135 [R/S]	*QI N Q T NP*	%AI printed
[D]	*T=?*	
1 [R/S]	*QI N Q T NP*	QO and NP printed

```
DECLINE

HYPERBOLIC

QO1=240.0000 BBL/DAY
T1=6.0000 MO
QO2=135.0000 BBL/DAY
T2=1.5000 YR
N=0.2263
%AI=45.9626

QO=2.0000 BBL/DAY
T=14.5559 YR
NP=231,188.7966 BBL

NEW QO=135.0000 BBL/DA\
%AI=39.8912

T=1.0000 YR
QO=178.3234 BBL/DAY
NP=89,309.7955 BBL
```

Example 6

Using the default method for calculating %AI, evaluate remaining reserves as of July 1, 1979 (T = 1.5 YR) for the data in Example 4. Input the values for %AI and N calculated in Example 4, predicting reserves to the economic limit of 2 BBL/DAY.

Keystrokes	Display	Comments
[XEQ] [ALPHA] DECLINE [ALPHA]	*OIL? Y/N:Y*	
[R/S]	*EXP HAR HYP*	
[E]	*QOI=?*	
135 [R/S]	*N Q*	
[B]	*N=?*	
[R/S]	*%AI=?*	
41.6952 [R/S]	*QI N Q T NP*	

Keystrokes	Display	Comments
[C]	*QO=?*	
2 [R/S]	*QI N Q T NP*	T and NP printed

```
DECLINE

HYPERBOLIC

QOI=135.0000 BBL/DAY
%AI=41.6952

QO=2.0000 BBL/DAY
T=13.0559 YR
NP=113,513.0787 BBL
```

Example 7

Gas well C has an initial producing rate of 2.1 MMCF/DAY. Based on similar well performance, it should have a hyperbolic decline with N = 0.35. Calculate the ultimate reserves to an economic limit of 100 MCF/DAY in 20 years. Determine the initial decline slope, reserves, and annual production for the first three years. Notice how the prompt names and units change to gas rates and productions.

Keystrokes	Display	Comments
[XEQ] [ALPHA] DECLINE [ALPHA]	*OIL? Y/N:Y*	
N [R/S]	*EXP HAR HYP*	
[E]	*QGI=?*	
2.1 [ALPHA] MMCF/DAY [R/S]	*N Q*	
[C]	*QG1=?*	
100 [ALPHA] MCF/DAY [R/S]	*T1=?*	
20 [R/S]	*N Q GP*	
[B]	*N=?*	
.35 [R/S]	*QI N Q T GP*	%AI printed
[H]	*DTI=?*	
[R/S]	*DTS=?*	
[R/S]	*TF=?*	
3 [R/S]	*GPI=?*	
[R/S]	*X=?*	
[R/S]	*QI N Q T GP*	Forecast printed

```
DECLINE

OIL: NO

HYPERBOLIC

QGI=2.1000 MMCF/DAY
QG1=100.0000 MCF/DAY
T1=20.0000 YR
N=0.3500
%AI=23.7986


SINGLE WELL

TF=3.0000 YR

QGI=2,100.0000 MCF/DAY

T=1.0000 YR
QG=1,619.8104 MCF/DAY
GP=673,752.5395 MCF
DGP=673,752.5395 MCF

T=2.0000 YR
QG=1,276.7534 MCF/DAY
GP=1,199,026.482 MCF
DGP=525,273.9425 MCF

T=3.0000 YR
QG=1,024.9462 MCF/DAY
GP=1,616,817.251 MCF
DGP=417,790.7690 MCF
```

Example 8

A certain well has an initial producing rate of 30 M3/DAY. If this well were to have an economic life of 20 years to reach 1 M3/DAY, what hyperbolic decline exponent and initial decline rate would be required? Use the SI option of the Unit Management system by setting flag 09 ([//] [SF] 09).

Keystrokes	Display	Comments
[XEQ] [ALPHA] DECLINE [ALPHA]	*OIL? Y/N:N*	
Y [R/S]	*EXP HAR HYP*	
[//] [SF] 09 [E]	*QOI=?*	Select the SI option
30 [R/S]	*N Q*	
[C]	*QO1=?*	
1 [R/S]	*T1=?*	
20 [R/S]	*N Q NP*	
[E]	*NP=?*	
60000 [R/S]	*QI N Q T NP*	N and %AI printed
[//] [CF] 09	*16.7716*	Back to English units

```
DECLINE

OIL: YES

HYPERBOLIC

QOI=30.0000 M3/DAY
QO1=1.0000 M3/DAY
T1=20.0000 YR
NP=60,000.0000 M3
N=0.0444
%AI=16.7716
```

Example 9: Multiple Well Forecast

Predict the total production for a certain field with two oil producing zones. Anticipated typical well performance is shown in Table 7–1, and the development schedule and a set of risk factors is shown in Table 7–2. Net interest is 82%. Predict the total gross and net production for zone A and the combined zones A and B.

Table 7–1

Zone	*A*	*B*
QOI (BBL/DAY)	100	90
Decline type	Exponential	Hyperbolic
%AI	30	65
Ending condition	QO = 4 BBL/DAY	T = 11 YR
Hyperbolic exponent	–	0.4501

Table 7–2

		Zone A		*Zone B*	
Date	*Time (YR)*	*Wells*	*Factor*	*Wells*	*Factor*
1-1-81	0.0	6	1.0	2	1.0
4-1-81	0.25	1	1.0	–	–
7-1-81	0.5	1	0.8	–	–
10-1-81	0.75	1	0.6	–	–
1-1-82	1.0	1	0.4	2	0.75
7-1-82	1.5	–	–	1	0.5
1-1-83	2.0	–	–	1	0.5

Keystrokes	Display	Comments
[XEQ] [ALPHA] DECLINE [ALPHA]	*OIL? Y/N:Y*	
[R/S]	*EXP HAR HYP*	
[A]	*QOI=?*	
100 [R/S]	*AI Q*	
[B]	*%AI=?*	
30 [R/S]	*Q T NP*	
[C]	*QO=?*	
4 [R/S]	*QI AI Q T NP*	T and NP printed
[I]	*CLEAR? Y/N:*	Multiple well input. Last character is Y or N
Y [R/S]	*BEG T=?*	
0 [R/S]	*NWELL=?*	
6 [R/S]	*X=?*	
1 [R/S]	*QI AI Q T NP*	
[I]	*CLEAR? Y/N:Y*	
N [R/S]	*BEG T=?*	
.25 [R/S]	*NWELL=?*	
1 [R/S]	*X=?*	
[R/S]	*QI AI Q T NP*	
[I]	*CLEAR? Y/N:N*	
[R/S]	*BEG T=?*	
.5 [R/S]	*NWELL=?*	
[R/S]	*X=?*	
.8 [R/S]	*QI AI Q T NP*	
[I]	*CLEAR? Y/N:N*	
[R/S]	*BEG T=?*	
.75 [R/S]	*NWELL=?*	
[R/S]	*X=?*	
.6 [R/S]	*QI AI Q T NP*	
[I]	*CLEAR? Y/N:N*	
[R/S]	*BEG T=?*	
1 [R/S]	*NWELL=?*	
[R/S]	*X=?*	
.4 [R/S]	*QI AI Q T NP*	
[J]	*X=?*	Multiple well output
.82 [R/S]	*QI AI Q T NP*	Annual gross and net production for zone A printed
[//] [E]	*EXP HAR HYP*	Select new decline type
[E]	*QOI=?*	
90 [R/S]	*N Q*	
[B]	*N=?*	
.4501 [R/S]	*%AI=?*	
65 [R/S]	*QI N Q T NP*	
[I]	*CLEAR? Y/N:N*	
[R/S]	*BEG T=?*	
0 [R/S]	*NWELL=?*	
2 [R/S]	*X=?*	
1 [R/S]	*QI N Q T NP*	
[I]	*CLEAR? Y/N:N*	
[R/S]	*BEG T=?*	
1 [R/S]	*NWELL=?*	
[R/S]	*X=?*	
.75 [R/S]	*QI N Q T NP*	
[I]	*CLEAR? Y/N:N*	
[R/S]	*BEG T=?*	
1.5 [R/S]	*NWELL=?*	
1 [R/S]	*X=?*	
.5 [R/S]	*QI N Q T NP*	
[I]	*CLEAR? Y/N:N*	
[R/S]	*BEG T=?*	
2 [R/S]	*NWELL=?*	
[R/S]	*X=?*	
.5 [R/S]	*QI N Q T NP*	
[J]	*X=?*	
.82 [R/S]	*QI N Q T NP*	Annual gross and net production for zones A and B printed

```
DECLINE

EXPONENTIAL

QOI=100.0000 BBL/DAY
%AI=30.0000
QO=4.0000 BBL/DAY
T=9.0247 YR
NP=98,240.7106 BBL

CLEAR: YES
BEG T=0.0000 YR
NWELL=6.0000
X=1.0000

CLEAR: NO
BEG T=0.2500 YR
NWELL=1.0000

BEG T=0.5000 YR
X=0.8000

BEG T=0.7500 YR
X=0.6000

BEG T=1.0000 YR
X=0.4000
```

```
MULTIWELL

X=0.8200

T=1.0000 YR
PROD=226,830.8082 BBL
XPROD=186,001.2628 BBL

T=2.0000 YR
PROD=202,112.7118 BBL
XPROD=165,732.4237 BBL

T=3.0000 YR
PROD=141,478.8983 BBL
XPROD=116,012.6966 BBL

T=4.0000 YR
PROD=99,035.2287 BBL
XPROD=81,208.8876 BBL

T=5.0000 YR
PROD=69,324.6601 BBL
XPROD=56,846.2212 BBL

T=6.0000 YR
PROD=48,527.2620 BBL
XPROD=39,792.3549 BBL

T=7.0000 YR
PROD=33,969.0835 BBL
XPROD=27,854.6484 BBL

T=8.0000 YR
PROD=23,778.3584 BBL
XPROD=19,498.2539 BBL

T=9.0000 YR
PROD=16,644.8509 BBL
XPROD=13,648.7777 BBL

T=10.0000 YR
PROD=2,801.9170 BBL
XPROD=2,297.5719 BBL

T=11.0000 YR
PROD=14.4744 BBL
XPROD=11.8690 BBL

ΣWELL=10.0000
ΣPROD=864,518.2533 BBL
ΣXPROD=708,904.9677 BBL

HYPERBOLIC

QOI=90.0000 BBL/DAY
N=0.4501
%AI=65.0000

BEG T=0.0000 YR
NWELL=2.0000
X=1.0000

BEG T=1.0000 YR
X=0.7500

BEG T=1.5000 YR
NWELL=1.0000
X=0.5000

BEG T=2.0000 YR
X=0.5000

MULTIWELL

X=0.8200

T=1.0000 YR
PROD=269,705.5353 BBL
XPROD=221,158.5389 BBL

T=2.0000 YR
PROD=254,648.0081 BBL
XPROD=208,811.3666 BBL

T=3.0000 YR
PROD=174,468.6118 BBL
XPROD=143,064.2617 BBL

T=4.0000 YR
PROD=112,417.5573 BBL
XPROD=92,182.3970 BBL

T=5.0000 YR
PROD=76,082.0908 BBL
XPROD=62,387.3144 BBL

T=6.0000 YR
PROD=52,407.4929 BBL
XPROD=42,974.1442 BBL

T=7.0000 YR
PROD=36,397.1248 BBL
XPROD=29,845.6423 BBL

T=8.0000 YR
PROD=25,394.9067 BBL
XPROD=20,823.8235 BBL

T=9.0000 YR
PROD=17,773.0091 BBL
XPROD=14,573.8674 BBL

T=10.0000 YR
PROD=3,341.4692 BBL
XPROD=2,740.0048 BBL

T=11.0000 YR
PROD=161.2668 BBL
XPROD=132.2388 BBL

T=12.0000 YR
PROD=1.9312 BBL
XPROD=1.5836 BBL

ΣWELL=16.0000
ΣPROD=1,022,799.004 BBL
ΣXPROD=838,695.1833 BBL
```

User Instructions

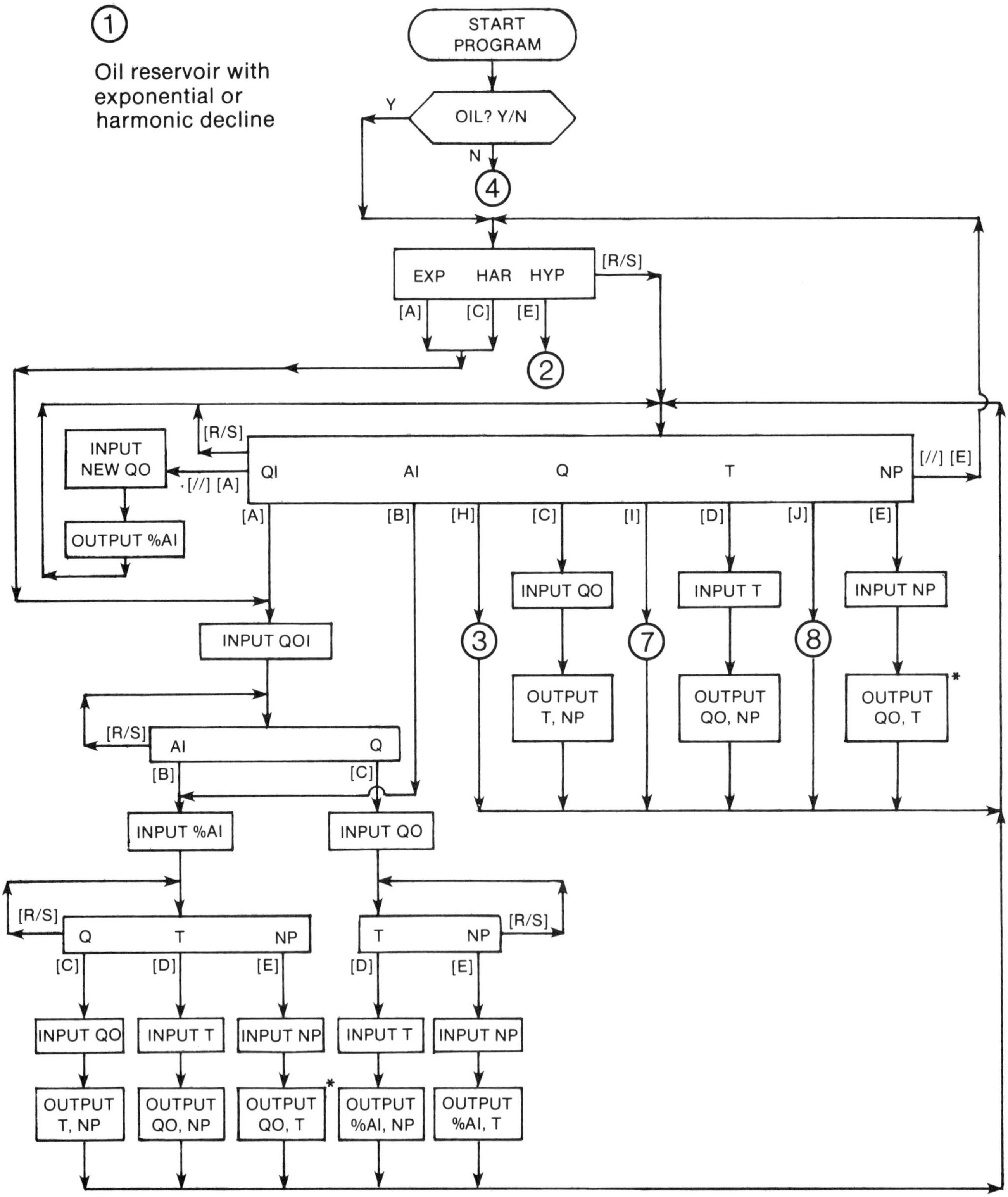

User Instructions (cont.)

(2)

Oil reservoir with hyperbolic decline

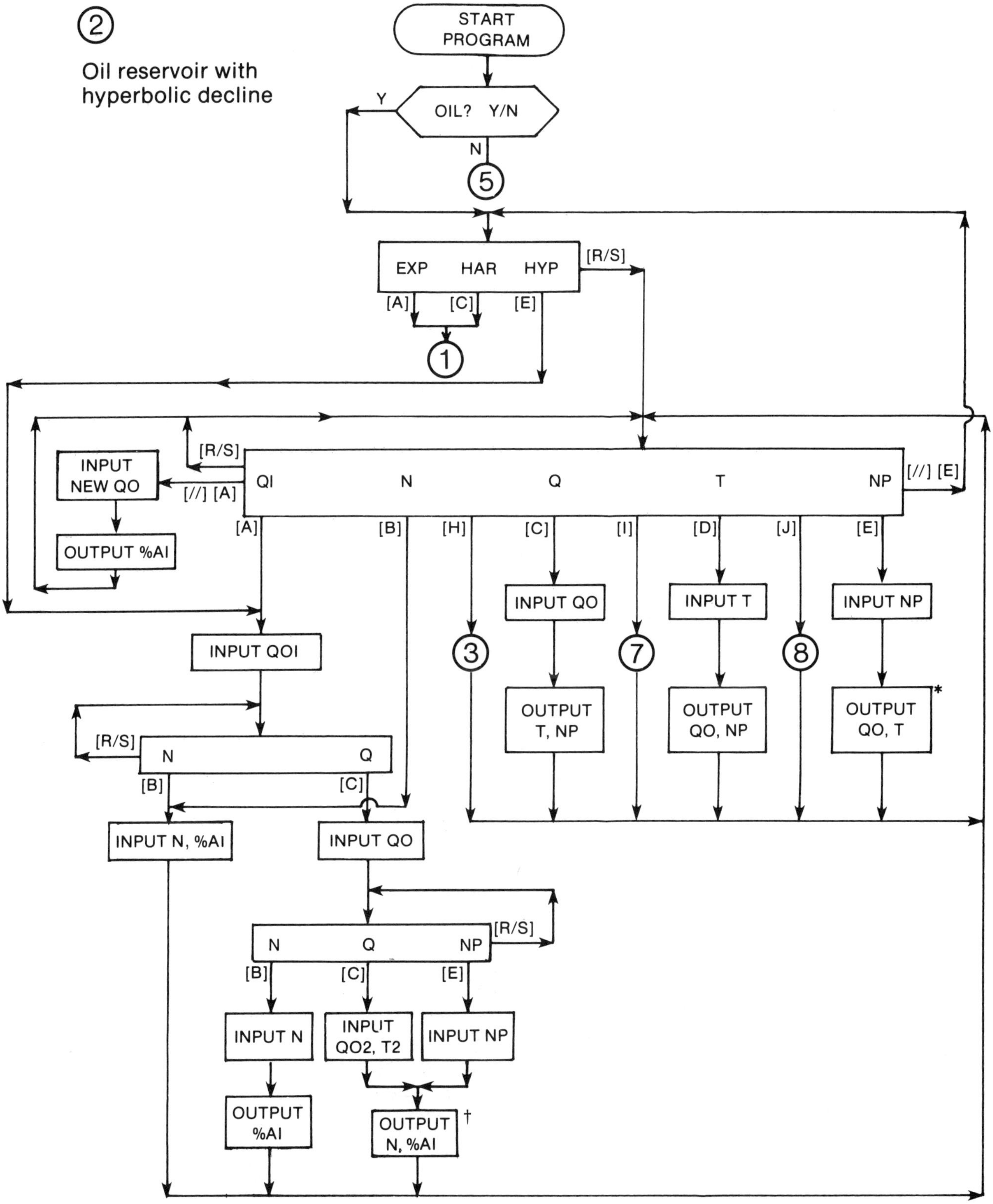

User Instructions (cont.)

③

Oil reservoir single well forecast

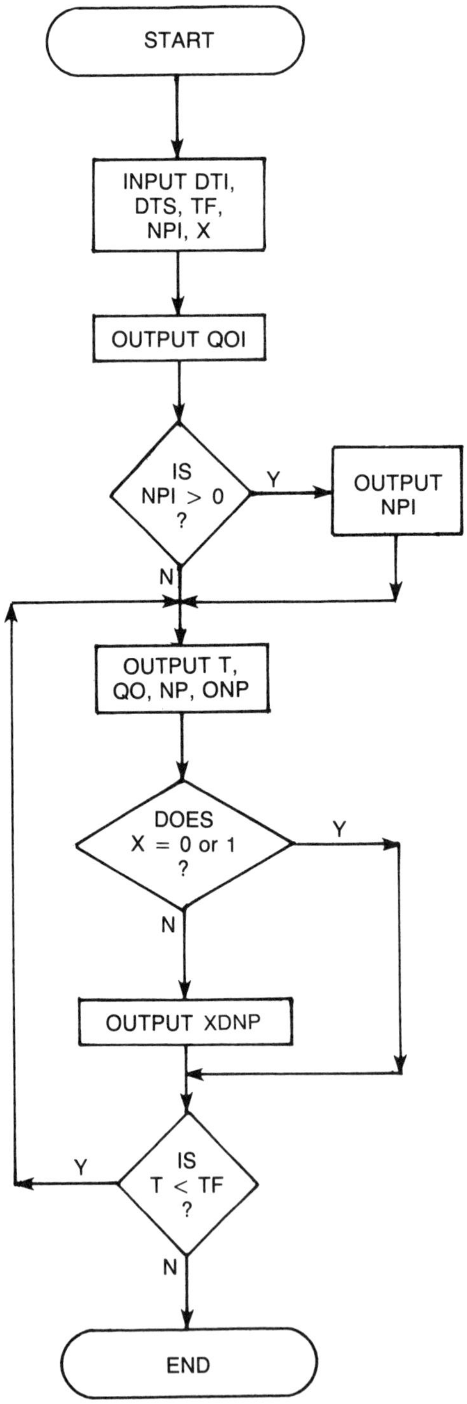

User Instructions (cont.)

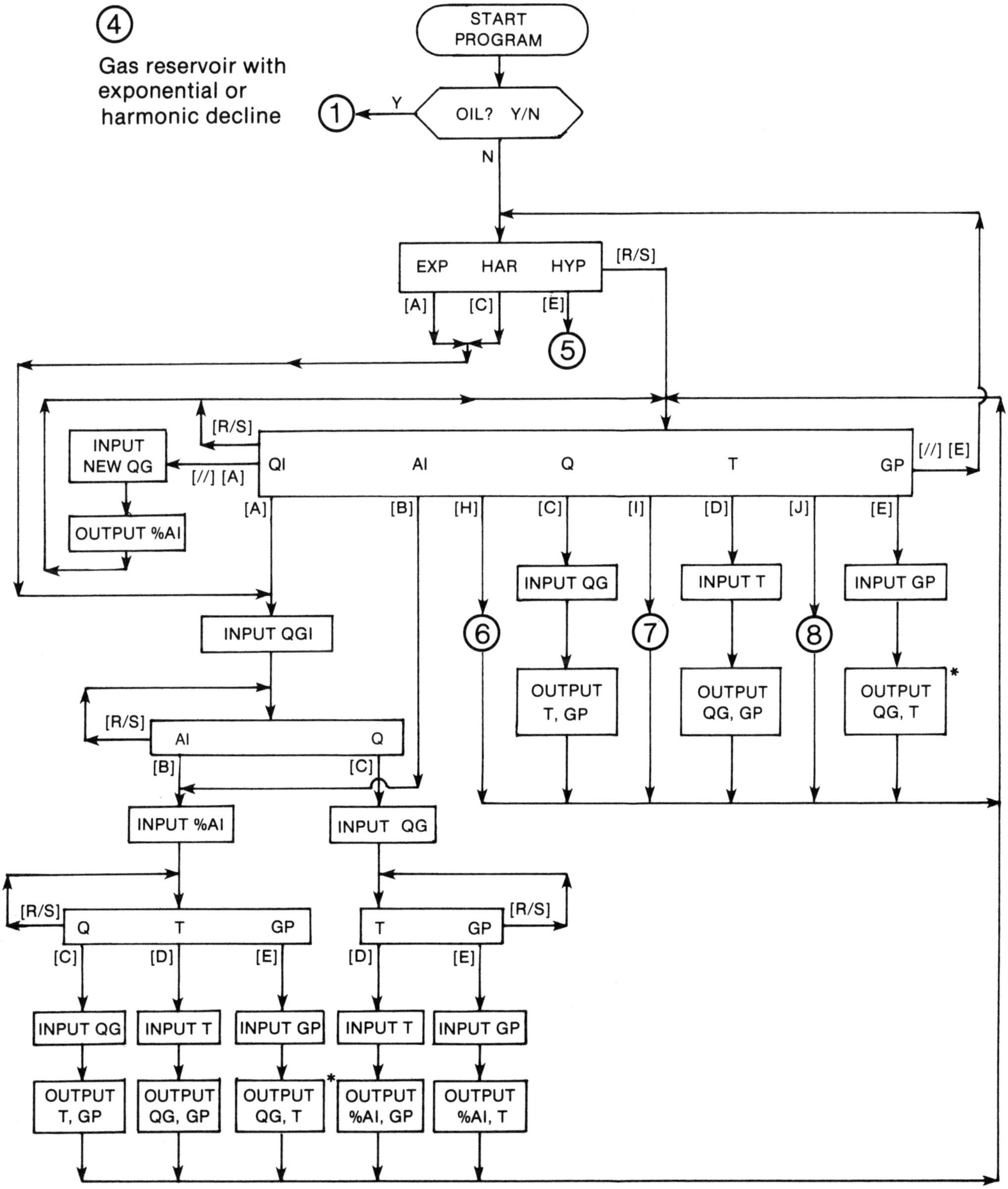

User Instructions (cont.)

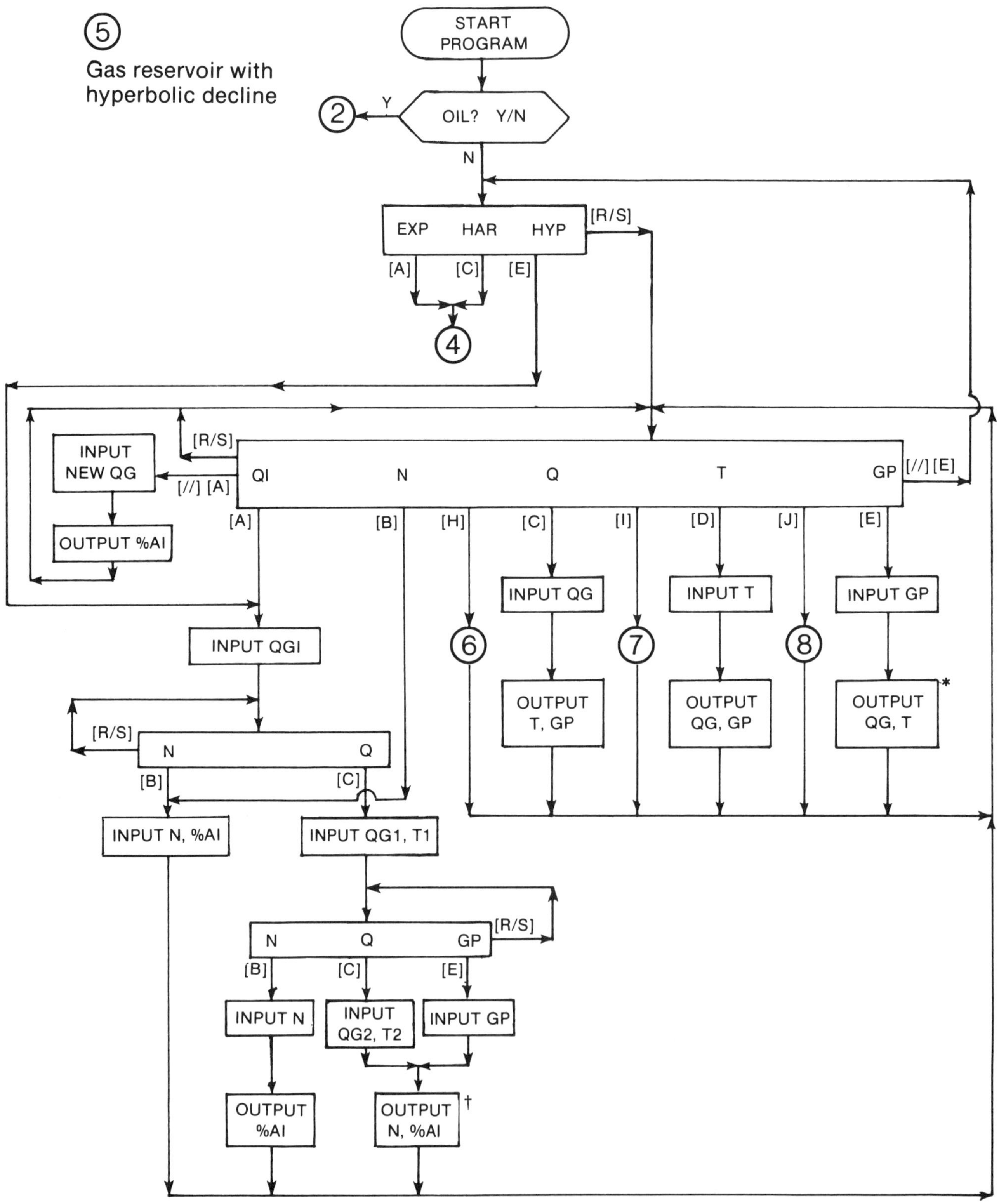

User Instructions (cont.)

⑥ Gas reservoir single well forecast

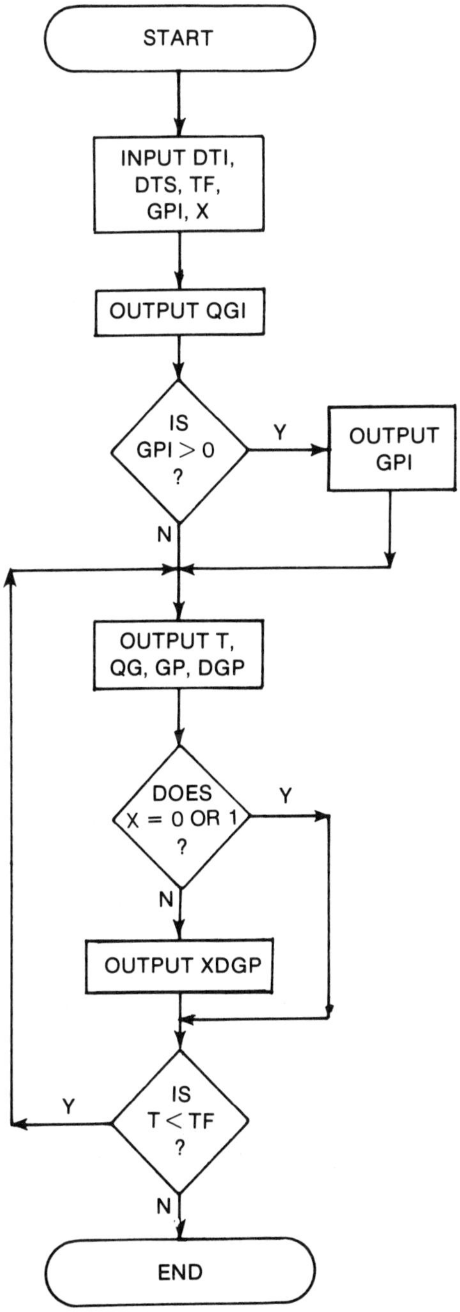

⑦ Multiple well input

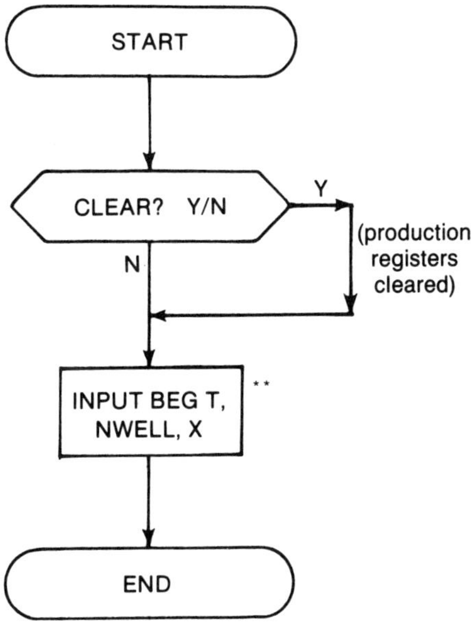

**Tones will sound while the production at each time is calculated.

User Instructions (cont.)

⑧ Multiple well output

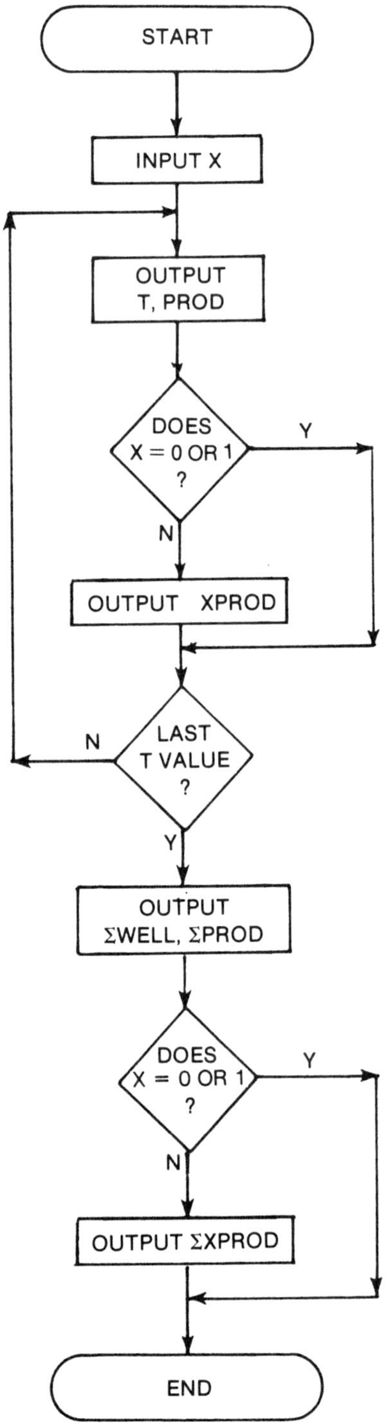

Notes on the User Instructions

*If the input value of NP (or GP) is so large that it cannot be produced at the current QOI (or QGI) and %AI, the program will halt with an error message (either NP BAD or GP BAD). To recover from the error, run the program again and change the input to represent a physically and mathematically realizable situation.

†Tones will sound while N is calculated iteratively. If the input data do not correspond to a true hyperbolic decline, the program will halt with an error message (either N < 0 or N > 1). To recover from the error, run the program again and change the input to represent a hyperbolic decline.

General Information

Memory Requirements

Program length:	1871 bytes (9 cards)
Minimum size:	049
Minimum hardware:	41C + quad memory module or 41CV

DECLINE requires a total of 2,214 bytes of the 2,237 available, leaving you with no more than 6 key assignments other than the DECLINE assignment. If you do not plan to use the multiple well input and output routines, you can remove them to save space. Simply delete all of the program steps beginning at line 823. This will yield a version of DECLINE that is 1,596 bytes long and requires size 023. This shortened version can be run in a 41C + 3 memory modules but will not leave a port available for a printer.

Hidden Options

When DECLINE is run, the default definition for %AI will be selected automatically. To change to the optional definition, set flag 00 ([//] [SF] 00) before the QOI prompt appears. To change back to the default definition, clear flag 00 ([//] [CF] 00) or run the program again. The flag 00 annunciator in the display will be off if the default definition is being used and on if the optional definition is being used.

Pac Subroutines Called

TITLE, Y/N?, OUT, INK, OUTK, IN

Registers

03 Oil or gas producing rate units
04 Oil or gas producing rate units
06 Time units
07 Time units
08 Oil or gas production units
09 Oil or gas production units
10 QGI (MCF/DAY), QOI (BBL/DAY)
11 %AI
12 QG, QG1(MCF/DAY), QO, QO1 (BBL/DAY)
13 T, T1 (YR), scratch
14 GP (MCF), NP (BBL)
15 D
16 QG2 (MCF/DAY), QO2 (BBL/DAY), DTI (YR)
17 DTS, T2, max T (YR)
18 TF (YR), scratch
19 GPI (MCF), NPI (BBL), NWELL, ΣPROD (BBL or MCF), scratch
20 X, scratch
21 Last GP (MCF), last NP (BBL), last QG (MCF/DAY), last QO (BBL/DAY)
22 N
23 ΣWELL
24–48 PROD (BBL or MCF)

Flags

00 Set: Use optional %AI definition.
Clear: Use default %AI definition.

01 Set: Oil or gas producing rate units not yet input.
Clear: Oil or gas producing rate units have been input.

02 Set: Time units not yet input.
Clear: Time units have been input.

03 Set: Oil or gas production units not yet input.
Clear: Oil or gas production units have been input.

04 Set: Clear production registers.
Clear: Leave production registers unchanged.

05 Set: Hyperbolic decline.
Clear: Exponential or harmonic decline.

06 Set: Hyperbolic case 1 (flag 05 set); exponential (flag 05 clear).
Clear: Hyperbolic case 2 (flag 05 set); harmonic (flag 05 clear).

07 Set: Oil reservoir.
Clear: Gas reservoir.

Flag 02 is also used in multiple well input to lump together all remaining production past year 25 into year 25.

Program Listing

```
 01♦LBL "DECLINE"
"DECLINE" 49
XROM "TITLE" FC?C 25
PROMPT CF 00 SF 01
SF 02 SF 03 SF 27
"OIL" 7 XROM "Y/N?"
"M3" FC? 07 "SCM"
ASTO 08 "⊢/DAY"
ASTO 03 ASHF ASTO 04
CLA ASTO 09 ASTO 07
"YR" ASTO 06

 28♦LBL 15
"EXP HAR HYP" PROMPT
GTO 17

 32♦LBL B
 33♦LBL D
GTO 15

 35♦LBL A
CF 05 SF 06
"EXPONENTIAL" GTO 08

 40♦LBL C
CF 05 CF 06 "HARMONIC"
GTO 08

 45♦LBL E
SF 05 "HYPERBOLIC"

 48♦LBL 08
ADV FS? 55 PRA ADV
GTO A

 54♦LBL 00
RCL 11 STO 20 RCL 15
STO 21 18 "NEW Q"
FS? 07 "⊢O" FC? 07
"⊢G" XEQ 35 RCL 10 /
FS? 05 RCL 22 FS? 05
Y↑X ST* 15 XEQ 32
RCL 21 STO 15 RCL 20
STO 11
```

Program Listing (cont.)

```
 78♦LBL 16
ADV

 80♦LBL 17
"QI N " FC? 05 "QI AI"
"⊢ Q T " FS? 07 "⊢NP"
FC? 07 "⊢GP" PROMPT
GTO 17

 91♦LBL e
GTO 15

 93♦LBL a
GTO 00

 95♦LBL H
GTO 22

 97♦LBL I
GTO 23

 99♦LBL J
GTO 24

101♦LBL 18
"     Q T " FS? 07
"⊢NP" FC? 07 "⊢GP"
PROMPT GTO 18

109♦LBL E
XEQ 41 XEQ 26 XEQ 36
XEQ 27 XEQ 40 GTO 16

116♦LBL D
XEQ 38 XEQ 25 XEQ 36
XEQ 29 XEQ 43 GTO 16

123♦LBL C
XEQ 34 XEQ 27 XEQ 40
XEQ 29 XEQ 43 GTO 16

130♦LBL A
9 "QOI" FC? 07 "QGI"
XEQ 35

136♦LBL 01
FS? 05 "  N Q"
FC? 05 "  AI Q"
PROMPT GTO 01

143♦LBL B
FS? 05 XEQ 46 10
STO 00 "%AI" XROM "IN"
XEQ 31 FC? 05 GTO 18
GTO 16

154♦LBL C
FS? 05 GTO 21 XEQ 34

158♦LBL 02
"        T " FS? 07
"⊢NP" FC? 07 "⊢GP"
PROMPT GTO 02

166♦LBL D
XEQ 38 XEQ 28 XEQ 32
XEQ 29 XEQ 43 GTO 16

173♦LBL E
XEQ 41 XEQ 30 XEQ 32
XEQ 27 XEQ 40 GTO 16

180♦LBL 21
11 "QO1" FC? 07 "QG1"
XEQ 35 "T1" XEQ 39

188♦LBL 03
"  N  Q   " FS? 07
"⊢NP" FC? 07 "⊢GP"
PROMPT GTO 03

196♦LBL B
XEQ 46 GTO 10

199♦LBL C
SF 06 15 "QO2" FC? 07
"QG2" XEQ 35 "T2"
XEQ 39 GTO 09

209♦LBL E
CF 06 XEQ 41

212♦LBL 09
XEQ 12 CLD TONE 9
RCL 22 "N" XROM "OUT"

219♦LBL 10
XEQ 28 XEQ 32 GTO 16

223♦LBL 12
CLX STO 00 1 STO 01
RCL 10 RCL 12 /
STO 19 FS? 06 GTO 13
1/X - RCL 19 LN
STO Z / RCL 14 RCL 10
/ RCL 13 / 365 /
STO 20 - STO 02
"N < 0" X<=0? GTO 20
X<>Y RCL 19 1 - /
RCL 20 GTO 14

260♦LBL 13
LN RCL 13 / RCL 10
RCL 16 / STO 20 LN
RCL 17 / - STO 02
"N < 0" X<=0? GTO 20
RCL 19 1 - RCL 13 /
RCL 20 1 - RCL 17 /

286♦LBL 14
- STO 15 "N > 1" X>0?
GTO 20 4 STO 18

294♦LBL 19
DSE 18 GTO 08 RCL 00
RCL 15 * RCL 01
RCL 02 * - RCL 15
RCL 02 - GTO 09

308♦LBL 08
RCL 00 RCL 01 + 2

313♦LBL 09
/ STO 22 TONE 5
FS? 06 GTO 10 1
RCL 19 RCL 22 Y↑X -
LASTX RCL 19 / 1 -
X<>Y / RCL 22 * 1
RCL 22 - / RCL 20 -
GTO 11

340♦LBL 10
RCL 19 RCL 22 Y↑X 1
- RCL 13 / RCL 20
RCL 22 Y↑X 1 -
RCL 17 / - RCL 22 /
```

Program Listing (cont.)

```
358♦LBL 11
ENTER↑  ABS  1 E-4  X>Y?
RTN  RCL 02  R↑  *  X>0?
GTO 12  LASTX  STO 15
RCL 22  STO 01  GTO 19

374♦LBL 12
LASTX  STO 02  RCL 22
STO 00  GTO 19

380♦LBL 22
1  STO 16  STO 17
"SINGLE WELL"  ADV
FS? 55  PRA  ADV  CLX
STO 19  STO 21  RCL 13
STO 18  15  STO 00
"DTI"  XEQ 39  "DTS"
XEQ 39  "TF"  XEQ 39
"NPI"  FC? 07  "GPI"
XEQ 42  XEQ 45  ADV
RCL 10  "QOI"  FC? 07
"QGI"  XEQ 37  RCL 19
"NPI"  FC? 07  "GPI"
X>0?  XEQ 44  CLX

420♦LBL 04
ADV  ST+ 16  RCL 16
RCL 18  X<=Y?  X<>Y
X<>Y  STO 13  XEQ 40
XEQ 25  XEQ 36  RCL 20
1  X=Y?  CLX  *  *  "X"
ARCL 05  X>0?  XEQ 37
XEQ 29  RCL 19  ST+ 14
RCL 14  XEQ 43  X<> 21
RCL 21  X<>Y  -  "D"
ARCL 05  XEQ 44  RCL 20
1  X=Y?  CLX  *  *  "X"
ARCL 05  X>0?  XEQ 44
RCL 13  RCL 18  X=Y?
GTO 16  RCL 17  GTO 04

470♦LBL 25
RCL 10  RCL 13  RCL 15
*  E↑X  FC? 05  FC? 06
XEQ 08  /  STO 12  RTN
```

```
482♦LBL 08
X<>Y  LASTX  1  +
FC? 05  RTN  1  -
RCL 22  *  1  +  RCL 22
1/X  Y↑X  RTN

499♦LBL 26
RCL 14  RCL 15  *  365
/  CHS  RCL 10  FC? 05
FS? 06  GTO 09  /  E↑X
RCL 10  *  GTO 10

515♦LBL 09
FC? 05  +  FC? 05
GTO 10  /  1  RCL 22  -
STO T  *  1  +  R↑  1/X
Y↑X  RCL 10  *

533♦LBL 10
X>0?  STO 12  X>0?  RTN
"N"  FC? 07  "G"
"⊦P BAD"

542♦LBL 20
TONE 3  PROMPT  GTO 20

546♦LBL 27
XEQ 11  RCL 15  /
STO 13  RTN

552♦LBL 28
XEQ 11  RCL 13  /
STO 15  RTN

558♦LBL 11
RCL 10  RCL 12  /  LN
FC? 05  FC? 06  GTO 12
RTN

567♦LBL 12
LASTX  1  -  FC? 05  RTN
1  +  RCL 22  Y↑X  1  -
RCL 22  /  RTN

582♦LBL 29
RCL 15  XEQ 13  *
STO 14  RTN
```

```
588♦LBL 30
RCL 14  XEQ 13  *
STO 15  RTN

594♦LBL 13
365  X<>Y  /  FS? 05
GTO 14  RCL 10  RCL 12
FS? 06  -  FS? 06  RTN
/  LN  *  RCL 10  RTN

611♦LBL 14
1  RCL 22  -  /  RCL 12
LASTX  Y↑X  RCL 10
RCL 22  Y↑X  *  RCL 10
X<>Y  -  RTN

627♦LBL 31
1  RCL 11  100  /  -
FC? 00  GTO 08  FC? 05
FS? 06  GTO 08  LASTX
X<>Y  /  STO 15  RTN

643♦LBL 08
1/X  LN  STO 15  FS? 00
FC? 05  RTN  LASTX
RCL 22  Y↑X  1  -
RCL 22  /  STO 15  RTN

659♦LBL 32
100  ENTER↑  1  RCL 15
FC? 00  GTO 10  FS? 05
GTO 09  FS? 06  GTO 10
+  /  RCL 15  GTO 14

674♦LBL 09
RCL 22  *  1  +  RCL 22
CHS  1/X  Y↑X  -  GTO 14

685♦LBL 10
CHS  E↑X  -

689♦LBL 14
*  STO 11  "%AI"
XROM "OUT"  RTN

695♦LBL 34
11  "QO"  FC? 07  "QG"
```

Program Listing (cont.)

```
700♦LBL 35
STO 00  XEQ 11
XROM "INK"  GTO 10

705♦LBL 36
"QO"  FC? 07  "QG"

709♦LBL 37
XEQ 11  XROM "OUTK"

712♦LBL 10
RDN  STO 03  X<>Y
STO 04  R↑  CF 08  RTN

720♦LBL 11
FS?C 01  SF 08  ASTO T
"BBL"  FC? 07  "MCF"
"⊦/DAY"  ASTO 01  ASHF
ASTO 02  CLA  ARCL T
RCL 04  RCL 03  RCL Z
RTN

737♦LBL 38
12  STO 00  "T"

741♦LBL 39
XEQ 12  XROM "INK"
GTO 10

745♦LBL 40
"T"  XEQ 12  XROM "OUTK"

749♦LBL 10
RDN  STO 06  X<>Y
STO 07  R↑  CF 08  RTN

757♦LBL 12
FS?C 02  SF 08  ASTO T
"YR"  ASTO 01  CLA
ASTO 02  ARCL T  RCL 07
RCL 06  RCL Z  RTN

770♦LBL 41
13  STO 00  "NP"  FC? 07
"GP"

776♦LBL 42
XEQ 13  XROM "INK"
GTO 10

780♦LBL 43
"NP"  FC? 07  "GP"

784♦LBL 44
XEQ 13  XROM "OUTK"

787♦LBL 10
RDN  STO 08  X<>Y
STO 09  R↑  CF 08  RTN

795♦LBL 13
FS?C 03  SF 08  ASTO T
"BBL"  FC? 07  "MCF"
ASTO 01  CLA  ASTO 02
ARCL T  RCL 09  RCL 08
RCL Z  RTN

810♦LBL 45
1  STO 20  19  STO 00
"X"  XROM "IN"  RTN

818♦LBL 46
21  STO 00  "N"
XROM "IN"  RTN

824♦LBL 23
"CLEAR"  4  XROM "Y/N?"
23.048  FC? 04  GTO 13
0

832♦LBL 05
STO IND Y  ISG Y  GTO 05

836♦LBL 13
CLX  STO 18  RCL 13
STO 16  RCL 10  STO 21
17  STO 00  "BEG T"
XEQ 39  "NWELL"
XROM "IN"  ST+ 23
XEQ 45  RCL 18  INT  23
+  STO 00  RCL 16
RCL 18  +  23  +  STO 17

862♦LBL 06
TONE 5  ISG 00  CLD
RCL 00  48  X<=Y?  SF 02
X<Y?  STO 00  X<>Y
RCL 17  X>Y?  X<>Y
FS? 02  X<>Y  RCL 18  -
23  -  STO 13  XEQ 25
RCL 10  X<> 21  STO 10
XEQ 29  RCL 19  *
RCL 20  *  ST+ IND 00
RCL 12  X<> 21  STO 10
FS?C 02  GTO 14  RCL 17
RCL 00  X<Y?  GTO 06

902♦LBL 14
RCL 16  STO 13  TONE 9
GTO 16

907♦LBL 24
"MULTIWELL"  ADV  FS? 55
PRA  ADV  XEQ 45  CLX
STO 19  24.048  STO 18

918♦LBL 07
RCL IND 18  X=0?  GTO 14
ADV  RCL 18  INT  23  -
XEQ 40  RCL IND 18
ST+ 19  "PROD"  XEQ 44
RCL 20  1  X=Y?  CLX  *
*  "X"  ARCL 05  X>0?
XEQ 44

942♦LBL 14
ISG 18  GTO 07  ADV
RCL 23  "ΣWELL"
XROM "OUT"  RCL 19
"ΣPROD"  XEQ 44  RCL 20
1  X=Y?  CLX  *  *
"ΣXPROD"  X>0?  XEQ 44
GTO 16  END
```

Section 3
Volumetric Calculations and Reserves

8. OIP — Calculating Oil in Place and Reserves Volumetrically

Calculates volumetric oil in place and estimates reserves from API correlations for water-drive and solution-gas-drive reservoirs.

9. GIP — Calculating Gas in Place and Reserves Volumetrically

Calculates volumetric gas in place and estimates reserves for water-drive and volumetric expansion.

8. OIP — Calculating Oil in Place and Reserves Volumetrically

Calculating oil in place volumetrically is a relatively simple chore since most reservoir engineers know the volumetric equation by heart. When using the equation, the ability to compute the oil formation volume factor (BOI) might be helpful. OIP allows the user to input or calculate PVT properties, as well as estimate recoveries using the API correlations (API Bull. D14). Frankly, the API correlations will rarely be better than an "experience factor" or "analogy" guess. This is particularly true of water-drive reservoirs. The API correlations for water drive should not be misused as estimates for waterflood recoveries. Also, the gas-drive correlations assume the reservoir is at the bubble point. Although the correlations have enormous estimation errors, they are made much simpler to use because of the Pac correlations for the PVT properties.

Volumetric estimates are generally used only at early times in the life of a field for reserve purposes. Reservoir heterogeneities are a commonly overlooked reason that volumetric estimates often exceed the results obtained by performance.

Equations

$$N = \frac{7758\ \text{POR}\ (1 - \text{SW})\ \text{AREA H}}{\text{BOI}}$$

$$\%\text{OILRF(gas drive)} = 41.815 \times \left[\frac{\text{POR}(1 - \text{SW})}{\text{BOBP}}\right]^{0.1611} \times \left[\frac{K'}{\text{UOBP}}\right]^{0.0979} \times \text{SW}^{0.3722}\left[\frac{\text{PBP}}{\text{P ABAN}}\right]^{0.1741}$$

$$\text{NPGD} = \text{OILRF (gas drive) N}$$

$$\%\text{OILRF (water drive)} = 54.898 \times \left[\frac{\text{POR}(1 - \text{SW})}{\text{BOI}}\right]^{0.0422} \times \left[\frac{K'}{\text{UOI}}\ \text{UWI}\right]^{0.077} \times \text{SW}^{-0.1903}\left[\frac{\text{PI}}{\text{P ABAN}}\right]^{-0.2159}$$

$$\text{NPWD} = \text{OILRF (water drive) N}$$

$$K' = \text{K in D}$$

$$\text{POR} = \frac{\%\text{POR}}{100}$$

$$\text{SW} = \frac{\%\text{SW}}{100}$$

$$\text{OILRF} = \frac{\%\text{OILRF}}{100}$$

Nomenclature

Symbol	*Variable Name*	*Input or Output*	*English Units*	*SI Units*
AREA	Reservoir area	I	ACRE	M2
BOI	Initial oil-formation volume factor	I	–	–
H	Formation thickness†	I	FT	M
K	Permeability	I	MD	MD
N*	Initial oil in place	O	BBL	M3
NPGD*	Cumulative oil production (gas drive)	O	BBL	M3
NPWD*	Cumulative oil production (water drive)	O	BBL	M3
P ABAN	Abandonment pressure	I	PSI	KPA
PI	Initial pressure	I	PSI	KPA
UOI	Initial oil viscosity	I	CP	PA*S
UWI	Initial water viscosity	I	CP	PA*S
%OILRF	Volume percent gas-drive or water-drive oil recovery factor	O	–	–

*The units for these variables are saved by the program.

†In the case of deviated wells or slanted beds, use the true stratigraphic thickness instead of the measured thickness of the formation.

Yes/No Questions

CORR?	Yes: Use Pac correlations to estimate PVT properties. No: Input PVT properties.
SKIP?	Yes: Skip input of PVT data. No: Allow input of PVT data.

Example

The Little Creek oil field has three producing wells offset by two dry holes, reasonably delineating the field. An isopachous map of the Woodbine pay indicates a total area of 200 acres with an average thickness of 24 FT. An average effective porosity of 14.3% and water saturation of 32.5% were calculated from well-log analysis. The reservoir was discovered at an initial pressure of 2,650 PSI; additional PVT data are as follows:

SEP T = 110 F
SEP P = 125 PSI
OIL G = 37.4 API
GAS G = 0.678
T = 155 F
RSI = 480 SCF/BBL

A very early pressure buildup was conducted for one of the wells, indicating an effective oil permeability of 2.2 MD. Estimate the original oil in place and ultimate recovery. Since this is a solution-gas-drive reservoir, the water-drive recovery should be ignored.

Keystrokes (SIZE> =041)	Display	Comments
[XEQ] [ALPHA] OIP [ALPHA]	*CORR? Y/N:*	Last character is Y or N
Y [R/S]	*SKIP? Y/N:*	Last character is Y or N
N [R/S]	*SEP T=?*	
110 [R/S]	*SEP P=?*	
125 [R/S]	*OIL G=?*	
37.4 [R/S]	*GAS G=?*	
.678 [R/S]	*T=?*	GAS GS printed
155 [R/S]	*RSI=?*	
480 [R/S]	*%POR=?*	PBP printed
14.3 [R/S]	*%SW=?*	
32.5 [R/S]	*PI=?*	
2650 [R/S]	*AREA=?*	
200 [R/S]	*H=?*	
24 [R/S]	*K=?*	N printed
2.2 [R/S]	*P ABAN=?*	
200 [R/S]	*%NACL=?*	NPGD and %OILRF (gas-drive) printed

Since this is a solution-gas-drive reservoir, it is not necessary to calculate the water-drive recovery.

```
OIL IN PLACE

CORR: YES
SKIP: NO
SEP T=110.0000 F
SEP P=125.0000 PSI
OIL G=37.4000 API
GAS G=0.6780
GAS GS=0.6842
T=155.0000 F
RSI=480.0000 SCF/BBL
PBP=2,182.2913 PSI
%POR=14.3000
%SW=32.5000
PI=2,650.0000 PSI
AREA=200.0000 ACRE
H=24.0000 FT

N=2,814,521.845 BBL

K=2.2000 MD
P ABAN=200.0000 PSI

NPGD=440,292.8451 BBL
%OILRF=15.6436
```

General Information

Memory Requirements

Program length:	608 bytes (3 cards)
Minimum size:	041
Minimum hardware:	41C + 2 memory modules

Hidden Options

None

Pac Subroutines Called

TITLE, Y/N?, W8, %POR, IN, INU, CBOb, CBO, CON, CUOd, OUT, %NACL, CRSb, CUOb, CUO, CUW, OUTK

User Instructions

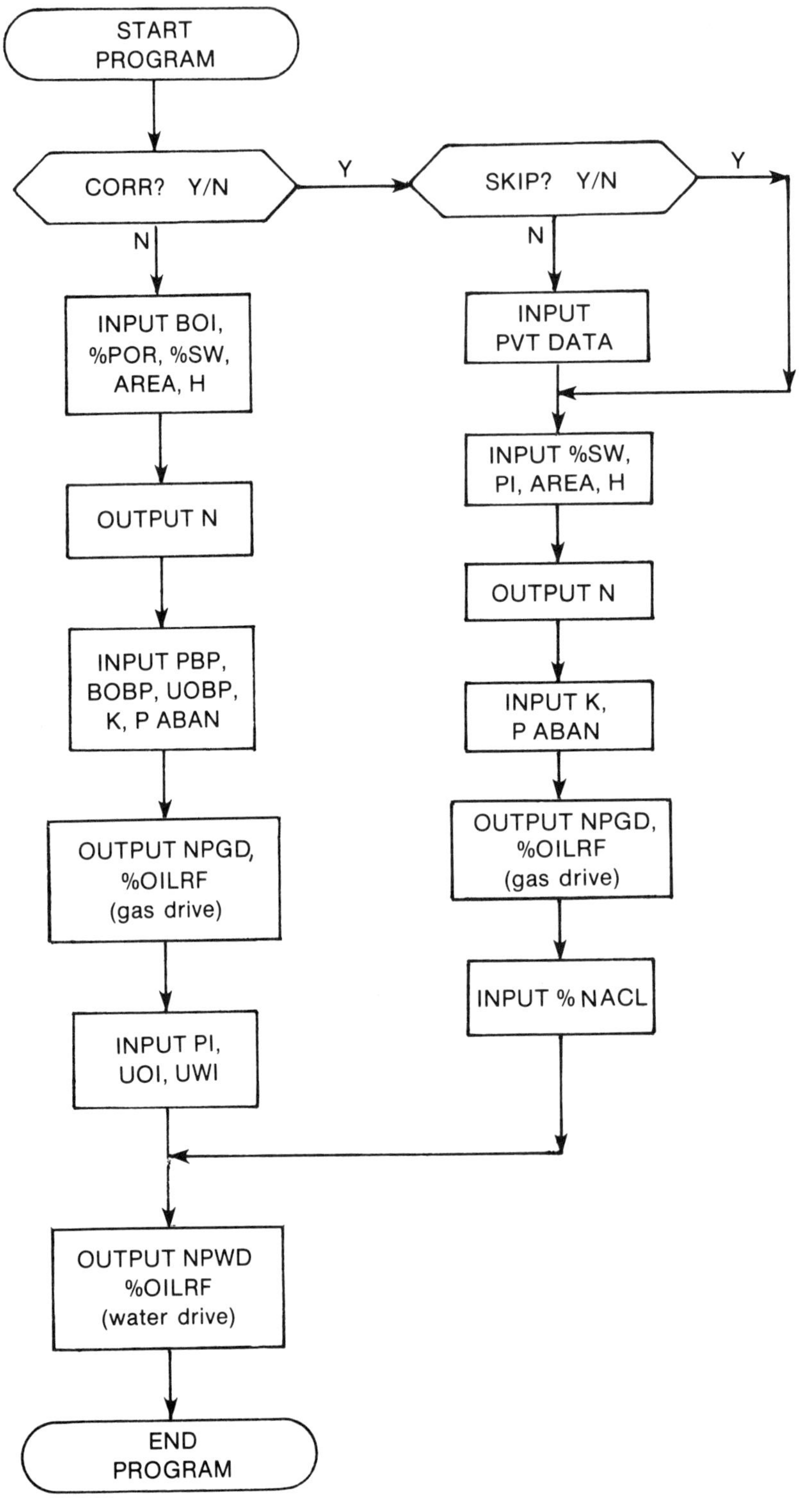

Registers

03	Oil production units
04	Oil production units
06	%OILRF
07	Scratch
08	UOI (CP)
09	UWI (CP)
17	PI (PSI)
26	BOI
27	AREA (ACRE)
28	H (FT)
29	K(MD)
30	BOBP
32	UOd (CP)
33	UOBP (CP)
34	Scratch
35	PI (PSI)
37	NPGD (BBL)
38	P ABAN (PSI)
40	N (BBL)

Registers 10, 11, 20, 23, 31, 36, and 39 unused

Flags

01 Set: Use Pac correlations to estimate PVT properties.
Clear: Input PVT properties.

02 Set: Skip input of PVT data.
Clear: Allow input of PVT data.

Program Listing

```
 01♦LBL "OIP"
"OIL IN PLACE" 41
XROM "TITLE" FC?C 25
PROMPT "M3" ASTO 03
CLA ASTO 04 "CORR" 1
XROM "Y/N?" FC? 01
GTO 00 "SKIP" 2
XROM "Y/N?" FC? 02
XROM "W8" FC? 02
XROM "%POR" GTO 01

 24♦LBL 00
25 STO 00 "BOI"
XROM "IN"

 29♦LBL 01
FC? 01 XROM "%POR" 20
STO 00 "%SW" XROM "IN"
FS? 01 XEQ 09 26
STO 00 "ACRE" ASTO 01
CLA ASTO 02 ASTO Z
"M2" ASTO Y "AREA"
XROM "INU" "FT"
ASTO 01 CLA ASTO 02
ASTO Z "M" ASTO Y "H"
XROM "INU" FC? 01
GTO 02 RCL 17 RCL 14
CF 03 X<=Y? SF 03
FC? 03 X<>Y
XROM "CBOb" STO 30
FS? 03 XROM "CBO"
STO 26

 72♦LBL 02
RCL 18 1 RCL 21 100
ST/ T / - * STO 07
RCL 26 / RCL 27 *
RCL 28 * "ACRE*FT-BBL"
CON STO 40 "N" XEQ 12
CF 08 ADV FS? 01
GTO 03 13 "PBP"
XEQ 10 29 STO 00
"BOBP" XROM "IN" 32
"UOBP" XEQ 11

107♦LBL 03
28 STO 00 "MD"
ASTO 01 ASTO Y CLA
ASTO 02 ASTO Z "K"
XROM "INU" 37 "P ABAN"
XEQ 10 FC? 01 GTO 04
XROM "CUOd" STO 32
RCL 13 X<>Y
XROM "CUOb" STO 33
FS? 03 GTO 04 RCL 14
XROM "CBOb" STO 30

134♦LBL 04
RCL 33 RCL 29 1 E3 /
STO 34 X<>Y / .0979
Y↑X RCL 21 100 /
.3722 Y↑X * RCL 14
RCL 38 / .1741 Y↑X *
RCL 07 RCL 30 / .1611
Y↑X * 41.815 *
STO 06 1 % RCL 40 *
STO 37 "NPGD" XEQ 12
RCL 06 "%OILRF"
XROM "OUT" ADV FC? 01
GTO 06 XROM "%NACL"
RCL 13 FC? 03 RCL 17
FC? 03 XROM "CRSb"
RCL 32 XROM "CUOb"
FS? 03 XROM "CUO"
STO 08 XROM "CUW"
GTO 07

191♦LBL 06
XEQ 09 7 "UOI" XEQ 11
8 "UWI" XEQ 11

199♦LBL 07
RCL 08 / RCL 34 *
.077 Y↑X RCL 21 100
/ -.1903 Y↑X *
RCL 17 RCL 38 /
-.2159 Y↑X * RCL 07
RCL 26 / .0422 Y↑X *
54.898 * STO 06 1 %
RCL 40 * "NPWD"
XEQ 12 RCL 06 "%OILRF"
XROM "OUT" ADV

237♦LBL 08
STOP GTO 08

240♦LBL 09
16 "PI"
```

Program Listing (cont.)

```
243♦LBL 10
STO 00  ASTO T  "PSI"
ASTO 01  "KPA"  ASTO Y
CLA  ASTO 02  ASTO Z
ARCL T  XROM "INU"  RTN
```

```
256♦LBL 11
STO 00  ASTO T  "CP"
ASTO 01  "PA*S"  ASTO Y
CLA  ASTO 02  ASTO Z
ARCL T  XROM "INU"  RTN
```

```
269♦LBL 12
ADV  ASTO T  "BBL"
ASTO 01  CLA  ASTO 02
ARCL T  RCL 04  RCL 03
RCL Z  XROM "OUTK"  RDN
STO 03  X<>Y  STO 04  R↑
END
```

9. GIP — Calculating Gas in Place and Reserves Volumetrically

Gas reserves can generally be calculated volumetrically with somewhat less error than oil reserves. However, low permeability and heterogeneous reservoirs require the same precautions as the oil volumetric reserve calculations. GIP calculates the original gas in place from input volumetric parameters, and recovery factors for volumetric and water-drive reservoirs. The residual gas saturation can be input or will be estimated by the program.

Equations

$$G = \frac{43.56\ POR\ (1 - SW)\ AREA\ H}{BGI}$$

$$GASRF\ (gas\ drive) = \frac{\frac{PI}{ZI} - \frac{P\ ABAN}{Z\ ABAN}}{\frac{PI}{ZI}}$$

$$GPGD = GASRF\ (gas\ drive)\ G$$

$$GASRF\ (water\ drive) = \frac{(1 - SW)\ \frac{PI}{ZI} - SGR\ \frac{P\ ABAN}{Z\ ABAN}}{(1 - SW)\ \frac{PI}{ZI}}$$

$$GPWD = GASRF\ (water\ drive)\ G$$

$$BGI = \frac{ZI\ T'\ STD\ P}{PI\ STD\ T'}$$

$$\%SGR = 62.5 - 1.3125\ \%POR$$

$$T' = T\ in\ R$$

$$STD\ T' = STD\ T\ in\ R$$

$$POR = \frac{\%POR}{100}$$

$$SW = \frac{\%SW}{100}$$

$$GASRF = \frac{\%GASRF}{100}$$

$$SGR = \frac{\%SGR}{100}$$

Nomenclature

Symbol	*Variable Name*	*Input or Output*	*English Units*	*SI Units*
AREA	Reservoir area	I	ACRE	M2
G*	Initial gas in place	O	MCF	SCM
GPGD*	Cumulative gas production (gas drive)	O	MCF	SCM
GPWD*	Cumulative gas production (water drive)	O	MCF	SCM
H	Formation thickness†	I	FT	M
P ABAN	Abandonment pressure	I	PSI	KPA
PI	Initial pressure	I	PSI	KPA
Z ABAN	Abandonment Z factor	I	–	–
ZI	Initial Z factor	I	–	–
%GASRF	Volume percent expansion-drive or water-drive gas recovery factor	O	–	–
%SGR	Volume percent residual gas saturation	I,O	–	–

*The units for these variables are saved by the program.
†In the case of deviated wells or slanted beds, use the true stratigraphic thickness instead of the measured thickness of the formation.

Yes/No Questions

CORR? Yes: Use Pac correlations to estimate PVT properties.
No: Input PVT properties.

SKIP? Yes: Skip input of PVT data.
No: Allow input of PVT data.

Example 1: Expansion Drive

Calculate the gas in place and estimated ultimate recovery for the South Texas Vicksburg reservoir (Example 2 of GASPVT). The estimated area is 460 ACRE with an average net thickness of 58 FT. Estimated abandonment pressure is 600 PSI. The porosity is 22.5% with a 37.8% water saturation. The remaining fluid data can be found in the GASPVT example. By running that example, one can answer Y to the SKIP? Y/N prompt.

Keystrokes (SIZE>=040)	Display	Comments
[XEQ] [ALPHA] GIP [ALPHA]	*CORR? Y/N:*	Last character is Y or N
Y [R/S]	*SKIP? Y/N:*	Last character is Y or N
N [R/S]	*Tc=?*	
441.8952 [R/S]	*Pc=?*	
661.5656 [R/S]	*STD T=?*	
60 [R/S]	*STD P=?*	
14.65 [R/S]	*T=?*	
180 [R/S]	*%POR=?*	
22.5 [R/S]	*%SW=?*	
37.8 [R/S]	*AREA=?*	
460 [R/S]	*H=?*	
58 [R/S]	*PI=?*	
6200 [R/S]	*P ABAN=?*	G printed
600 [R/S]	*%SGR=?*	GPGD %GASRF (gas drive), and %SGR printed

Since this is an expansion-drive reservoir, it is not necessary to calculate the water-drive recovery.

```
GAS IN PLACE

CORR: YES
SKIP: NO
Tc=441.8952 R
Pc=661.5656 PSI
STD T=60.0000 F
STD P=14.6500 PSI
T=180.0000 F
%POR=22.5000
%SW=37.8000
AREA=460.0000 ACRE
H=58.0000 FT
PI=6,200.0000 PSI

G=51,579,018.04 MCF

P ABAN=600.0000 PSI

GPGD=45,561,566.91 MCF
%GASRF=88.3335

%SGR=32.9688
```

Example 2: Water Drive

Use Example 1 but assume that the reservoir has a water drive, leaving a residual gas saturation of 22%. Estimated abandonment pressure is 3500 PSI based on analogy with nearby reservoirs. How is the recovery affected by using the calculated %SGR?

Keystrokes	Display	Comments
[XEQ] [ALPHA] GIP [ALPHA]	*CORR? Y/N:Y*	
[R/S]	*SKIP? Y/N:N*	
Y [R/S]	*%SW=?*	
[R/S]	*AREA=?*	
[R/S]	*H=?*	
[R/S]	*PI=?*	
[R/S]	*P ABAN=?*	G printed
3500 [R/S]	*%SGR=?*	GPGD %GASRF (gas drive), and %SGR printed
22 [R/S]	*P ABAN=?*	GPWD and %GASRF (water drive) printed
[R/S]	*%SGR=?*	GPGD, %GASRF (gas drive), and %SGR printed
[R/S]	*P ABAN=?*	GPWD and %GASRF (water drive) printed

```
GAS IN PLACE

SKIP: YES

G=51,579,018.04 MCF

P ABAN=3,500.0000 PSI

GPGD=12,290,566.96 MCF
%GASRF=23.8286

%SGR=32.9688
%SGR=22.0000

GPWD=37,682,781.32 MCF
%GASRF=73.0584

GPGD=12,290,566.96 MCF
%GASRF=23.8286

%SGR=32.9688

GPWD=30,754,401.93 MCF
%GASRF=59.6258
```

User Instructions

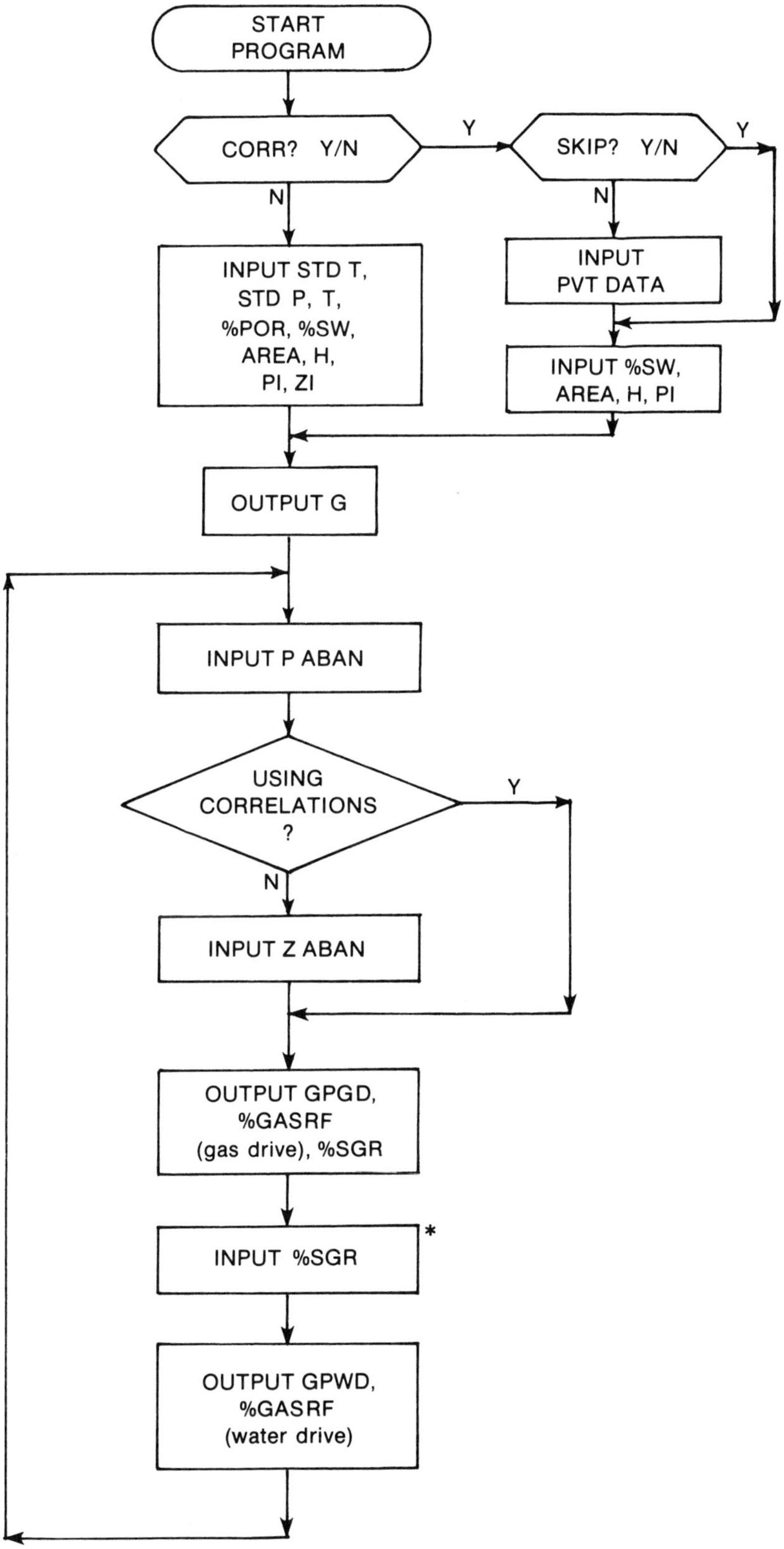

* This option is provided in case you know the value for %SGR.

General Information

Memory Requirements

Program length: 439 bytes (2 cards)
Minimum size: 040
Minimum hardware: 41C + 1 memory module

Hidden Options

None

Pac Subroutines Called

TITLE, Y/N?, ITcPc, STDTP, T, %POR, IN, INU, CZ, CON, OUT, OUTK

Registers

03	%GASRF
04	Scratch
06	Gas production units
07	Gas production units
08	ZI
09	Z ABAN
17	PI (PSI)
26	%SGR
27	AREA (ACRE)
28	H (FT)
35	PI (PSI)
36	GPGD (MCF)
38	P ABAN (PSI)
39	G (MCF)

Registers 12–15, 19, 20, 24, 25, 29–34, and 37 unused

Flags

01	Set:	Use Pac correlations to estimate PVT properties.
	Clear:	Input PVT properties.
02	Set:	Skip input of PVT data.
	Clear:	Allow input of PVT data.

Program Listing

```
 01♦LBL "GIP"
"GAS IN PLACE"  40
XROM "TITLE"  FC?C 25
PROMPT  "SCM"  ASTO 06
CLA  ASTO 07  "CORR"  1
XROM "Y/N?"  FC? 01
GTO 00  "SKIP"  2
XROM "Y/N?"  FS? 02
GTO 01  XROM "ITcPc"

 22♦LBL 00
XROM "STDTP"  XROM "T"
XROM "%POR"

 26♦LBL 01
20  STO 00  "%SW"
XROM "IN"  26  STO 00
"ACRE"  ASTO 01  CLA
ASTO 02  ASTO Z  "M2"
ASTO Y  "AREA"
XROM "INU"  "FT"
ASTO 01  CLA  ASTO 02
ASTO Z  "M"  ASTO Y  "H"
XROM "INU"  16  "PI"
XEQ 07  STO 35  FC? 01
GTO 02  XEQ 09  CZ
STO 08  GTO 03

 61♦LBL 02
7  STO 00  "ZI"
XROM "IN"

 66♦LBL 03
RCL 17  /  RCL 23  *
RCL 22  "F-R"  CON  /
RCL 16  CON  *  RCL 18
X<>Y  /  1  RCL 21  100
ST/ T  /  -  STO 04  *
RCL 27  *  RCL 28  *
43.56  *  STO 39  "G"
XEQ 08  CF 08

 99♦LBL 15
ADV  37  "P ABAN"
XEQ 07  FC? 01  GTO 04
RCL 38  XEQ 09  CZ
GTO 05

110♦LBL 04
8  STO 00  "Z ABAN"
XROM "IN"

115♦LBL 05
RCL 08  X<>Y  /  RCL 38
*  RCL 17  /  1  X<>Y  -
STO 03  RCL 39  *
STO 36  "GPGD"  XEQ 08
RCL 03  100  *  "%GASRF"
XROM "OUT"  ADV  62.5
RCL 18  1.3125  *  -
STO 26  "%SGR"
XROM "OUT"  25  STO 00
"%SGR"  XROM "IN"  100
/  1  RCL 03  -  *
RCL 04  /  1  X<>Y  -
STO 03  RCL 39  *
"GPWD"  XEQ 08  RCL 03
100  *  "%GASRF"
XROM "OUT"  GTO 15

172♦LBL 07
STO 00  ASTO T  "PSI"
ASTO 01  "KPA"  ASTO 02
CLA  ASTO 02  ASTO Z
ARCL T  XROM "INU"  RTN

185♦LBL 08
ADV  ASTO T  "MCF"
ASTO 01  CLA  ASTO 02
ARCL T  RCL 07  RCL 06
RCL Z  XROM "OUTK"  RDN
STO 06  X<>Y  STO 07  R↑
RTN

203♦LBL 09
RCL 16  "F-R"  CON
RCL 10  /  X<>Y  RCL 11
/  END
```

Section 4
Material Balance

10. OILMBE — Calculating N or We from Material Balance

Calculates initial oil in place or cumulative water influx above or below the bubble point. PVT properties can be input by the user or calculated from the Pac subroutines.

11. KG/KO — Gas-Oil Relative-Permeability Ratio

A utility program to calculate gas-oil relative permeability ratio from well performance or from two different correlations. The input coefficients for either set of correlations can be varied to calculate KG/KO curves that match desired values.

12. OILPRED — Predicting Solution-Gas-Drive Performance

Predicts solution gas-drive performance above and below the bubble point using Pac PVT subroutines.

13. QOVST — Rate-Time Forecast from Material Balance

Predicts rate-time performance for solution-gas-drive reservoirs above and below the bubble point.

14. INFCOEF — Calculating the Water Influx Coefficient

Calculates the water influx coefficient from performance or theory. Valid for both oil and gas reservoirs.

15. INFLUX — Predicting Water Influx

Predicts water influx for future pressure decrements for any value of Re/RW. Valid for both oil and gas reservoirs.

16. GASMBE — Gas Material Balance

Calculates initial gas in place or cumulative water influx from performance. Other flexible material balance capabilities are provided.

10. OILMBE — Calculating N or We from Material Balance

Practically all reservoir engineering texts present stirring derivations of the general material-balance equation. All of the programs in this section will assume that the user is familiar with material-balance equation concepts. OILMBE is particularly helpful in calculating the original oil in place (N) based on performance. It can also calculate water influx at any point knowing N. It is flexible enough to be used in calculating MBE terms for graphical estimates. For example, N can be estimated as a function of the original free gas in place (G) from the cumulative production history of saturated reservoirs. Water influx can be calculated for several values of N to provide sensitivities for predictive purposes.

One point is worthy of note. Calculating N from pressures and cumulative production above the bubble point is generally much more successful than below the bubble point for similar quality data. Below the bubble point, the compressibility of the free gas begins to dominate the total reservoir compressibility. Consequently, small errors for saturated reservoirs may lead to large errors in estimating oil in place.

Also, if the initial reservoir pressure is below the bubble point, you will be prompted for both the initial free gas in place (G) and the initial gas saturation (%SGI). Only G is required for the material-balance calculation. %SGI is used to calculate the oil saturation (%SO) and will not affect the material balance. If you are not concerned about %SO, you can enter 0 for %SGI.

Equations

Above bubble point:

$$N = \frac{NP\,BO - (We - BW\,WP)}{BO - BOI + (CFR + CWI\,SWC)(PI - P)BOI/(1 - SWC)}$$

Below bubble point:

$$N = \frac{NP\,BOb + BG'(GP - RSb\,NP) - G(BG' - BGI') - (We - BW\,WP)}{BOb - BOI + BG'(RSI - RSb) + (CFR + CWI\,SWC)(PI - P)BOI/(1 - SWC)}$$

$$\%SO = \frac{BOb}{BOI}\ \frac{N - NP}{N}\,(100 - \%SWC - \%SGI)$$

$$BG' = BG,\ BBL/SCF$$

$$BGI' = BGI,\ BBL/SCF$$

$$SWC = \frac{\%SWC}{100}$$

Nomenclature

Symbol	*Variable Name*	*Input or Output*	*English Units*	*SI Units*
BGI	Initial gas formation volume factor	I	FT3/SCF	M3/SCM
BOI	Initial oil formation volume factor	I	-	-
CWI	Initial water isothermal compressibility	I	1/PSI	1/KPA
G*	Initial free gas in place	I	MCF	SCM
GP*	Cumulative gas production	I	MCF	SCM
N*	Initial oil in place	I,O	BBL	M3
NP*	Cumulative oil production	I	BBL	M3
P*	Pressure	I	PSI	KPA
PI*	Initial pressure	I	PSI	KPA
WP*	Cumulative water production	I	BBL	M3
We*	Cumulative water influx	I,O	BBL	M3
%SGI	Volume percent initial gas saturation	I	-	-
%SWC	Volume percent connate water saturation	I	-	-

*The units for these variables are saved by the program.

Yes/No Questions

CORR? Yes: Use Pac correlations to estimate PVT properties.
No: Input PVT properties.

SKIP? Yes: Skip input of PVT data.
No: Allow input of PVT data.

Example 1

The Little Creek oil reservoir used in Example 1 of OIP has had historical production as presented in Table 10-1. Calculate the initial oil in place (N) from this performance using the Pac correlations and assuming no water influx or water production. Required data are as follows:

Tc = 377.8726 R
Pc = 665.8449 PSI
STD T = 60 F
STD P = 14.65 PSI
SEP T = 110 F
SEP P = 125 PSI
OIL G = 37.4 API
GAS G = 0.678
Reservoir temperature = 155 F
RSI = 480 SCF/BBL
Water salinity = 1.9%
Porosity = 14.3%
Water saturation = 32.5%
Initial pressure = 2,650 PSI

The calculated values of N are graphed in Figure 10-1. A good approximate value for N is $3.25(10^6)$ BBL. This is the average of all the values except the one at 2,108 PSI, which is excessively high to be considered valid. Note that even though the pressure points were entered in descending order, OILMBE allows any order.

Table 10-1

Date	*P (PSI)*	*NP (BBL)*	*GP (MCF)*
9-76	2,650	-	-
1-77	2,532	7,070	3,394
2-77	2,428	13,621	6,552
3-77	2,301	22,088	10,580
5-77	2,198	29,606	14,300
6-78	2,108	73,571	34,848
1-80	1,996	129,670	59,865
12-81	1,905	189,574	88,984

Keystrokes (SIZE> =044)	**Display**	**Comments**
[XEQ] [ALPHA] OILMBE [ALPHA]	*CORR? Y/N:*	Last character is Y or N
Y [R/S]	*SKIP? Y/N:*	Last character is Y or N
N [R/S]	*Tc=?*	
377.8726 [R/S]	*Pc=?*	
665.8449 [R/S]	*STD T=?*	
60 [R/S]	*STD P=?*	
14.65 [R/S]	*SEP T=?*	
110 [R/S]	*SEP P=?*	
125 [R/S]	*OIL G=?*	
37.4 [R/S]	*GAS G=?*	

Keystrokes (SIZE> =044)	**Display**	**Comments**
.678 [R/S]	*T=?*	GAS GS printed
155 [R/S]	*RSI=?*	
480 [R/S]	*%NACL=?*	PBP printed
1.9 [R/S]	*%POR=?*	
14.3 [R/S]	*%SWC=?*	
32.5 [R/S]	*PI=?*	
2650 [R/S]	*P=?*	
2532 [R/S]	*NP=?*	
7070 [R/S]	*GP=?*	
3394 [R/S]	*WP=?*	
0 [R/S]	*N We*	
[E]	*We=?*	We known
0 [R/S]	*N We*	N printed
[R/S]	*P=?*	
2428 [R/S]	*NP=?*	
13621 [R/S]	*GP=?*	
6552 [R/S]	*WP=?*	
[R/S]	*N We*	
[E]	*We=?*	
[R/S]	*N We*	N printed
[R/S]	*P=?*	
2301 [R/S]	*NP=?*	
22088 [R/S]	*GP=?*	
10580 [R/S]	*WP=?*	
[R/S]	*N We*	
[E]	*We=?*	
[R/S]	*N We*	N printed
[R/S]	*P=?*	
2198 [R/S]	*NP=?*	
29606 [R/S]	*GP=?*	
14300 [R/S]	*WP=?*	
[R/S]	*N We*	
[E]	*We=?*	
[R/S]	*N We*	N printed
[R/S]	*P=?*	
2108 [R/S]	*NP=?*	
73571 [R/S]	*GP=?*	
34848 [R/S]	*WP=?*	
[R/S]	*N We*	
[E]	*We=?*	
[R/S]	*N We*	%SO and N printed (P<PBP)
[R/S]	*P=?*	
1996 [R/S]	*NP=?*	
129670 [R/S]	*GP=?*	
59865 [R/S]	*WP=?*	
[R/S]	*N We*	
[E]	*We=?*	
[R/S]	*N We*	%SO and N printed
[R/S]	*P=?*	
1905 [R/S]	*NP=?*	
189574 [R/S]	*GP=?*	
88984 [R/S]	*WP=?*	
[R/S]	*N We*	
[E]	*We=?*	
[R/S]	*N We*	%SO and N printed

```
OIL MATL BAL

CORR: YES
SKIP: NO
Tc=377.8726 R
Pc=665.8449 PSI
STD T=60.0000 F
STD P=14.6500 PSI
SEP T=110.0000 F
SEP P=125.0000 PSI
OIL G=37.4000 API
GAS G=0.6780
GAS GS=0.6842
T=155.0000 F
RSI=480.0000 SCF/BBL
PBP=2,182.2913 PSI
%NACL=1.9000
%POR=14.3000
%SWC=32.5000
PI=2,650.0000 PSI

P=2,532.0000 PSI
NP=7,070.0000 BBL
GP=3,394.0000 MCF
WP=0.0000 BBL
We=0.0000 BBL
N=3,249,834.778 BBL

P=2,428.0000 PSI
NP=13,621.0000 BBL
GP=6,552.0000 MCF
N=3,249,905.859 BBL

P=2,301.0000 PSI
NP=22,088.0000 BBL
GP=10,580.0000 MCF
N=3,250,022.138 BBL

P=2,198.0000 PSI
NP=29,606.0000 BBL
GP=14,300.0000 MCF
N=3,274,166.268 BBL

P=2,108.0000 PSI
NP=73,571.0000 BBL
GP=34,848.0000 MCF
%SO=65.9837
N=3,477,801.532 BBL

P=1,996.0000 PSI
NP=129,670.0000 BBL
GP=59,865.0000 MCF
%SO=63.9371
N=3,146,151.077 BBL

P=1,905.0000 PSI
NP=189,574.0000 BBL
GP=88,984.0000 MCF
%SO=62.2193
N=3,226,794.884 BBL
```

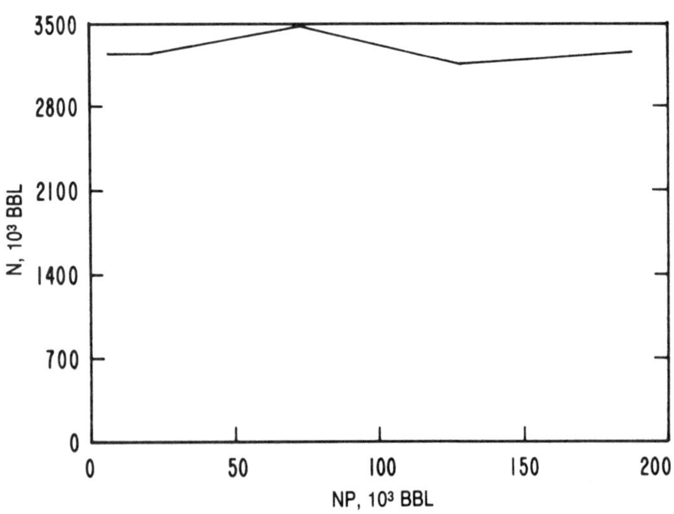

Figure 10–1

Example 2

The NW Franklin field in East Texas has an estimated original oil in place of 6,500,000 BBL (see the example for OILPVT). Its historical production is given in Table 10–2. Other pertinent data include:

Water salinity = 17,500 PPM
Porosity = 12.5%
Water saturation = 40.2%

Calculate the water influx at each point shown in Table 10–2. Although some water production has occurred, the water support is not very strong with little water influx to date.

Table 10–2

P (PSI)	*NP (BBL)*	*GP (MCF)*	*WP (BBL)*
2,850	–	–	–
2,723	30,704	15,674	0
2,605	64,120	32,637	150
2,475	103,300	52,802	9,420

Keystrokes	**Display**	**Comments**
[XEQ] [ALPHA] OILMBE [ALPHA]	*CORR? Y/N:Y*	
[R/S]	*SKIP? Y/N:N*	
[R/S]	*Tc=?*	
396.944 [R/S]	*Pc=?*	
660.1595 [R/S]	*STD T=?*	
[R/S]	*STD P=?*	
[R/S]	*SEP T=?*	
95 [R/S]	*SEP P=?*	
[R/S]	*OIL G=?*	
31.7 [R/S]	*GAS G=?*	
.762 [R/S]	*T=?*	GAS GS printed
[R/S]	*RSI=?*	
510 [R/S]	*%NACL=?*	PBP printed
[R/S]	*PPM=?*	
17500 [R/S]	*%POR=?*	
12.5 [R/S]	*%SWC=?*	
40.2 [R/S]	*PI=?*	
2850 [R/S]	*P=?*	
2723 [R/S]	*NP=?*	
30704 [R/S]	*GP=?*	
15674 [R/S]	*WP=?*	
0 [R/S]	*N We*	
[A]	*N=?*	N known
6.5 [EEX] 6 [R/S]	*N We*	We printed
[R/S]	*P=?*	
2605 [R/S]	*NP=?*	
64120 [R/S]	*GP=?*	
32637 [R/S]	*WP=?*	
150 [R/S]	*N We*	
[A]	*N=?*	
[R/S]	*N We*	We printed
[R/S]	*P=?*	
2475 [R/S]	*NP=?*	
103300 [R/S]	*GP=?*	

Keystrokes	Display	Comments
52802 [R/S]	*WP=?*	
9420 [R/S]	*N We*	
[A]	*N=?*	
[R/S]	*N We*	%SO and We printed

```
OIL MATL BAL

Tc=396.9440 R
Pc=660.1595 PSI
SEP T=95.0000 F
OIL G=31.7000 API
GAS G=0.7620
GAS GS=0.7671
RSI=510.0000 SCF/BBL
PBP=2,514.4215 PSI
PPM=17,500.0000
%POR=12.5000
%SWC=40.2000
PI=2,850.0000 PSI

P=2,723.0000 PSI
NP=30,704.0000 BBL
GP=15,674.0000 MCF
WP=0.0000 BBL
N=6,500,000.000 BBL
We=18,034.0875 BBL

P=2,605.0000 PSI
NP=64,120.0000 BBL
GP=32,637.0000 MCF
WP=150.0000 BBL
We=40,319.9839 BBL

P=2,475.0000 PSI
NP=103,300.0000 BBL
GP=52,802.0000 MCF
WP=9,420.0000 BBL
%SO=58.8713
We=48,656.1191 BBL
```

Example 3

Repeat Example 2 at the first pressure using the following PVT properties as input:

P (PSI)	2,850	2,723
BO	1.280	1.281
BG (FT3/SCF)	0.005	–
BW	–	1.019
CWI (1/PSI)	$3.05\ (10^{-6})$	–

Bubble-point pressure = 2,552 PSI

Formation compressibility = $4.4\ (10^{-6})$ 1/PSI

Notice that the calculated water influx is slightly higher using these PVT properties than that calculated using the Pac correlations.

Keystrokes	Display	Comments
[XEQ] [ALPHA] OILMBE [ALPHA]	*CORR? Y/N:Y*	
N [R/S]	*BOI=?*	
1.28 [R/S]	*BGI=?*	
.005 [R/S]	*RSI=?*	
[R/S]	*PBP=?*	
2552 [R/S]	*CWI=?*	
3.05 [EEX] [CHS] 6 [R/S]	*CFR=?*	
4.4 [EEX] [CHS] 6 [R/S]	*%SWC=?*	
[R/S]	*PI=?*	
[R/S]	*P=?*	
2723 [R/S]	*NP=?*	
30704 [R/S]	*GP=?*	
15674 [R/S]	*WP=?*	
0 [R/S]	*BO=?*	
1.281 [R/S]	*BW=?*	
1.019 [R/S]	*N We*	
[A]	*N=?*	
[R/S]	*N We*	We printed

```
OIL MATL BAL

CORR: NO
BOI=1.2800
BGI=0.0050 FT3/SCF
PBP=2,552.0000 PSI
CWI=3.0500E-6 1/PSI
CFR=4.4000E-6 1/PSI

P=2,723.0000 PSI
NP=30,704.0000 BBL
GP=15,674.0000 MCF
WP=0.0000 BBL
BO=1.2810
BW=1.0190
We=22,890.7500 BBL
```

User Instructions

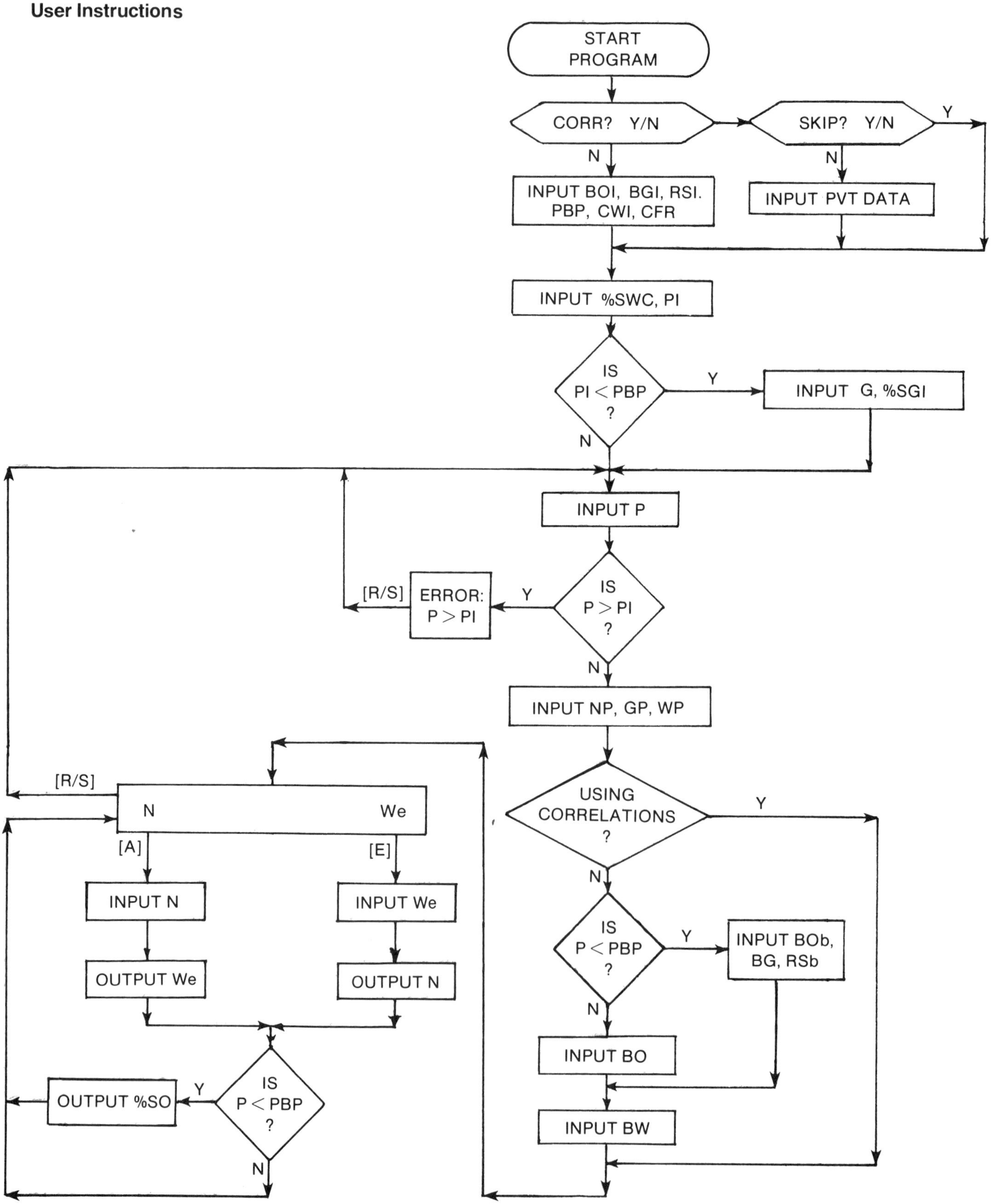

General Information

Memory Requirements

Program length: 896 bytes (4 cards)
Minimum size: 044
Minimum hardware: 41C + 2 memory modules

Hidden Options

None

Pac Subroutines Called

TITLE, Y/N?, W7, %NACL, %POR, IN, INU, RSI, CBOb, CBO, CBG, CCW, CCFR, OUTK, OUT, INK, CON, CRSb, CBW

Registers

03	Oil and water production units
04	Oil and water production units
06	Gas production units
07	Gas production units
08	Pressure units
09	Pressure units
20	%SGI
21	%SWC
26	BOI
27	BO, BOb
28	BGI (FT3/SCF)
29	BG (FT3/SCF)
30	RSb (SCF/BBL)
31	BW
32	CWI (1/PSI)
33	CFR (1/PSI)
34	Scratch
35	PI (PSI)
36	GP (MCF)
37	NP (BBL)
38	WP (BBL)
39	G (MCF)
40	N (BBL)
41	We (BBL)
42	BOBP
43	Scratch

Flags

00	Set:	Oil production units not yet input.
	Clear:	Oil production units have been input.
01	Set:	Use Pac correlations to estimate PVT properties.
	Clear:	Input PVT properties.
02	Set:	Skip input of PVT data.
	Clear:	Allow input of PVT data.
03	Set:	Current pressure ≥ PBP.
	Clear:	Current pressure < PBP.
04	Set:	Gas production units not yet input.
	Clear:	Gas production units have been input.

Program Listing

```
 01♦LBL "OILMBE"
"OIL MATL BAL"  44
XROM "TITLE"  FC?C 25
PROMPT  SF 00  SF 04
SF 06  SF 27  "M3"
ASTO 03  "SCM"  ASTO 06
"KPA"  ASTO 08  CLA
ASTO 04  ASTO 07
ASTO 09  "CORR"  1
XROM "Y/N?"  FC? 01
GTO 00  "SKIP"  2
XROM "Y/N?"  FS? 02
GTO 01  XROM "W7"
XROM "%NACL"
XROM "%POR"  GTO 01

 35♦LBL 00
25  STO 00  "BOI"
XROM "IN"  27  STO 00
"FT3/SCF"  ASTO 01  ASHF
ASTO 02  "M3/SCM"
ASTO Y  CLA  ASTO Z
"BGI"  XROM "INU"
XROM "RSI"  "PSI"
ASTO 01  CLA  ASTO 02
ASTO Z  "KPA"  ASTO Y
"PBP"  XROM "INU"  31
STO 00  "1/PSI"  ASTO 01
CLA  ASTO 02  ASTO Z
"1/KPA"  ASTO Y  "CWI"
XROM "INU"  "1/KPA"
ASTO Y  CLA  ASTO Z
"CFR"  XROM "INU"

 79♦LBL 01
20  STO 00  "%SWC"
XROM "IN"  34  "PI"
XEQ 06  RND  RCL 14  RND
CF 03  X<=Y?  SF 03  0
FS? 03  GTO 02  38  "G"
XEQ 07  19  STO 00
"%SGI"  XROM "IN"  CF 04

104♦LBL 02
STO 20  ADV  FC? 01
GTO 15  RCL 35  X<> 17
STO 35  RCL 14
XROM "CBOb"  STO 42
FS? 03  XROM "CBO"
FC? 03  RCL 17  FC? 03
XROM "CBOb"  STO 26
XEQ 00  XROM "CBG"
STO 28  XROM "CCW"
STO 32  RCL 35  X<> 17
STO 35  XROM "CCFR"
STO 33

132♦LBL 15
CF 08  16  "P"  XEQ 06
RCL 35  X<>Y  X<=Y?
GTO 02  TONE 3  "P > PI"
PROMPT  GTO 15

145♦LBL 02
```

Program Listing (cont.)

```
RND  RCL 14  RND  CF 03
X<=Y?  SF 03  FS?C 00
SF 08  36  "NP"  XEQ 08
XEQ 09  FC?C 04  CF 08
35  "GP"  XEQ 07  CF 08
37  "WP"  XEQ 08  XEQ 09
FS? 01  GTO 04  26
STO 00  "BO"  FC? 03
"⊦b"  XROM "IN"  FS? 03
GTO 03  28  STO 00
"FT3/SCF"  ASTO 01  ASHF
ASTO 02  "M3/SCM"
ASTO Y  CLA  ASTO Z
"BG"  XROM "INU"
"SCF/BBL"  ASTO 01  ASHF
ASTO 02  "SCM/M3"
ASTO Y  CLA  ASTO Z
"RSb"  XROM "INU"

200♦LBL 03
30  STO 00  "BW"
XROM "IN"

205♦LBL 04
" N        We"  PROMPT
GTO 15

209♦LBL A
39  "N"  XEQ 08  XEQ 09
XEQ 11  X<>Y  RCL 40  *
-  STO 41  XEQ 01  "We"
GTO 05

223♦LBL E
40  "We"  XEQ 08  XEQ 09
XEQ 11  RCL 41  -  X<>Y
/  STO 40  XEQ 01  "N"

236♦LBL 05
XEQ 08  XROM "OUTK"
XEQ 10  ADV  GTO 04

242♦LBL 01
FS? 03  RTN  STO 34  100
RCL 20  -  RCL 21  -
RCL 26  /  RCL 27  *  1
RCL 37  RCL 40  /  -  *
"%SO"  XROM "OUT"
RCL 34  RTN

265♦LBL 06
STO 00  ASTO T  "PSI"
ASTO 01  CLA  ASTO 02
ARCL T  RCL 09  RCL 08
RCL Z  XROM "INK"  RDN
STO 08  X<>Y  STO 09  R↑
RTN

283♦LBL 07
STO 00  ASTO T  "MCF"
ASTO 01  CLA  ASTO 02
ARCL T  RCL 07  RCL 06
RCL Z  XROM "INK"  RDN
STO 06  X<>Y  STO 07  R↑
RTN

301♦LBL 08
STO 00  ASTO T  "BBL"
ASTO 01  CLA  ASTO 02
ARCL T  RCL 04  RCL 03
RCL Z  RTN

313♦LBL 09
XROM "INK"

315♦LBL 10
RDN  STO 03  X<>Y
STO 04  R↑  RTN

322♦LBL 11
0  STO 34  STO 43
FS? 03  GTO 13  RCL 29
FC? 01  GTO 12  XEQ 00
XROM "CBG"  STO 29

334♦LBL 12
RCL 28  X<>Y  -
"FT3-BBL"  CON  RCL 39
*  1 E3  *  STO 34
RCL 17  FS? 01
XROM "CRSb"  FC? 01
RCL 30  STO 30  RCL 13
X<>Y  -  RCL 29
"FT3-BBL"  CON  *
STO 43  RCL 36  1 E3  *
RCL 37  RCL 30  *  -
RCL 29  CON  *  ST+ 34
RCL 27  FC? 01  GTO 14
RCL 17  XROM "CBOb"
GTO 14

376♦LBL 13
RCL 27  FC? 01  GTO 14
RCL 42  XROM "CBO"

382♦LBL 14
STO 27  ST+ 43  RCL 37
*  ST+ 34  FS? 01
XROM "CBW"  FC? 01
RCL 31  STO 31  RCL 38
*  ST+ 34  RCL 32
RCL 21  100  /  STO Z  *
RCL 33  +  RCL 35
RCL 17  -  *  1  R↑  -
/  1  -  RCL 26  *
RCL 43  +  RCL 34  RTN

420♦LBL 00
RCL 16  "F-R"  CON
RCL 10  /  RCL 17
RCL 11  /  END
```

11. KG/KO — Gas-Oil Relative-Permeability Ratio

A key factor in predicting the behavior of solution-gas-drive reservoirs is predicting the behavior of the gas-oil ratio. Above the bubble point, the producing gas-oil ratio (RP) is equal to the dissolved gas-oil ratio (RS=RSI). As the pressure drops below the bubble point and after enough free gas has formed, an additional component will be added due to free gas flow in the reservoir. Based on the radial flow equation, it is not difficult to derive the dynamic equation:

$$RP = RSb + \frac{KG}{KO}\,\frac{UOb}{UG}\,\frac{BOb}{BG'}$$

The gas and oil viscosities and formation-volume factors can be calculated easily. The ratio of the effective permeabilities (KG/KO) is equal to the ratio of the relative permeabilities (KRG/KRO). Since the water saturation of a solution-gas-drive reservoir is (neglecting expansion) constant, the only variable that determines KG/KO at a given pressure is the oil saturation.

KG/KO does two different types of calculations:

1. Given the initial oil in place, the cumulative oil and gas production, and the producing gas-oil ratio, it calculates the KG/KO term.
2. Using one of two correlations and two user-input coefficients, it calculates KG/KO for an array of oil saturations. This is useful to select coefficients to use for predictive purposes in OILPRED.

The two correlations used by the program originally were developed to represent trends observed in laboratory data but do not, of course, duplicate all possible KG/KO curves. Although some KG/KO curves may not be characterized by one of these correlations, most curves can be represented.

The default correlation can calculate values of KRG and KRO and, consequently, KRG/KRO (which equals KG/KO). Typical values for the coefficients a and b are given in Table 11-1. The values of a = b = 2 represent good-quality consolidated sandstones with very favorable KG/KO characteristics.

The optional correlation was developed to represent typical KG/KO curves primarily as a function of connate water saturation, with typical values of %SGc = 5 and c = 25. These coefficients represent the critical gas saturation and minimum oil saturation, respectively. Unlike the default correlation, the optional correlation provides for the formation of a critical gas saturation, since KG/KO will be zero until the gas saturation exceeds %SGc. However, the optional correlation cannot provide estimates of KRG or KRO. An estimate of KRO is required in the QOVST program.

While using the KG/KO program to select the different coefficients, the user frequently will alter the coefficients to fit field- or laboratory-measured KG/KO values. Since field-generated points are usually fairly sparse, it is recommended that the user try to select coefficients that honor the field data points and reflect the shape of the laboratory (core) KG/KO points.

Equations

P and RP known:

$$KGKO = (RP - RSb)\,\frac{UG}{UOb}\,\frac{BG'}{BOb}$$

$BG' = BG$ in BBL/SCF

%SO known:

Default KG/KO correlation:

$$KGKO_i = \frac{(1 - S_*)^a\,(1 - S_*{}^b)}{S_*{}^{a+b}}$$

$$S_* = \frac{\%SO_i}{100 - \%SWC}$$

Optional KG/KO correlation:

$$KGKO_i = A(0.04335 + 0.4556A)$$

$$A = \frac{100 - \%SGc - \%SWC - \%SO_i}{\%SO_i - c}$$

$$\%SL_i = \%SO_i + \%SWC$$

Nomenclature

Symbol	Variable Name	Input or Output	English Units	SI Units
a	First coefficient of default KG/KO correlation	I	-	-
b	Second coefficient of default KG/KO correlation	I	-	-

Symbol	Variable Name	Input or Output	English Units	SI Units
c	Second coefficient of optional KG/KO correlation, minimum oil saturation	I	–	–
KGKO*	Gas-oil relative permeability ratio	O	–	–
$KGKO_i$*	Gas-oil relative permeability ratio at $\%SO_i$	O	–	–
RP†	Producing gas-oil ratio	I,O	SCF/BBL	SCM/M3
%SGc	First coefficient of optional KG/KO correlation, volume percent critical gas saturation	I	–	–
$\%SL_i$	Volume percent liquid saturation at $\%SO_i$	O	–	–
$\%SO_i$	Volume percent oil saturation	I,O	–	–
%SWC	Volume percent connate water saturation	I	–	–

Note: For $KGKO_i$, $\%SL_i$, and $\%SO_i$, i = 1, 2, 3, . . . , n, where n is the number of %SO values input by the user.

*The program uses the symbol KGKO. The documentation often uses the symbol KG/KO because it is more readily recognized.
†The units for these variables are saved by the program.

Yes/No Questions

CORR? Yes: Use Pac correlations to estimate PVT properties.
No: Input PVT properties.

EDIT? Yes: Allow editing of %SO values.
No: No editing necessary.

SKIP? Yes: Skip input of PVT data.
No: Allow input of PVT data.

Table 11-1 (after Slider)

Formation	a	b
Unconsolidated sandstones (well sorted)	3	0
Unconsolidated sandstones (poorly sorted)	2	1.5
Cemented sandstones, oolitic limestones, and vugular limestones	2	2

Example 1

In the first example for OILMBE, we calculated the estimated oil in place from performance of an oil reservoir. Three of the seven data points were below the bubble point. In this example, we will calculate the KG/KO values for those data points. Besides the initial PVT data, the required information is as follows:

P (PSI)	RP (SCF/BBL)
2,108	458
1,996	434
1,905	539

Note that the calculated KG/KO value at 2,108 PSI is negative because the observed value of RP (458) is less than the solution value (460.7). This is, of course, either an error in the measured producing GOR or a difference between the correlations and the actual PVT performance.

Keystrokes (SIZE> =041)	Display	Comments
[/] [FIX] 6		
[XEQ] [ALPHA] KG/KO [ALPHA]	*P,RP %SO*	P and RP known
[A]	*CORR? Y/N:*	Last character is Y or N
Y [R/S]	*SKIP? Y/N:*	Last character is Y or N
N [R/S]	*Tc=?*	
377.8726 [R/S]	*Pc=?*	
665.8449 [R/S]	*STD T=?*	
60 [R/S]	*STD P=?*	
14.65 [R/S]	*SEP T=?*	
110 [R/S]	*SEP P=?*	
125 [R/S]	*OIL G=?*	
37.4 [R/S]	*GAS G=?*	
.678 [R/S]	*T=?*	GAS GS printed
155 [R/S]	*RSI=?*	
480 [R/S]	*P=?*	PBP printed
2108 [R/S]	*RP=?*	
458 [R/S]	*P=?*	RSb and KGKO printed
1996 [R/S]	*RP=?*	
434 [R/S]	*P=?*	RSb and KGKO printed
1905 [R/S]	*RP=?*	
539 [R/S]	*P=?*	RSb and KGKO printed

```
KG/KO

CORR: YES
SKIP: NO
Tc=377.872600 R
Pc=665.844900 PSI
STD T=60.000000 F
STD P=14.650000 PSI
SEP T=110.000000 F
SEP P=125.000000 PSI
OIL G=37.400000 API
GAS G=0.678000
GAS GS=0.684159
T=155.000000 F
RSI=480.000000 SCF/BBL
PBP=2,182.291272 PSI

P=2,108.000000 PSI
RP=458.000000 SCF/BBL
RSb=460.666099 SCF/BBL
KGKO=-0.000018

P=1,996.000000 PSI
RP=434.000000 SCF/BBL
RSb=431.759996 SCF/BBL
KGKO=0.000016

P=1,905.000000 PSI
RP=539.000000 SCF/BBL
RSb=408.495414 SCF/BBL
KGKO=0.001739
```

Example 2

Figure 11–1 is a graph of core-analysis results for KG/KO with the field performance points plotted. The object of this example is to calculate coefficients for a correlation equation that will be similar in shape to the core data but match the field performance as well.

The first attempt was to use the default correlation with a = 2 and b = 2. This shape is so dissimilar from the core data that the optional correlation with %SGc = 5 and c = 25 was tried. Based on the field performance, it is clear that the critical gas saturation is less than five percent but probably greater than 2 percent. After some experimentation with different values for the coefficients, %SGc = 3.825 and c = 10 appear to match both the lab shape and the field performance points. These different KG/KO curves are shown in Figure 11–2.

Note that as many %SO points as desired can be input (see the "Memory Requirements" section for details). The value of %SL (= %SO + %SWC) is printed for convenience since it is used often to plot KG/KO data. The program contains a feature enabling the user to edit the input %SO data as well. This is shown in the user instructions.

Keystrokes	Display	Comments
[XEQ] [ALPHA] KG/KO [ALPHA]	*P,RP %SO*	
[E]	*%SWC=?*	
32.5 [R/S]	*%SO1=?*	
65.5 [R/S]	*%SO2=?*	
63.65 [R/S]	*%SO3=?*	
61.89 [R/S]	*%SO4=*	
60 [R/S]	*%SO5=?*	
55 [R/S]	*%SO6=?*	
50 [R/S]	*%SO7=?*	
45 [R/S]	*%SO8=?*	
[R/S]	*a=?*	
2 [R/S]	*b=?*	
2 [R/S]	*EDIT? Y/N:N*	%SO, %SL, and KG/KO values printed

If you made any errors while inputing the %SO values, the edit option will allow you to go back and change the incorrect values.

Keystrokes	Display	Comments
[ALPHA] [//] [SF] 00 [R/S]	*%SGc=?*	Select optional KG/KO correlation
5 [R/S]	*c=?*	
25 [R/S]	*EDIT? Y/N:N*	%SO, %SL, and KG/KO values printed
[R/S]	*%SGc=?*	
3.825 [R/S]	*c=?*	
10 [R/S]	*EDIT? Y/N:N*	%SO, %SL, and KG/KO values printed
[ALPHA] [//] [FIX] 4		Back to FIX 4

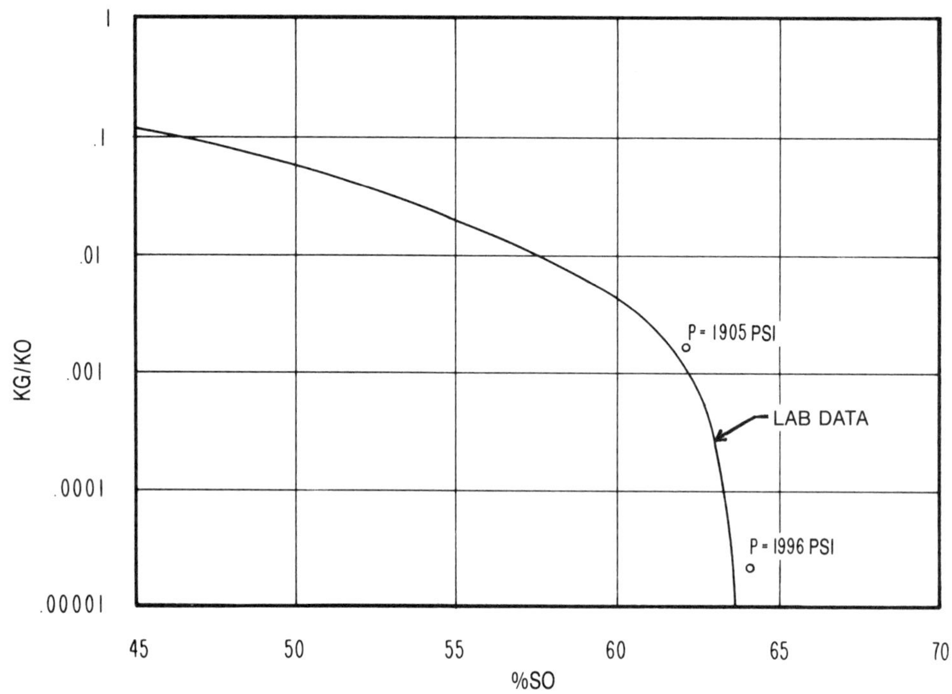

Figure 11–1

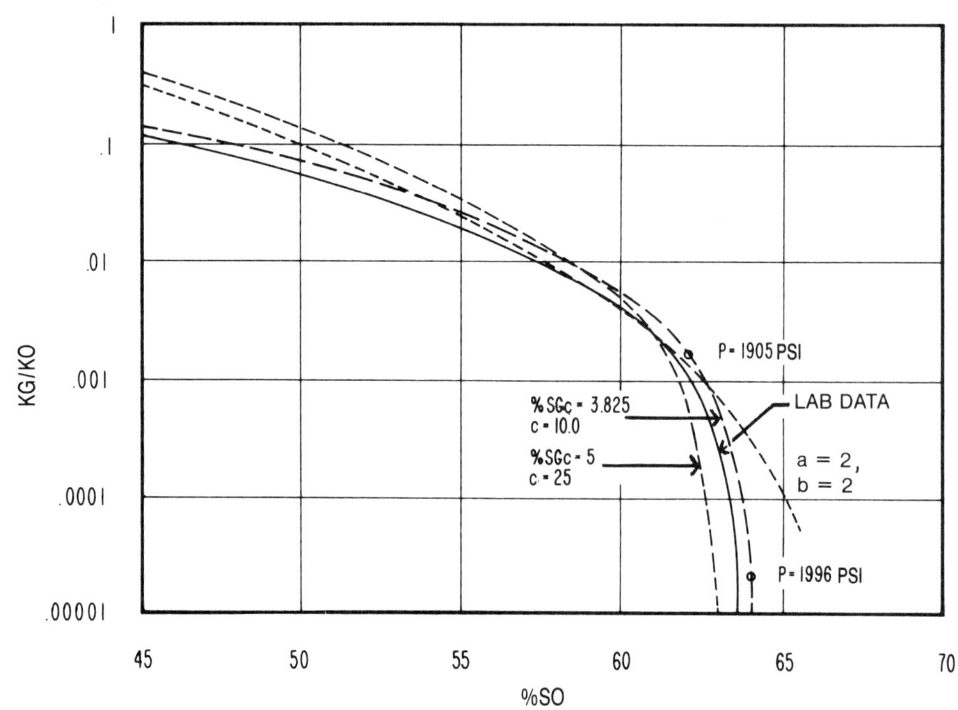

Figure 11–2

```
KG/KO

%SWC=32.500000
%SO1=65.500000
%SO2=63.650000
%SO3=61.890000
%SO4=60.000000
%SO5=55.000000
%SO6=50.000000
%SO7=45.000000

a=2.000000
b=2.000000

%SO1=65.500000
%SL1=98.000000
KGKO1=0.000058

%SO2=63.650000
%SL2=96.150000
KGKO2=0.000456

%SO3=61.890000
%SL3=94.390000
KGKO3=0.001557

%SO4=60.000000
%SL4=92.500000
KGKO4=0.004150

%SO5=55.000000
%SL5=87.500000
KGKO5=0.026147

%SO6=50.000000
%SL6=82.500000
KGKO6=0.100756

%SO7=45.000000
%SL7=77.500000
KGKO7=0.312500

%SGc=5.000000
c=25.000000

%SO1=65.500000
%SL1=98.000000
KGKO1=0.000000

%SO2=63.650000
%SL2=96.150000
KGKO2=0.000000

%SO3=61.890000
%SL3=94.390000
KGKO3=0.000841

%SO4=60.000000
%SL4=92.500000
KGKO4=0.005421

%SO5=55.000000
%SL5=87.500000
KGKO5=0.039313

%SO6=50.000000
%SL6=82.500000
KGKO6=0.135575

%SO7=45.000000
%SL7=77.500000
KGKO7=0.386750

%SGc=3.825000
c=10.000000

%SO1=65.500000
%SL1=98.000000
KGKO1=0.000000

%SO2=63.650000
%SL2=96.150000
KGKO2=0.000020

%SO3=61.890000
%SL3=94.390000
KGKO3=0.002030

%SO4=60.000000
%SL4=92.500000
KGKO4=0.005647

%SO5=55.000000
%SL5=87.500000
KGKO5=0.025289

%SO6=50.000000
%SL6=82.500000
KGKO6=0.068070

%SO7=45.000000
%SL7=77.500000
KGKO7=0.152839
```

User Instructions

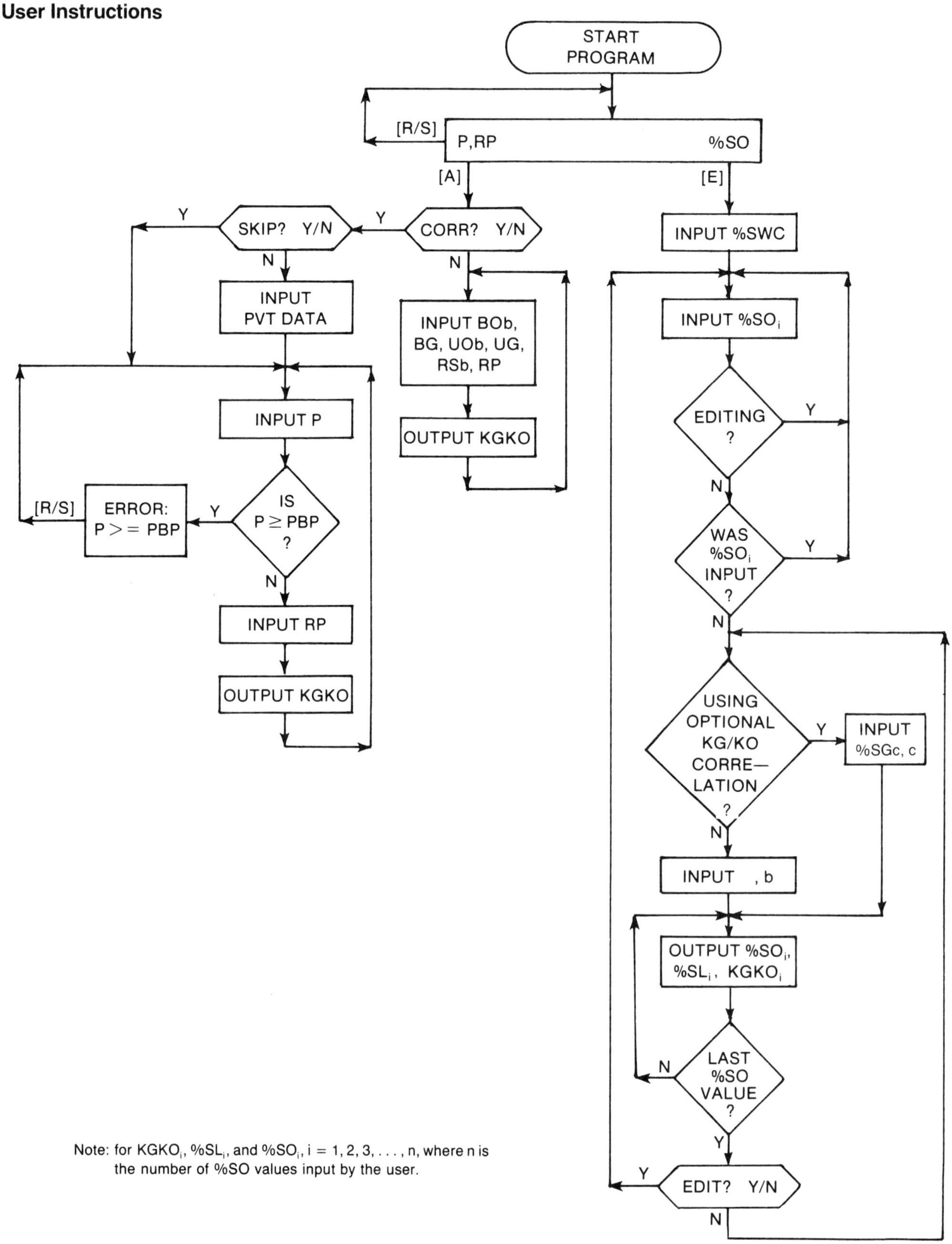

Note: for KGKO$_i$, %SL$_i$, and %SO$_i$, i = 1, 2, 3, . . . , n, where n is the number of %SO values input by the user.

General Information

Memory Requirements

Program length:	661 bytes (3 cards)
Minimum size:	041*
Minimum hardware:	41C + 2 memory modules

*This size will allow up to 10 %SO values. To accommodate v %SO values, use size 31 + v.

Hidden Options

When KG/KO is run, the default KG/KO correlation will be selected automatically. To change to the optional correlation, set flag 00 ([//] [SF] 00) any time before the two coefficient prompts appear. To change back to the default correlation, clear flag 00 ([//] [CF] 00) or run the program again. The flag 00 annunciator in the display will be off if the default correlation is being used and on if the optional correlation is being used.

Pac Subroutines Called

TITLE, Y/N?, W7, CUOd, P, CRSb, X8, CUOb, CUG, CBOb, CBG, CON, OUT, IN, INU, INK

Registers

03	Producing gas-oil ratio units
04	Producing gas-oil ratio units
06	UOd, UOb (CP)
07	TR, UG (CP)
08	PR
09	Scratch
21	%SWC
26	RP (SCF/BBL)
27	Pointer, BOb
28	a, %SGc
29	b, c, BG (FT3/SCF)
30	%SO1, RSb (SCF/BBL)
31	%SO2
32	%SO3
33	%SO4
34	%SO5
35	%SO6
36	%SO7
37	%SO8
38	%SO9
39	%SO10

Registers 18–20 unused

Flags

00	Set:	Use optional KG/KO correlation.
	Clear:	Use default KG/KO correlation.
01	Set:	Use Pac correlations to estimate PVT properties.
	Clear:	Input PVT properties.
02	Set:	Skip input of PVT data.
	Clear:	Allow input of PVT data.
03	Set:	Allow editing of %SO values.
	Clear:	No editing necessary.

Program Listing

```
 01♦LBL "KG/KO"
"KG/KO" 41
XROM "TITLE" FC?C 25
PROMPT CF 00 CF 03
SF 27 "SCM/M3" ASTO 03
CLA ASTO 04

 14♦LBL 00
"P,RP      %SO" PROMPT
GTO 00

 18♦LBL A
 19♦LBL B
"CORR" 1 XROM "Y/N?"
FC? 01 GTO 15 "SKIP"
2 XROM "Y/N?" FC? 02
XROM "W7" XROM "CUOd"
STO 06 ADV

 33♦LBL 01
XROM "P" RND RCL 14
RND X>Y? GTO 10
TONE 3 "P >= PBP"
PROMPT GTO 01

 44♦LBL 10
R↑ STO 07 R↑ STO 08
XEQ 02 RCL 17
XROM "CRSb" XROM "X8"
RCL 26 X<>Y - STO 09
RCL 06 XROM "CUOb"
ST/ 09 RCL 07 RCL 08
XROM "CUG" ST* 09
RCL 17 XROM "CBOb"
ST/ 09 RCL 07 RCL 08
XROM "CBG" RCL 09
XEQ 11 GTO 01

 73♦LBL 15
26 STO 00 "BOb"
XROM "IN" 28 STO 00
"FT3/SCF" ASTO 01 ASHF
ASTO 02 "M3/SCM"
ASTO Y CLA ASTO Z
"BG" XROM "INU" 5
STO 00 "CP" ASTO 01
CLA ASTO 02 ASTO Z
"PA*S" ASTO Y "UOb"
XROM "INU" "PA*S"
ASTO Y CLA ASTO Z
"UG" XROM "INU" 29
STO 00 "SCF/BBL"
ASTO 01 ASHF ASTO 02
"SCM/M3" ASTO Y CLA
ASTO Z "RSb"
XROM "INU" XEQ 02
```

Program Listing (cont.)

```
RCL 30  -  RCL 07  *
RCL 06  /  RCL 27  /
RCL 29  XEQ 11  GTO 15

131♦LBL 11
"FT3-BBL"  CON  *
"KGKO"  XROM "OUT"  ADV
RTN

139♦LBL 02
"SCF/BBL"  ASTO 01  ASHF
ASTO 02  RCL 04  RCL 03
25  STO 00  "RP"
XROM "INK"  RDN  STO 03
X<>Y  STO 04  R↑  CF 08
RTN

157♦LBL D
158♦LBL E
20  STO 00  "%SWC"
XROM "IN"  30  STO 27

165♦LBL 16
29  STO 00

168♦LBL 04
"%SO"  XEQ 09  XROM "IN"
FC? 03  GTO 05  ISG 27
GTO 04  GTO 17

177♦LBL 05
ISG 27  CLD  FS? 22
GTO 04  RCL 00  1  -
1 E3  /  30  +  STO 09

190♦LBL 17
ADV  27  STO 00  "a"
FS? 00  "%SGc"
XROM "IN"  "b"  FS? 00
"c"  XROM "IN"  RCL 09
STO 27  ADV

205♦LBL 18
RCL IND 27  "%SO"
XEQ 09  XROM "OUT"
RCL 21  +  "%SL"  XEQ 09
XROM "OUT"  FC? 00
GTO 07  RCL 28  +  100
-  X>0?  0  RCL 29
RCL IND 27  -  X=0?
GTO 06  /  X<0?  GTO 06
.4556  RCL Y  *  .04335
+  *  GTO 08

238♦LBL 06
TONE 3  "c >= %SO"
PROMPT  GTO 06

243♦LBL 07
RCL IND 27  100  RCL 21
-  /  1  RCL Y  RCL 29
Y↑X  -  1  RCL Z  -
RCL 28  Y↑X  *  R↑
RCL 28  RCL 29  +  Y↑X
/

266♦LBL 08
"KGKO"  XEQ 09
XROM "OUT"  ADV  ISG 27
GTO 18  "EDIT"  3
XROM "Y/N?"  FC? 03
GTO 17  ADV  RCL 09
STO 27  GTO 16

282♦LBL 09
STO 05  CLST  FS? 41  1
+  FS? 40  2  +  5  /
FS? 39  1  +  FS? 38  2
+  FS? 37  4  +  FS? 36
8  +  FS? 29  CHS
RCL 27  29  -  FIX 0
CF 29  ARCL X  X<>Y
X<0?  SF 29  ENTER↑  FRC
5  *  FIX IND Y  X=0?
SCI IND Y  1  X=Y?
ENG IND Z  RCL 05  END
```

12. OILPRED — Predicting Solution-Gas-Drive Performance

It often is useful to predict reservoir performance from the material-balance equation. There is some question whether material-balance forecasts are substantially better than conventional decline-curve analysis. This is due to the problems in determining gas-oil relative-permeability performance. Core measurements of relative permeability usually are inadequate to predict performance; it is generally necessary to calculate KG/KO data from performance. By the time sufficient data has been acquired to do so, it may well be the case that decline-curve analysis may yield just as good forecasts. However, the material-balance method may be more useful, especially with the GOR and pressure-depletion forecasts it provides.

The method used in these programs for predicting the performance is basically similar to that described by Slider. The program KG/KO is used to determine relative-permeability data from performance as well as to determine coefficients for one of two KG/KO correlations available.

OILPRED uses the material-balance equation to calculate directly the oil and gas production for pressures above the bubble point. Below the bubble point, the Tarner method is used. This method involves an iterative procedure to calculate RP at each pressure step. Since the Tarner method (unlike some others) tends to be self-correcting, pressure steps of less than 200 PSI generally add little significance to the forecast. Unlike the Slider example, the effect of rock and water compressibilities are included below the bubble point.

Equations

Above bubble point:

$$NP = \frac{N[BO - BOI + (CFR + CWI\ SWC)(PI - P)BOI/(1 - SWC)]}{BO}$$

$$GP = RSI\ NP$$

$$\%SO = \frac{BO}{BOI}\ \frac{N - NP}{N}\ (100 - \%SWC)$$

Below bubble point: NP is solved iteratively at the ith pressure step as follows:

$$\begin{aligned} NP_i = \{\ & N[BOb - BOI + (CFR + CWI\ SWC) \\ & (PI - P)BOI/(1 - SWC) + (RSI - RSb)BG'] \\ & + G(BG' - BGI') - BG'[GP_{i-1} \\ & - (RP_i + RP_{i-1})NP_{i-1}/2]\ \} \div [BOb - BG'RSb \\ & + (RP_i + RP_{i-1})BG'/2] \end{aligned}$$

$$\%SO_i = \frac{BOb}{BOI}\ \frac{N - NP_i}{N}\ (100 - \%SWC - \%SGI)$$

$(KG/KO)_i$ is calculated from SO_i using the KG/KO correlations described in the KG/KO program.

$$RP_i = RSb + \left(\frac{KG}{KO}\right)_i \frac{UOb}{UG}\ \frac{BOb}{BG'}$$

This is repeated until the RP_i is within 0.01% of the RP_i used in the NP_i equation.

NP_{i-1} at next pressure step $= NP_i$

GP_{i-1} at next pressure step $=$
$GP_{i-1} + (NP_i - NP_{i-1})(RP_i + RP_{i-1})/2$

RP_i at next pressure step $= RP_i + (RP_i - RP_{i-1})$

RP_{i-1} at next pressure step $= RP_i$

Initial guesses are $NP_{i-1} = GP_{i-1} = 0$ and $RP = RP_{i-1}$
$= RSI$ (if BEG $P \geq PBP$)
or RPI (if BEG $P < PBP$).

If $PI \geq PBP$ and BEG $P < PBP$,

$$\%SGI = 100 - \%SWC - \left(\frac{BOb}{BOI}\ \frac{N - NPI}{N}\right) \times (100 - \%SWC)$$

$$G = N\ RSI - GPI - (N - NP)RSb$$

$$NR = N - NP$$

$$\%OILRF = \frac{NP}{NR}(100)$$

$$GR = G + N\ RS - GP$$

$$\%GASRF = \frac{GP}{GR}(100)$$

$$BG' = BG,\ BBL/SCF$$

$$BGI' = BGI,\ BBL/SCF$$

$$SWC = \frac{\%SWC}{100}$$

Nomenclature

Symbol	Variable Name	Input or Output	English Units	SI Units
a	First coefficient of default KG/KO correlation	I	–	–
BEG P*	Beginning pressure of forecast	I	PSI	KPA
b	Second coefficient of default KG/KO correlation	I	–	–
c	Second coefficient of optional KG/KO correlation, minimum oil saturation	I	–	–
DGP*	Delta GP, change in cumulative gas production	O	MCF	SCM
DNP*	Delta NP, change in cumulative oil production	O	BBL	M3
END P*	Ending pressure of forecast	I	PSI	KPA
G*	Initial free gas in place	I	MCF	SCM
GP*	Cumulative gas production	O	MCF	SCM
GPBP*	Cumulative gas production up to PBP	O	MCF	SCM
GPI*	Initial cumulative gas production	I,O	MCF	SCM
GR*	Remaining total gas in place at end of forecast	O	MCF	SCM
N*	Initial oil in place	I	BBL	M3
NP*	Cumulative oil production	O	BBL	M3
NPBP*	Cumulative oil production up to PBP	O	BBL	M3
NPI*	Initial cumulative oil production	I,O	BBL	M3
NR*	Remaining oil in place at the end of forecast	O	BBL	M3
P*	Pressure	O	PSI	KPA
PI*	Initial pressure	I	PSI	KPA
P DEC*	Pressure decrement of forecast	I	PSI	KPA
RP*	Producing gas-oil ratio	O	SCF/BBL	SCM/M3
RPI*	Initial producing gas-oil ratio	I	SCF/BBL	SCM/M3
%GASRF	Volume percent gas recovery factor	O	–	–
%OILRF	Volume percent oil recovery factor	O	–	–
%SGc	First coefficient of optional KG/KO correlation, volume percent critical gas saturation	I	–	–
%SGI	Volume percent initial gas saturation	I,O	–	–
%SWC	Volume percent connate water saturation	I	–	–

*The units for these variables are saved by the program.

Yes/No Questions

SKIP? Yes: Skip input of PVT data.
No: Allow input of PVT data.

Example 1

This example will predict the performance of the oil field used in Example 1 of OILMBE and Examples 1 and 2 of KG/KO. Our producing history for OILMBE ended at a pressure of 1,905 PSI with cumulative production of 189,574 BBL and 88,984 MCF. The original value of N was estimated to be 3,250,000 BBL with an initial dissolved gas-oil ratio of 480 SCF/BBL. Using the initial pressure and oil in place, a starting pressure of 1,905 PSI, and the cumulative production, forecast the reservoir production.

Figures 12–1 and 12–2 present the historical and calculated values. Notice that %SO at 1,905 PSI is 100 − %SWC − %SGI = 62.2471, which is almost identical to the %SO value at 1,905 PSI that was calculated by OILMBE. The difference is that N in this example is the approximate value of $3.25(10^6)$ BBL, while N in the OILMBE example is $3.227(10^6)$ BBL, calculated at 1,905 PSI.

Keystrokes (SIZE>=063)	Display	Comments
[XEQ] [ALPHA] OILPRED [ALPHA]	*SKIP? Y/N:*	Last character is Y or N

Keystrokes (SIZE> =063)	Display	Comments
N [R/S]	*Tc=?*	
[/] [SF] 00		Select optional KG/KO corre-lation
377.8726 [R/S]	*Pc=?*	
665.8449 [R/S]	*STD T=?*	
60 [R/S]	*STD P=?*	
14.65 [R/S]	*SEP T=?*	
110 [R/S]	*SEP P=?*	
125 [R/S]	*OIL G=?*	
37.4 [R/S]	*GAS G=?*	
.678 [R/S]	*T=?*	GAS GS printed
155 [R/S]	*RSI=?*	
480 [R/S]	*%NACL=?*	PBP printed
1.9 [R/S]	*%POR=?*	
12.5 [R/S]	*%SWC=?*	
32.5 [R/S]	*%SGc=?*	Notice prompts for optional KG/KO corre-lation
3.825 [R/S]	*c=?*	
10 [R/S]	*PI=?*	
2650 [R/S]	*N=?*	
3.25 [EEX] 6 [R/S]	*BEG P=?*	
1905 [R/S]	*END P=?*	
200 [R/S]	*P DEC=?*	
200 [R/S]	*NPI=?*	

Since PI is above the bubble point and BEG P is below the bubble point, you must provide historical production data.

Keystrokes	Display	Comments
189574 [R/S]	*GPI=?*	
88984 [R/S]	*RPI=?*	
539 [R/S]	*BEG P=?*	%SGI printed, followed by the produc-tion forecast

```
OIL MBE PRED

SKIP: NO
Tc=377.8726 R
Pc=665.8449 PSI
STD T=60.0000 F
STD P=14.6500 PSI
SEP T=110.0000 F
SEP P=125.0000 PSI
OIL G=37.4000 API
GAS G=0.6780
GAS GS=0.6842
T=155.0000 F
RSI=480.0000 SCF/BBL
PBP=2,182.2913 PSI
%NACL=1.9000
%POR=12.5000
%SWC=32.5000
%SGc=3.8250
c=10.0000
PI=2,650.0000 PSI
N=3,250,000.000 BBL

BEG P=1,905.0000 PSI
END P=200.0000 PSI
P DEC=200.0000 PSI
NPI=189,574.0000 BBL
GPI=88,984.0000 MCF
RPI=539.0000 SCF/BBL
%SGI=5.2529

P=1,705.0000 PSI
NP=314,009.4626 BBL
DNP=124,435.4625 BBL
GP=169,732.8381 MCF
DGP=80,748.8381 MCF
%SO=58.5758
RP=758.8429 SCF/BBL

P=1,505.0000 PSI
NP=424,326.9298 BBL
DNP=110,317.4674 BBL
GP=281,027.9210 MCF
DGP=111,295.0828 MCF
%SO=55.3013
RP=1,258.8808 SCF/BBL

P=1,305.0000 PSI
NP=512,930.4402 BBL
DNP=88,603.5103 BBL
GP=420,238.8873 MCF
DGP=139,210.9665 MCF
%SO=52.5529
RP=1,883.4544 SCF/BBL

P=1,105.0000 PSI
NP=584,787.7414 BBL
DNP=71,857.3012 BBL
GP=579,444.4440 MCF
DGP=159,205.5566 MCF
%SO=50.2135
RP=2,547.7044 SCF/BBL

P=905.0000 PSI
NP=645,160.1232 BBL
DNP=60,372.3817 BBL
GP=751,936.9436 MCF
DGP=172,492.4996 MCF
%SO=48.1695
RP=3,166.5807 SCF/BBL

P=705.0000 PSI
NP=698,234.2109 BBL
DNP=53,074.0879 BBL
GP=932,385.1123 MCF
DGP=180,448.1688 MCF
%SO=46.3362
RP=3,633.2788 SCF/BBL

P=505.0000 PSI
NP=747,810.1483 BBL
DNP=49,575.9373 BBL
GP=1,116,316.938 MCF
DGP=183,931.8259 MCF
%SO=44.6443
RP=3,786.9269 SCF/BBL

P=305.0000 PSI
NP=799,127.2209 BBL
DNP=51,317.0726 BBL
GP=1,299,391.084 MCF
DGP=183,074.1459 MCF
%SO=43.0099
RP=3,348.0921 SCF/BBL

P=200.0000 PSI
NP=830,165.7521 BBL
DNP=31,038.5312 BBL
GP=1,393,397.113 MCF
DGP=94,006.0293 MCF
%SO=42.1240
RP=2,709.2840 SCF/BBL

NR=2,419,834.248 BBL
%OILRF=25.5436
GR=155,058.9970 MCF
%GASRF=89.9862
```

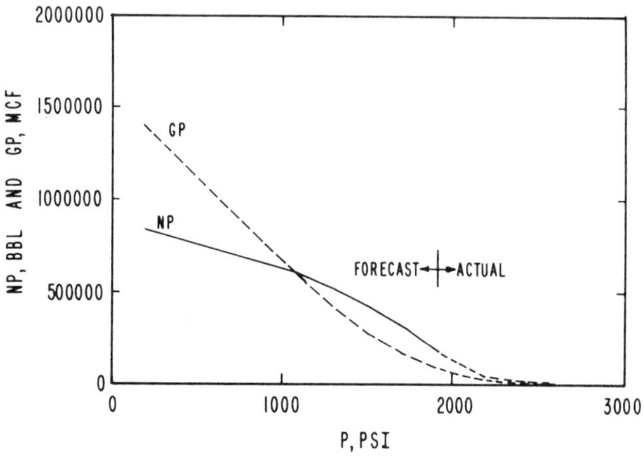

Figure 12–1

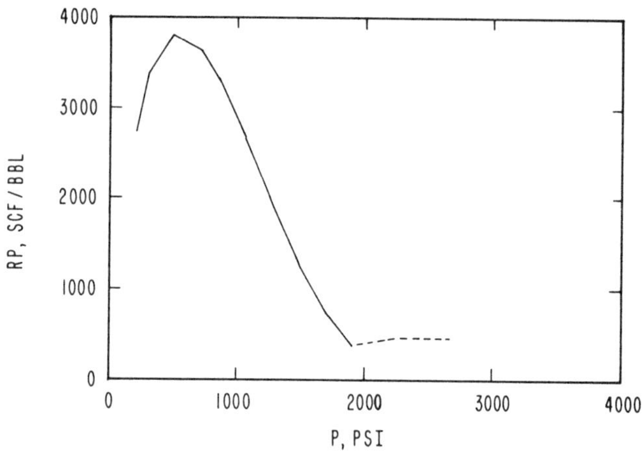

Figure 12–2

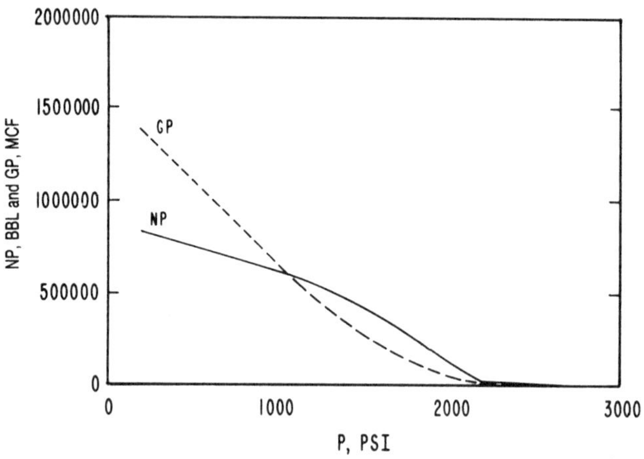

Figure 12–3

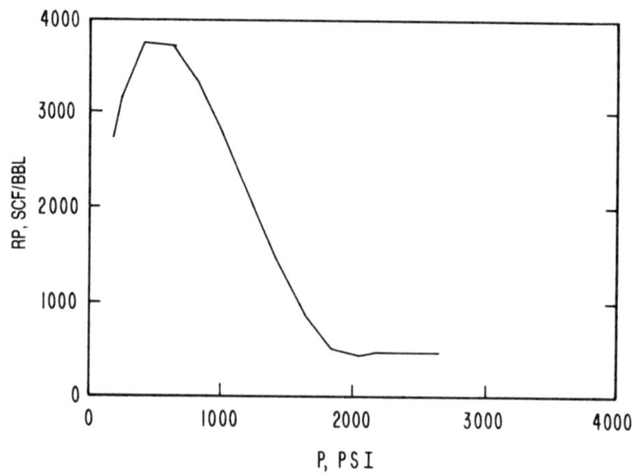

Figure 12–4

Example 2

Another method of forecasting the production would be to start at P = 2,650 PSI and N = 3.25(10^6)BBL, ignoring the actual historical performance. Figures 12–3 and 12–4 present this forecast. Note that the ultimate recoveries are, as expected, quite similar. This illustrates the procedure which would normally be used to forecast reservoir production for undersaturated reservoirs prior to any production.

Keystrokes	Display	Comments
2650 [R/S]	*END P=?*	
[R/S]	*P DEC=?*	
[R/S]	*BEG P=?*	NPI and GPI printed, followed by the production forecast

```
BEG P=2,650.0000 PSI
NPI=0.0000 BBL
GPI=0.0000 MCF

P=2,450.0000 PSI
NP=12,439.8705 BBL
DNP=12,439.8705 BBL
GP=5,971.1378 MCF
DGP=5,971.1378 MCF

P=2,250.0000 PSI
NP=26,111.2543 BBL
DNP=13,671.3839 BBL
GP=12,533.4021 MCF
DGP=6,562.2643 MCF

NPBP=31,077.4621 BBL
GPBP=14,917.1818 MCF

P=2,050.0000 PSI
NP=101,934.1052 BBL
DNP=75,822.8508 BBL
GP=47,711.7627 MCF
DGP=35,178.3606 MCF
%SO=65.1778
RP=445.6600 SCF/BBL

P=1,850.0000 PSI
NP=228,265.5473 BBL
DNP=126,331.4422 BBL
GP=108,524.5946 MCF
DGP=60,812.8320 MCF
%SO=61.3670
RP=517.0905 SCF/BBL

P=1,650.0000 PSI
NP=354,496.2214 BBL
DNP=126,230.6741 BBL
GP=196,016.5210 MCF
DGP=87,491.9264 MCF
%SO=57.6811
RP=869.1324 SCF/BBL

P=1,450.0000 PSI
NP=459,537.2851 BBL
DNP=105,041.0637 BBL
GP=315,901.9765 MCF
DGP=119,885.4554 MCF
%SO=54.5313
RP=1,413.5074 SCF/BBL

P=1,250.0000 PSI
NP=543,475.7937 BBL
DNP=83,938.5085 BBL
GP=461,665.0975 MCF
DGP=145,763.1210 MCF
%SO=51.8917
RP=2,059.5856 SCF/BBL

P=1,050.0000 PSI
NP=611,959.2148 BBL
DNP=68,483.4212 BBL
GP=625,432.9095 MCF
DGP=163,767.8120 MCF
%SO=49.6335
RP=2,723.1139 SCF/BBL

P=850.0000 PSI
NP=670,055.3091 BBL
DNP=58,096.0943 BBL
GP=800,827.7840 MCF
DGP=175,394.8745 MCF
%SO=47.6485
RP=3,314.9813 SCF/BBL

P=650.0000 PSI
NP=721,812.4584 BBL
DNP=51,757.1493 BBL
GP=982,814.2516 MCF
DGP=181,986.4676 MCF
%SO=45.8566
RP=3,717.3406 SCF/BBL

P=450.0000 PSI
NP=771,183.3424 BBL
DNP=49,370.8841 BBL
GP=1,167,027.606 MCF
DGP=184,213.3537 MCF
%SO=44.1879
RP=3,745.0882 SCF/BBL

P=250.0000 PSI
NP=824,621.0171 BBL
DNP=53,437.6746 BBL
GP=1,348,979.227 MCF
DGP=181,951.6212 MCF
%SO=42.5409
RP=3,064.7748 SCF/BBL

P=200.0000 PSI
NP=840,011.0536 BBL
DNP=15,390.0365 BBL
GP=1,393,447.156 MCF
DGP=44,467.9287 MCF
%SO=42.1119
RP=2,714.0196 SCF/BBL

NR=2,409,988.946 BBL
%OILRF=25.8465
GR=166,552.8441 MCF
%GASRF=89.3235
```

Example 3

For undersaturated reservoirs, the initial free gas in place (G) and the initial gas saturation must be input. If we had discovered this reservoir at 1,905 PSI, we could easily forecast future production. The total gas in place initially was 480 * 3.25(10^6) SCF, or 1.56 BCF. After the production to date, the remaining gas in place is 1.56 − 0.089 = 1.471 BCF. Based on the current dissolved gas-oil ratio of 408.5 SCF/BBL and the remaining oil in place of 3.25(10^6) − 189,574 = 3,060,426 BBL, there is 408.5(3,060,426)SCF, or 1.25 BCF gas in solution. Therefore, the free gas in place is 1.471 − 1.25 BCF or 221 MMCF.

See Figures 12-5 and 12-6 for the results of this forecast. Although these techniques theoretically are applicable, oil reservoirs with large gas caps often behave somewhat differently due to GOR control in actual operations.

Keystrokes	Display	Comments
[XEQ] [ALPHA] OILPRED [ALPHA]	*SKIP? Y/N:N*	
Y [R/S]	*%SWC=?*	
[//] [SF] 00 [R/S]	*%SGc=?*	
[R/S]	*c=?*	
[R/S]	*PI=?*	
1905 [R/S]	*N=?*	

Keystrokes	Display	Comments
3060426 [R/S]	*G=?*	
221 [ALPHA] MMCF [R/S]	*%SGI=?*	
[←]	*0.0000*	

At BEG P = 2,650 PSI (above the bubble point) in Example 2, there was no free gas in the reservoir, so %SGI was set to zero.

Keystrokes	Display	Comments
5.2529 [R/S]	*BEG P=?*	
1905 [R/S]	*END P=?*	
[R/S]	*P DEC=?*	
[R/S]	*NPI=?*	

With N and G relative to discovery at 1,905 PSI, NPI = GPI = 0.

Keystrokes	Display	Comments
0 [R/S]	*GPI=?*	
0 [ALPHA] MCF [R/S]	*RPI=?*	
[R/S]	*BEG P=?*	Production forecast printed

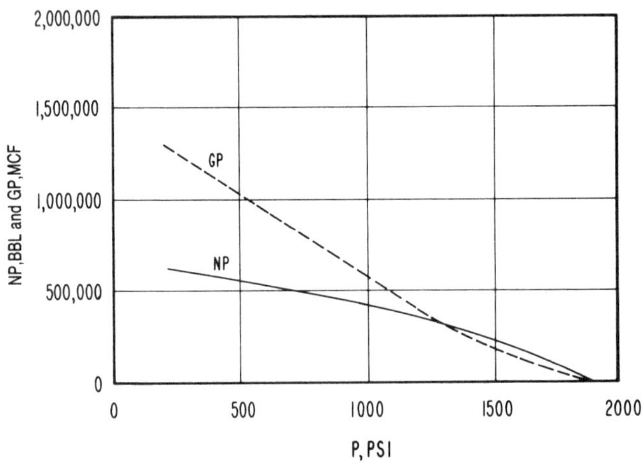

Figure 12–5

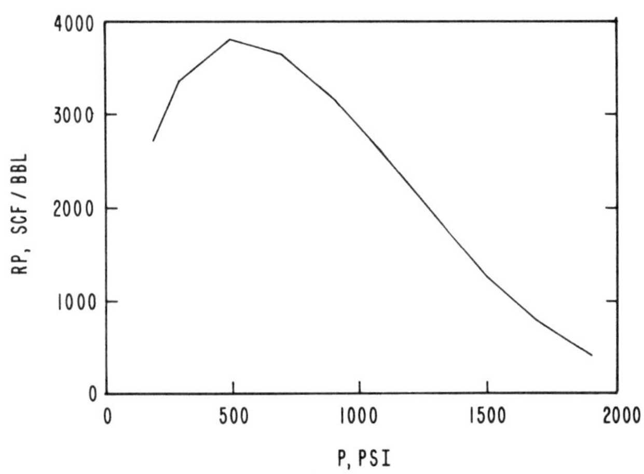

Figure 12–6

```
OIL MBE PRED

SKIP: YES
PI=1,905.0000 PSI
N=3,060,426.000 BBL
G=221.0000 MMCF
%SGI=5.2529

BEG P=1,905.0000 PSI
NPI=0.0000 BBL
GPI=0.0000 MCF

P=1,705.0000 PSI
NP=124,254.6468 BBL
DNP=124,254.6468 BBL
GP=80,601.5595 MCF
DGP=80,601.5595 MCF
%SO=58.5795
RP=758.3609 SCF/BBL

P=1,505.0000 PSI
NP=234,467.6746 BBL
DNP=110,213.0278 BBL
GP=191,695.6891 MCF

DGP=111,094.1296 MCF
%SO=55.3069
RP=1,257.6282 SCF/BBL

P=1,305.0000 PSI
NP=323,019.4382 BBL
DNP=88,551.7636 BBL
GP=330,678.1209 MCF
DGP=138,982.4318 MCF
%SO=52.5594
RP=1,881.3817 SCF/BBL

P=1,105.0000 PSI
NP=394,849.7526 BBL
DNP=71,830.3144 BBL
GP=489,647.6893 MCF
DGP=158,969.5682 MCF
%SO=50.2204
RP=2,544.8712 SCF/BBL

P=905.0000 PSI
NP=455,206.7084 BBL
DNP=60,356.9558 BBL
GP=661,906.0348 MCF
DGP=172,258.3455 MCF
%SO=48.1766
RP=3,163.1154 SCF/BBL

P=705.0000 PSI
NP=508,271.1142 BBL
DNP=53,064.4058 BBL
GP=842,126.2679 MCF
DGP=180,220.2331 MCF
%SO=46.3433
RP=3,629.3938 SCF/BBL

P=505.0000 PSI
NP=557,840.2255 BBL
DNP=49,569.1113 BBL
GP=1,025,838.413 MCF
DGP=183,712.1446 MCF
%SO=44.6514
RP=3,782.9700 SCF/BBL

P=305.0000 PSI
NP=609,151.3752 BBL
DNP=51,311.1497 BBL
GP=1,208,701.837 MCF
DGP=182,863.4250 MCF
%SO=43.0170
RP=3,344.6592 SCF/BBL

P=200.0000 PSI
NP=640,186.2668 BBL
DNP=31,034.8917 BBL
GP=1,302,600.736 MCF
DGP=93,898.8980 MCF
%SO=42.1311
RP=2,706.5234 SCF/BBL

NR=2,420,239.733 BBL
%OILRF=20.9182
GR=166,403.7440 MCF
%GASRF=88.6723
```

User Instructions

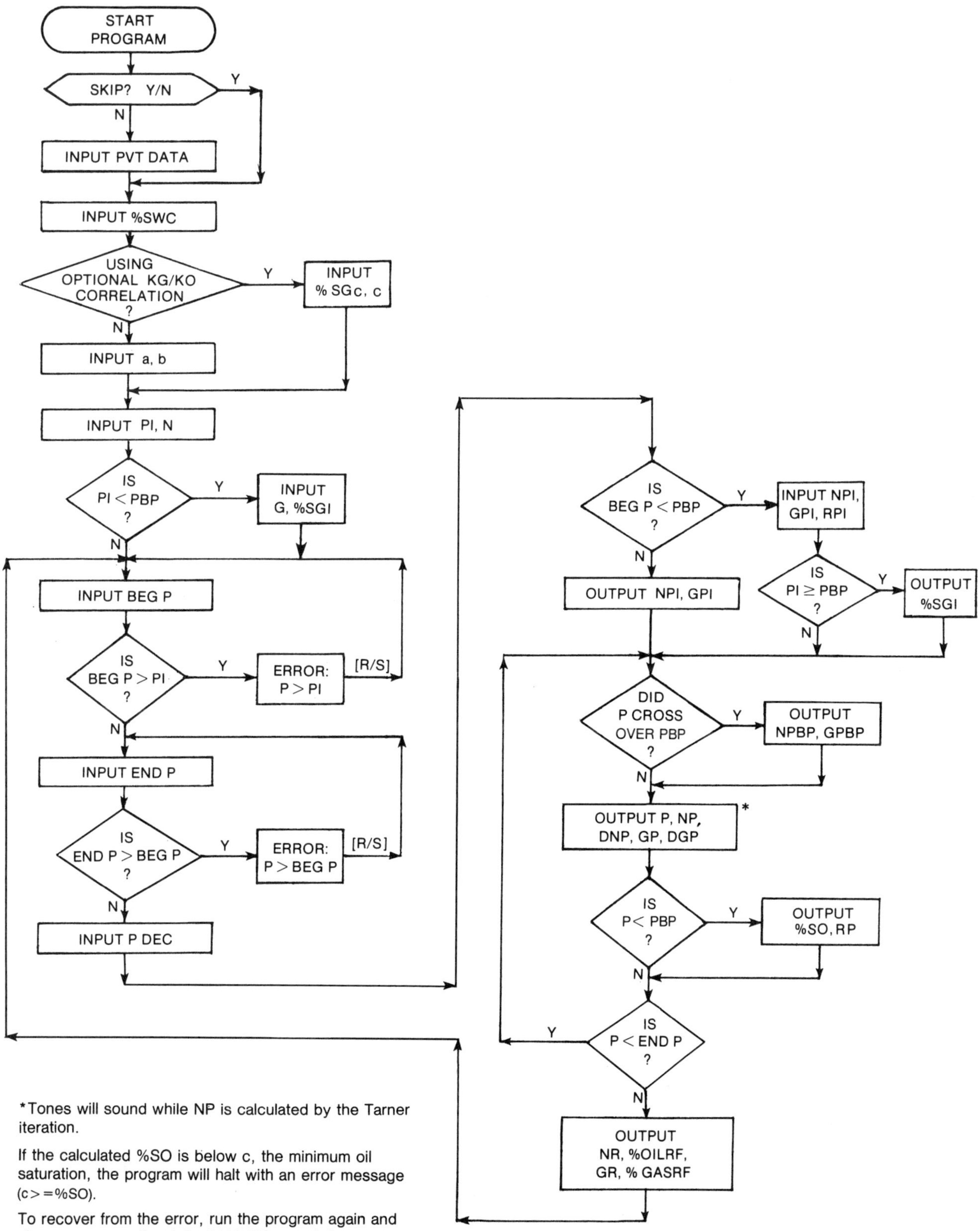

*Tones will sound while NP is calculated by the Tarner iteration.

If the calculated %SO is below c, the minimum oil saturation, the program will halt with an error message (c> =%SO).

To recover from the error, run the program again and change c or use the default KG/KO correlation.

General Information

Memory Requirements

Program length: 1508 bytes (7 cards)
Minimum size: 063
Minimum hardware: 41C + quad memory module or 41CV

Hidden Options

When OILPRED is run, the default KG/KO correlation will be selected automatically. To change to the optional correlation, set flag 00 ([//] [SF] 00) any time before the two coefficient prompts appear. To change back to the default correlation, clear flag 00 ([//] [CF] 00) or run the program again. The flag 00 annunciator in the display will be off if the default correlation is being used, and on if the optional correlation is being used.

Pac Subroutines Called

TITLE, Y/N?, W7, %NACL, %POR, CBOb, CCFR, IN, CBO, OUT, INK, OUTK, CRSb, CBG, CON, CUOd, CCW, CUG

Registers

03	Oil production units
04	Oil production units
06	Gas production units
07	Gas production units
08	Pressure units
09	Pressure units
17	BEG P, P (PSI)
20	%SGI
21	%SWC
26	BOI
27	BO
28	BGI' (BBL/SCF)
29	BG' (BBL/SCF)
30	RS, RSb (SCF/BBL), SO
31	Scratch
32	CWI (1/PSI)
33	CFR (1/PSI)
34	Scratch
35	PI (PSI)
36	GPI (MCF)
37	NPI (BBL)
38	END P (PSI)
39	G (MCF)
40	N (BBL)
41	P DEC (PSI)
42	BOBP
43	DGP (SCF), scratch
44	DNP (BBL), scratch
45	UOd (CP)
46	UO (CP), P (PSI)
47	UG(CP), scratch
48	a, %SGc
49	b, c
50	RP_i (SCF/BBL)
51	RP_{i-1} (SCF/BBL)
52	RS (SCF/BBL)
53	RPI (SCF/BBL)
54	Producing gas-oil ratio units
55	Producing gas-oil ratio units
56	G (SCF)
57	GP_{i-1} (SCF)
58	NP_{i-1} (BBL)
59	0, GPI, GPBP (MCF)
60	0, NPI, NPBP (BBL)
61	N, N − NPI, N − NPBP (BBL)
62	PI, BEG P, PBP (PSI)

Flags

00	Set:	Use optional KG/KO correlation.
	Clear:	Use default KG/KO correlation.
01	Set:	Producing gas-oil ratio units not yet input.
	Clear:	Producing gas-oil ratio units have been input.
02	Set:	Skip input of PVT data.
	Clear:	Allow input of PVT data.
03	Set:	Current pressure ≥ PBP.
	Clear:	Current pressure < PBP.
04	Set:	Gas production units not yet input.
	Clear:	Gas production units have been input.

Program Listing

```
 01♦LBL "OILPRED"
"OIL MBE PRED" 63
XROM "TITLE" FC?C 25
PROMPT CF 00 SF 01
SF 04 SF 06 SF 27
"M3" ASTO 03 "SCM"
ASTO 06 "⊦/M3" ASTO 54
"KPA" ASTO 08 CLA
ASTO 04 ASTO 07
ASTO 09 ASTO 55 "SKIP"
2 XROM "Y/N?" FS? 02
GTO 14 XROM "W7"
XROM "%NACL"
XROM "%POR"

 33♦LBL 14
RCL 14 XROM "CBOb"
STO 42 XROM "CCFR"
STO 33 20 STO 00
"%SWC" XROM "IN" 47
STO 00 "a" FS? 00
"%SGc" XROM "IN" "b"
FS? 00 "c" XROM "IN"
34 "PI" XEQ 24 RND
RCL 14 RND CF 03
X<=Y? SF 03 39 "N"
XEQ 30 FS? 03 ADV
FS? 03 GTO 15 38 "G"
XEQ 28 CF 04 19
STO 00 "%SGI"
XROM "IN" ADV

 78♦LBL 15
CF 08 16 "BEG P"
XEQ 24 RCL 35 X<>Y
X<=Y? GTO 00 TONE 3
"P > PI" PROMPT GTO 15

 91♦LBL 00
37 "END P" XEQ 24
RCL 17 X<>Y X<=Y?
GTO 03 TONE 3
"P > BEG P" PROMPT
GTO 00

103♦LBL 03
40 "P DEC" XEQ 24
RCL 17 RND RCL 14 RND
X>Y? GTO 04 RCL 35
X<> 17 STO 35 CLST
STO 20 RCL 17 XEQ 32
RCL 35 X<> 17 STO 35
XEQ 33 "NPI" XEQ 22
"GPI" XEQ 23 GTO 18

129♦LBL 04
36 "NPI" XEQ 30
FS?C 04 SF 08 35
"GPI" XEQ 28 CF 08
FS?C 01 SF 08 52
"RPI" XEQ 26 CF 08
RCL 35 RND RCL 14 RND
X>Y? GTO 05 RCL 35
X<> 17 STO 35 RCL 42
XROM "CBO" STO 31
RCL 17 X<> 35 STO 17
XROM "CBOb" RCL 31 /
1 RCL 37 RCL 40 / -
* 1 - RCL 21 100 /
1 - * 100 * STO 20
"%SGI" XROM "OUT"

182♦LBL 05
RCL 37 RCL 36 RCL 53
RCL 17 XEQ 32 RCL 39
1 E3 * STO 56 FS? 03
GTO 18 RCL 40 RCL 13
* RCL 59 1 E3 * -
RCL 61 RCL 52 * -
STO 56 GTO 18

207♦LBL 16
XEQ 33

209♦LBL 17
RCL 17 "P" XEQ 25
"NP" XEQ 22 CF 08
RCL 44 "DNP" XEQ 31
"GP" XEQ 23 RCL 43
1 E3 / "DGP" XEQ 29
FS? 03 GTO 18 RCL 30
100 * "%SO"
XROM "OUT" FS?C 01
SF 08 RCL 50 "RP"
XEQ 27 CF 08

239♦LBL 18
ADV RCL 38 RND RCL 17
RCL 41 - RND X>Y?
GTO 06 X<>Y RCL 17
RND X=Y? GTO 19
RCL 38 RND

256♦LBL 06
LASTX X<>Y RCL 14 RND
X<=Y? GTO 08 X<>Y CLX
RCL 17 RND X<Y?
GTO 08 X=Y? GTO 07
RCL Z STO 46 RCL 14
STO 17 XEQ 33 RCL 58
"NPBP" XEQ 31 RCL 57
1 E3 / "GPBP" XEQ 29
ADV RCL 44 STO 47
XEQ 09 XEQ 33 RCL 05
ST+ 44 RCL 13 *
ST+ 43 GTO 17

295♦LBL 07
RCL Z STO 46 XEQ 09
GTO 16

300♦LBL 08
RCL Z STO 17 GTO 16

304♦LBL 09
RCL 58 RCL 57 1 E3 /
RCL 13 RCL 14 XEQ 32
RCL 46 STO 17 RTN

315♦LBL 19
RCL 41 ST- 17 RCL 40
RCL 31 - "NR" XEQ 31
RCL 31 RCL 40 / 100
* "%OILRF" XROM "OUT"
RCL 56 RCL 40 RCL 52
* + 1 E3 / STO 31
RCL 34 - "GR" XEQ 29
RCL 34 RCL 31 / 100
* "%GASRF" XROM "OUT"
ADV GTO 15

351♦LBL 22
RCL 58 RCL 60 +
STO 31 XEQ 31 RTN

358♦LBL 23
RCL 57 1 E3 / RCL 59
+ STO 34 FS?C 04
SF 08 XEQ 29 CF 08
RTN

370♦LBL 24
STO 00 XEQ 10
XROM "INK" RDN STO 08
X<>Y STO 09 R↑ RTN
```

Program Listing (cont.)

```
380♦LBL 25
XEQ 10  XROM "OUTK"  RDN
STO 08  X<>Y  STO 09  R↑
RTN

389♦LBL 10
ASTO T  "PSI"  ASTO 01
CLA  ASTO 02  ARCL T
RCL 09  RCL 08  RCL Z
RTN

400♦LBL 26
STO 00  XEQ 11
XROM "INK"  RDN  STO 54
X<>Y  STO 55  R↑  RTN

410♦LBL 27
XEQ 11  XROM "OUTK"  RDN
STO 54  X<>Y  STO 55  R↑
RTN

419♦LBL 11
ASTO T  "SCF/BBL"
ASTO 01  ASHF  ASTO 02
CLA  ARCL T  RCL 55
RCL 54  RCL Z  RTN

431♦LBL 28
STO 00  XEQ 12
XROM "INK"  RDN  STO 06
X<>Y  STO 07  R↑  RTN

441♦LBL 29
XEQ 12  XROM "OUTK"  RDN
STO 06  X<>Y  STO 07  R↑
RTN

450♦LBL 12
ASTO T  "MCF"  ASTO 01
CLA  ASTO 02  ARCL T
RCL 07  RCL 06  RCL Z
RTN

461♦LBL 30
STO 00  XEQ 13
XROM "INK"  RDN  STO 03
X<>Y  STO 04  R↑  RTN

471♦LBL 31
XEQ 13  XROM "OUTK"  RDN
STO 03  X<>Y  STO 04  R↑
RTN

480♦LBL 13
ASTO T  "BBL"  ASTO 01
CLA  ASTO 02  ARCL T
RCL 04  RCL 03  RCL Z
RTN

491♦LBL 32
STO 62  RDN  STO 50
STO 51  RCL 40  R↑
STO 60  -  STO 61  0
STO 56  STO 57  STO 58
R↑  STO 59  RCL 17  RND
RCL 14  RND  CF 03
X<=Y?  SF 03  RCL 42
FS? 03  XROM "CBO"
FC? 03  RCL 17  FC? 03
XROM "CBOb"  STO 26
RCL 13  FC? 03  RCL 17
FC? 03  XROM "CRSb"
STO 52  XEQ 01
XROM "CBG"  "FT3-BBL"
CON  STO 28  XROM "CUOd"
STO 45  XROM "CCW"
STO 32  RTN

538♦LBL 33
CLX  STO 31  STO 43
STO 44  RCL 17  RND
RCL 14  RND  CF 03
X<=Y?  GTO 14  XEQ 01
XROM "CBG"  "FT3-BBL"
CON  STO 29  RCL 28  -
RCL 56  *  STO 44
RCL 17  XROM "CRSb"
STO 30  RCL 29  *
ST- 43  RCL 52  RCL 30
-  RCL 29  *  STO 31
RCL 17  XROM "CBOb"
GTO 03

575♦LBL 14
SF 03  RCL 42
XROM "CBO"

579♦LBL 03
STO 27  ST+ 43  RCL 32
RCL 21  100  /  STO Z  *
RCL 33  +  RCL 62
RCL 17  -  *  1  R↑  -
/  RCL 26  *  LASTX  -
RCL 27  +  RCL 31  +
RCL 61  *  ST+ 44
FC? 03  GTO 03  RCL 44
RCL 43  /  ENTER↑
X<> 58  -  STO 44
RCL 13  *  STO 43
RCL 58  LASTX  *  STO 57
TONE 9  RTN

627♦LBL 03
1  RCL 21  100  /  -
STO 34  RCL 27  RCL 29
RCL 30  *  -  STO 43
RCL 30  RCL 45
XROM "CUOb"  STO 46
XEQ 01  XROM "CUG"
X<> 47  STO 05  RCL 50
ENTER↑  X<> 51  -  X<0?
CLX  ST+ 50

655♦LBL 34
TONE 5  RCL 44  RCL 57
RCL 50  RCL 51  +  2  /
STO 31  RCL 58  *  -
RCL 29  *  -  RCL 31
RCL 29  *  RCL 43  +  /
STO 31  RCL 34  RCL 20
100  /  -  RCL 27
RCL 26  /  *  1  R↑
RCL 61  /  -  *  STO 00
FS? 00  GTO 04  RCL 34
/  1  RCL Y  RCL 49  Y↑X
-  1  RCL Z  -  RCL 48
Y↑X  *  R↑  RCL 48
RCL 49  +  Y↑X  /
GTO 05

716♦LBL 04
RCL 49  RCL 34  RCL 48
100  ST/ T  /  -  RCL 00
-  X<0?  0  LASTX  R↑  -
X=0?  GTO 02  /  X<0?
GTO 02  .4556  RCL Y  *
.04335  +  *

742♦LBL 05
RCL 46  *  RCL 47  /
RCL 27  *  RCL 29  /
RCL 30  +  ENTER↑
X<> 50  RCL 50  -  X<>Y
/  ABS  1 E-4  X<Y?
GTO 34  RCL 31  ENTER↑
X<> 58  -  STO 44
RCL 50  RCL 51  +  2  /
*  STO 43  ST+ 57
RCL 00  STO 30  TONE 9
RTN

780♦LBL 02
RCL 00  TONE 3
"c >= %SO"  PROMPT
GTO 02

786♦LBL 01
RCL 16  "F-R"  CON
RCL 10  /  RCL 17
RCL 11  /  END
```

13. QOVST — Rate-Time Forecast from Material Balance

The solution-gas-drive predictions of reservoir performance calculated from OILMBE are only a part of the information required by the reservoir engineer. Usually, a rate-time forecast also will be required. Once again, putting aside the issue of whether this procedure is better than decline-curve analysis, QOVST predicts the times and rates corresponding to information provided from an oil material balance. The method requires calculating the well productivity index and predicting its decline.

The productivity index (JO) is simply the oil producing rate divided by the pressure drawdown in the reservoir. In the case of pseudosteady-state radial flow, JO can be shown to be:

$$JO = \frac{0.00708\ KO\ H}{BO\ UO\ [\ln(Re/RW) - 1/2]}$$

In the case of steady-state radial flow:

$$JO = \frac{0.00708\ KO\ H}{BO\ UO\ \ln(Re/RW)}$$

where KO = K KRO.

The only terms in either of these equations that vary as the reservoir is depleted are KO, BO, and UO. QOVST uses the Pac correlations for BO and UO and user-input coefficients for the KRO correlation. If the initial oil-producing rate is unknown, enough information to calculate an initial JO must be provided from an early test or estimate of productivity.

Gas producing rates also can be forecast based on the instantaneous values of RP. For example, if QO is calculated as 68.7 BBL/DAY at 1,305 PSI where RP is 1881.4 SCF/BBL, QG = (68.7) (1881.4) SCF/DAY = 129.2 MCF/DAY.

For most reservoirs, the theoretical material-balance performance and rate-time prediction will yield hyperbolic declines. QOVST calculates the incremental time values with a series of exponential declines. The result generally approximates a hyperbolic decline over the life of the reservoir.

Equations

Above bubble point:

$$QO = QOI\ \frac{P - PWF}{PI - PWF}\ \frac{KRO}{KROI}\ \frac{BOI\ UOI}{BO\ UO}$$

Below bubble point:

$$QO = QOI\ \frac{P - PWF}{PI - PWF}\ \frac{KRO}{KROI}\ \frac{BOI\ UOI}{BOb\ UOb}$$

$$\%SO = \frac{BOb}{BOI}\ \frac{N - NP}{N}\ (100 - \%SWC - \%SGI)$$

$$KRO = \left|\frac{\%SO}{100 - \%SWC}\right|^{a+b}$$

KROI = KRO at the initial %SO (%SO at PI)

If QOI is unknown, it is calculated from the initial productivity, JOI:

$$QOI = JOI(PI - PWF) =$$

$$\frac{0.00708\ K\ KROI\ H}{BOI\ UOI\ \ln(Re/RW)}\ (PI - PWF)\ \text{(steady-state radial flow)}$$

TIME is calculated at the ith pressure step as follows:

$$TIME_i = TIME_{i-1} + \frac{\ln(QO_i/QO_{i-1})\ (NP_i - NP_{i-1})}{(QO_i - QO_{i-1})\ 365}$$

Nomenclature

Symbol	*Variable Name*	*Input or Output*	*English Units*	*SI Units*
a	First coefficient of KRO correlation	I	-	-
BOI	Initial oil-formation volume factor	I	-	-
b	Second coefficient of KRO correlation	I	-	-
DNP*	Delta NP, change in cumulative oil production	O	BBL	M3
H	Formation thickness†	I	FT	M
K	Permeability	I	MD	MD
N*	Initial oil in place	I	BBL	M3
NP*	Cumulative oil production	I	BBL	M3
P*	Pressure	O	PSI	KPA
PI*	Initial pressure	I	PSI	KPA
PWF*	Flowing bottom-hole pressure	I	PSI	KPA

Symbol	Variable Name	Input or Output	English Units	SI Units
QO*	Oil producing rate	O	BBL/ DAY	M3/ DAY
QOI*	Initial oil producing rate	I,O	BBL/ DAY	M3/ DAY
RW	Effective well-bore radius	I	FT	M
Re	Radius of drainage	I	FT	M
TIME*	Cumulative production time	O	YR	YR
UOI	Initial oil viscosity	I	CP	PA*S
%SGI	Initial volume percent gas saturation	I	–	–
%SWC	Volume percent connate water saturation	I	–	–

*The units for these variables are saved by the program.
†In the case of deviated wells or slanted beds, use the true vertical thickness instead of the measured thickness of the formation.

Yes/No Questions

CORR? Yes: Use Pac correlations to estimate PVT properties.
No: Input PVT properties.

SKIP? Yes: Skip input of PVT data.
No: Allow input of PVT data.

Example 1

To forecast the production at the Little Creek oil reservoir (see Example 3 of OILPRED), we need to know that the oil producing rate at the current (12-81) pressure of 1,905 PSI is 300 BBL/DAY. The estimated economic limit is 2 BBL/DAY per well or 6 BBL/DAY total. The historical and forecast production are shown in Figure 13-1. Use the production forecast from Example 3 of OILPRED to get the calculated production. The bottom-hole pressure is forecast to remain essentially constant at 600 PSI. As a separate case, assume that when the field producing rate drops below 70 BBL/DAY, lift equipment will be installed to lower the effective bottom-hole pressure to 50 PSI.

In reality, one could allocate the reserves (volumetrically, with consideration for individual well deliverabilities) between wells and forecast them individually. It is, of course, likely that the wells would require pumping units at separate times.

The value of N used in the second part of the forecast is the original N minus the production to 1,305 PSI (the crossover point). The current value of %SO is printed when the pressure is below the bubble point; %SGI is equal to 100 − %SO − %SWC, or 14.9406%. The current value of JO is QO/(P − PWF) = 68.7078/(1305 − 600) = 0.097. The new value of QO with PWF reduced to 50 PSI is 0.097(1305 − 50) = 121.7 BBL/DAY. The values for NP are incremental from the value at 1,705 PSI.

Note that the remaining life of the field in the second case is over 28 years to deplete the reservoir to 305 PSI. The oil producing rate is 6.4 BBL/DAY for all three wells at that point, with gas production of about 20 MCF/DAY (negligible). The minimum pressure used for the OILPRED forecast should not be so low as to produce below the economic limit. This abandonment rate indicates that the ultimate recovery will be to a reservoir of about 300 PSI, not the 200 PSI limit assumed in the OILPRED example. Once again, it should be pointed out that this forecast essentially is what would be done solely under primary depletion. Although the oil recovery factor is relatively high for primary depletion, waterflooding and other enhanced-recovery techniques should definitely be considered immediately.

Keystrokes (SIZE> =057)	Display	Comments
[XEQ] [ALPHA] QOVST [ALPHA]	*CORR? Y/N:*	Last character is Y or N
Y [R/S]	*SKIP? Y/N:*	Last character is Y or N
N [R/S]	*SEP T=?*	
110 [R/S]	*SEP P=?*	
125 [R/S]	*OIL G=?*	
37.4 [R/S]	*GAS G=?*	
.678 [R/S]	*T=?*	GAS GS printed
155 [R/S]	*RSI=?*	
480 [R/S]	*%SWC=?*	PBP printed
32.5 [R/S]	*a=?*	
2 [R/S]	*b=?*	
2 [R/S]	*PI=?*	
1905 [R/S]	*N=?*	
3060426 [R/S]	*%SGI=?*	
5.2529 [R/S]	*PWF=?*	
600 [R/S]	*QOI H,R,K*	
[A]	*QOI=?*	QOI known
300 [R/S]	*P=?*	
1705 [R/S]	*NP=?*	
124255 [R/S]	*P=?*	DNP, QO, %SO, and TIME printed
1505 [R/S]	*NP=?*	
234468 [R/S]	*P=?*	DNP, QO, %SO, and TIME printed

Keystrokes (SIZE> =057)	Display	Comments
1305 [R/S]	*NP=?*	
323019 [R/S]	*P=?*	DNP, QO, %SO, and TIME printed
1105 [R/S]	*NP=?*	
394850 [R/S]	*P=?*	DNP, QO, %SO, and TIME printed
905 [R/S]	*NP=?*	
455207 [R/S]	*P=?*	DNP, QO, %SO, and TIME printed
705 [R/S]	*NP=?*	
508271 [R/S]	*P=?*	DNP, QO, %SO, and TIME printed. The economic limit has been reached
[XEQ] [ALPHA] QOVST [ALPHA]	*CORR? Y/N:Y*	
[R/S]	*SKIP? Y/N:N*	
Y [R/S]	*%SWC=?*	
[R/S]	*a=?*	
[R/S]	*b=?*	
[R/S]	*PI=?*	
1305 [R/S]	*N=?*	
323019 [−]	*2,737,406.562*	Since the original value of N was in X when the prompt appeared, subtract the NP to 1,305 PSI
[R/S]	*%SGI=?*	
14.9406 [R/S]	*PWF=?*	
50 [R/S]	*QOI H,R,K*	
[A]	*QOI=?*	QOI known
121.7 [R/S]	*P=?*	
1105 [R/S]	*NP=?*	
71830 [R/S]	*P=?*	DNP, QO, %SO, and TIME printed

Keystrokes (SIZE> =057)	Display	Comments
905 [R/S]	*NP=?*	
60357 [+] [R/S]	*P=?*	DNP, QO, %SO, and TIME printed
705 [R/S]	*NP=?*	
53064 [+] [R/S]	*P=?*	DNP, QO, %SO, and TIME printed
505 [R/S]	*NP=?*	
49569 [+] [R/S]	*P=?*	DNP, QO, %SO, and TIME printed
305 [R/S]	*NP=?*	
51311 [+] [R/S]	*P=?*	DNP, QO, %SO, and TIME printed
200 [R/S]	*NP=?*	
31035 [+] [R/S]	*P=?*	DNP, QO %SO, and TIME printed

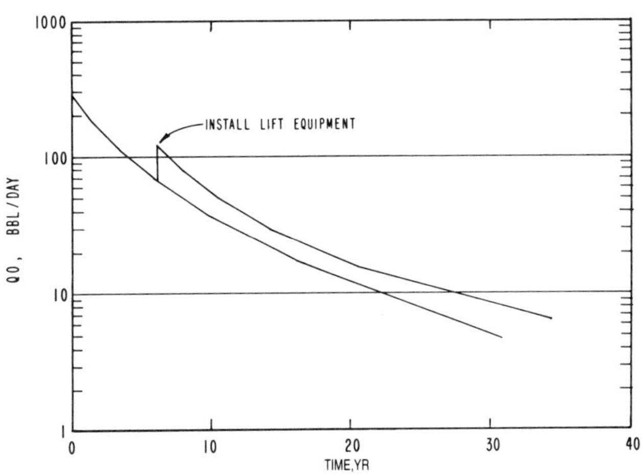

Figure 13–1

```
QO VS TIME

CORR: YES
SKIP: NO
SEP T=110.0000 F
SEP P=125.0000 PSI
OIL G=37.4000 API
GAS G=0.6780
GAS GS=0.6842
T=155.0000 F
RSI=480.0000 SCF/BBL
PBP=2,182.2913 PSI
%SWC=32.5000
a=2.0000
b=2.0000
PI=1,905.0000 PSI
N=3,060,426.000 BBL
%SGI=5.2529
PWF=600.0000 PSI
QOI=300.0000 BBL/DAY

P=1,705.0000 PSI
NP=124,255.0000 BBL
DNP=124,255.0000 BBL
QO=188.9658 BBL/DAY
%SO=58.5795
TIME=1.4171 YR

P=1,505.0000 PSI
NP=234,468.0000 BBL
DNP=110,213.0000 BBL
QO=115.8073 BBL/DAY
%SO=55.3069
TIME=3.4381 YR

P=1,305.0000 PSI
NP=323,019.0000 BBL
DNP=88,551.0000 BBL
QO=68.7078 BBL/DAY
%SO=52.5594
TIME=6.1272 YR

P=1,105.0000 PSI
NP=394,850.0000 BBL
DNP=71,831.0000 BBL
QO=37.9143 BBL/DAY
%SO=50.2204
TIME=9.9268 YR

P=905.0000 PSI
NP=455,207.0000 BBL
DNP=60,357.0000 BBL
QO=17.6981 BBL/DAY
%SO=48.1766
TIME=16.1586 YR

P=705.0000 PSI
NP=508,271.0000 BBL
DNP=53,064.0000 BBL
QO=4.6873 BBL/DAY
%SO=46.3433
TIME=31.0042 YR

QO VS TIME

SKIP: YES
PI=1,305.0000 PSI
N=2,737,407.000 BBL
%SGI=14.9406
PWF=50.0000 PSI
QOI=121.7000 BBL/DAY

P=1,105.0000 PSI
NP=71,830.0000 BBL
DNP=71,830.0000 BBL
QO=78.8123 BBL/DAY
%SO=50.2204
TIME=1.9937 YR

P=905.0000 PSI
NP=132,187.0000 BBL
DNP=60,357.0000 BBL
QO=49.3654 BBL/DAY
%SO=48.1765
TIME=4.6208 YR

P=705.0000 PSI
NP=185,251.0000 BBL
DNP=53,064.0000 BBL
QO=29.0943 BBL/DAY
%SO=46.3432
TIME=8.4126 YR

P=505.0000 PSI
NP=234,820.0000 BBL
DNP=49,569.0000 BBL
QO=15.3490 BBL/DAY
%SO=44.6514
TIME=14.7308 YR

P=305.0000 PSI
NP=286,130.9999 BBL
DNP=51,310.9999 BBL
QO=6.3816 BBL/DAY
%SO=43.0170
TIME=28.4891 YR

P=200.0000 PSI
NP=317,165.9999 BBL
DNP=31,035.0000 BBL
QO=3.1635 BBL/DAY
%SO=42.1311
TIME=47.0303 YR
```

Example 2

Calculate the average deliverability per well of the Little Creek reservoir based on steady-state flow theory and the buildup permeability estimate of 2.2 MD. The calculated value of Re is 961 FT based on the average area/well of 66.67 ACRE. RW is 0.422 FT for each well. The total field deliverability is estimated to be three times the individual well average.

This method should be used only when good values of deliverability are unavailable. Since the calculated deliverability (3 times 47.7 = 143.1 BBL/DAY) is less than the actual deliverability of 300 BBL/DAY, it is entirely likely that the permeability estimate is too low. A value of 4.6 MD would have matched the measured value of productivity.

Keystrokes	Display	Comments
[XEQ] [ALPHA] QOVST [ALPHA]	*CORR? Y/N:Y*	
[R/S]	*SKIP? Y/N:Y*	
[R/S]	*%SWC=?*	
[R/S]	*a=?*	
[R/S]	*b=?*	
[R/S]	*PI=?*	
1905 [R/S]	*N=?*	
3060426 [R/S]	*%SGI=?*	
5.2529 [R/S]	*PWF=?*	
600 [R/S]	*QOI H,R,K*	
[E]	*H=?*	H,Re,RW, and K known
24 [R/S]	*Re=?*	
961 [R/S]	*RW=?*	
.422 [R/S]	*K=?*	
2.2 [R/S]	*P=?*	QOI printed

```
QO VS TIME

PI=1,905.0000 PSI
N=3,060,426.000 BBL
%SGI=5.2529
PWF=600.0000 PSI
H=24.0000 FT
Re=961.0000 FT
RW=0.4220 FT
K=2.2000 MD
QOI=47.7259 BBL/DAY
```

General Information

Memory Requirements

Program length: 895 bytes (4 cards)
Minimum size: 057
Minimum hardware: 41C and 2 memory modules

Hidden Options

None

Pac Subroutines Called

TITLE, Y/N?, W8, CBOb, IN, INU, OUTK, INK, CUOd, CBO, CRSb, CUOb, CUO

Registers

03	Oil production units
04	Oil production units
06	Oil producing rate units
07	Oil producing rate units
08	Pressure units
09	Pressure units
20	%SGI
21	%SWC
26	BOI
27	BO, BOb
28	H (FT)
29	K (MD)
30	Re (FT)
31	RW (FT)
32	KROI, KRO
33	P_{i-1} (PSI)
34	Scratch
35	PI (PSI)
37	NP_i (BBL)
40	N (BBL)
42	BOBP
43	QO_i (BBL/DAY)
44	NP_{i-1} (BBL)
45	UOI (CP)
46	UOd, UO, UOb (CP)
47	PWF (PSI)
48	a
49	b
50	QOI (BBL/DAY)
51	QO_{i-1} (BBL/DAY)
52	$TIME_i$ (YR)
53	$TIME_{i-1}$ (YR)
54	Time units
55	Time units
56	SO

Registers 10, 11, 18, 19, 22, 23, 36, 38, 39, 41 unused

Flags

01	Set:	Use Pac correlations to estimate PVT properties.
	Clear:	Input PVT properties.
02	Set:	Skip input of PVT data.
	Clear:	Allow input of PVT data.
03	Set:	Current pressure ≥ PBP.
	Clear:	Current pressure < PBP.
04	Set:	Time units not yet input.
	Clear:	Time units have been input.

User Instructions

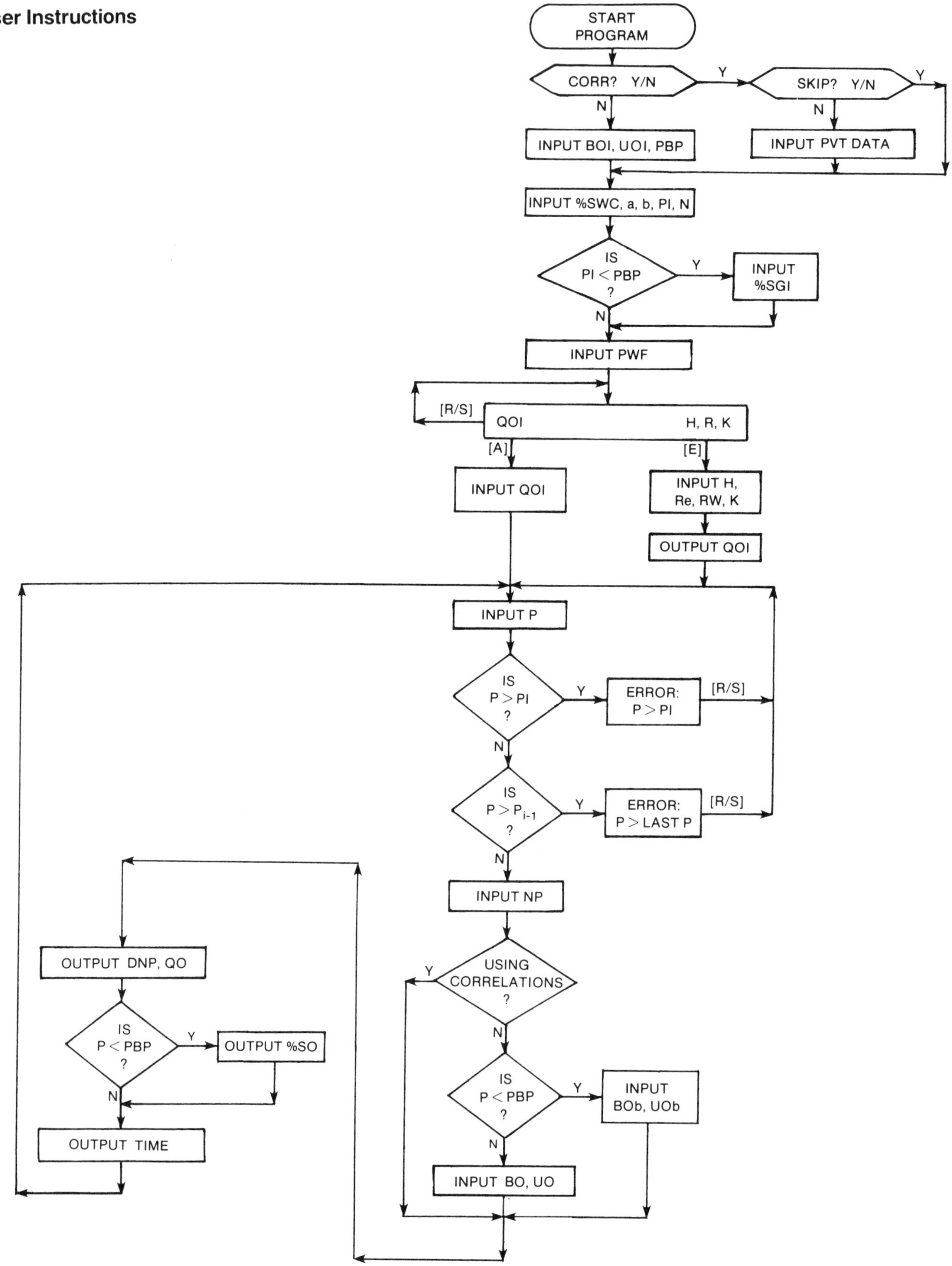

Program Listing

```
01♦LBL "QOVST"
"QO VS TIME" 57
XROM "TITLE" FC?C 25
PROMPT SF 04 SF 27
"M3" ASTO 03 "⊦/DAY"
ASTO 06 "YR" ASTO 54
"KPA" ASTO 08 CLA
ASTO 04 ASTO 07
ASTO 09 ASTO 55 CLX
STO 44 STO 53 "CORR"
1 XROM "Y/N?" FC? 01
GTO 00 "SKIP" 2
XROM "Y/N?" FC? 02
XROM "W8" RCL 14
XROM "CBOb" STO 42
GTO 01

39♦LBL 00
25 STO 00 "BOI"
XROM "IN" 44 STO 00
"CP" ASTO 01 CLA
ASTO 02 ASTO Z "PA*S"
ASTO Y "UOI"
XROM "INU" 13 STO 00
"PSI" ASTO 01 CLA
ASTO 02 ASTO Z "KPA"
ASTO Y "PBP"
XROM "INU"

66♦LBL 01
20 STO 00 "%SWC"
XROM "IN" 47 STO 00
"a" XROM "IN" "b"
XROM "IN" 34 "PI"
XEQ 16 STO 33 RND
RCL 14 RND CF 03
X<=Y? SF 03 39 "N"
XEQ 19 CF 08 CLX
FS? 03 STO 20 19
STO 00 "%SGI" FC? 03
XROM "IN" 46 "PWF"
XEQ 16 SF 08 1 XEQ 22
FC? 01 GTO 14 RCL 17
X<> 35 STO 17 XEQ 23
RCL 17 X<> 35 STO 17

114♦LBL 14
"QOI      H,R,K" PROMPT
GTO 14

118♦LBL A
49 "QOI" XEQ 17
STO 51 RCL 26 *
RCL 45 * RCL 32 /
RCL 35 RCL 47 - /
STO 34 ADV GTO 15

136♦LBL E
27 STO 00 "FT"
ASTO 01 CLA ASTO 02
ASTO Z "M" ASTO Y "H"
XROM "INU" 29 STO 00
"M" ASTO Y CLA ASTO Z
"Re" XROM "INU" "M"
ASTO Y CLA ASTO Z
"RW" XROM "INU" 28
STO 00 "MD" ASTO 01
ASTO Y CLA ASTO 02
ASTO Z "K" XROM "INU"
RCL 28 .00708 *
RCL 30 RCL 31 / LN /
STO 34 RCL 32 * *
RCL 26 / RCL 45 /
RCL 35 RCL 47 - *
STO 50 "QOI" XEQ 18
ADV

196♦LBL 15
CF 08 16 "P" XEQ 16
RCL 35 X<>Y X<=Y?
GTO 02 "P > PI" GTO 07

207♦LBL 02
RCL 33 X<>Y X<=Y?
GTO 03 "P > LAST P"

213♦LBL 07
TONE 3 PROMPT GTO 15

217♦LBL 03
36 "NP" XEQ 19 FS? 01
GTO 04 26 STO 00 "BO"
FC? 03 "⊦b" XROM "IN"
45 STO 00 "CP"
ASTO 01 CLA ASTO 02
ASTO Z "PA*S" ASTO Y
"UO" FC? 03 "⊦b"
XROM "INU"

242♦LBL 04
XEQ 20 RCL 37 ENTER↑
X<> 44 - "DNP" XEQ 24
RCL 43 "QO" XEQ 18
RCL 56 100 * "%SO"
FC? 03 XROM "OUT"
FS?C 04 SF 08 "YR"
ASTO 01 CLA ASTO 02
RCL 55 RCL 54 RCL 52
"TIME" XROM "OUTK" RDN
STO 54 X<>Y STO 55 R↑
ADV GTO 15

277♦LBL 16
STO 00 ASTO T "PSI"
ASTO 01 CLA ASTO 02
ARCL T RCL 09 RCL 08
RCL Z XROM "INK" RDN
STO 08 X<>Y STO 09 R↑
RTN

295♦LBL 17
STO 00 XEQ 05
XROM "INK" RDN STO 06
X<>Y STO 07 R↑ RTN

305♦LBL 18
XEQ 05 XROM "OUTK" RDN
STO 06 X<>Y STO 07 R↑
RTN

314♦LBL 05
ASTO T "BBL/DAY"
ASTO 01 ASHF ASTO 02
CLA ARCL T RCL 07
RCL 06 RCL Z RTN

326♦LBL 19
STO 00 XEQ 06
XROM "INK" RDN STO 03
X<>Y STO 04 R↑ RTN

336♦LBL 24
XEQ 06 XROM "OUTK" RDN
STO 03 X<>Y STO 04 R↑
RTN

345♦LBL 06
ASTO T "BBL" ASTO 01
CLA ASTO 02 ARCL T
RCL 04 RCL 03 RCL Z
RTN

356♦LBL 20
FS? 01 XEQ 23 FC? 03
XEQ 21 RCL 34 RCL 32
* RCL 46 / RCL 27 /
RCL 17 STO 33 RCL 47
- * STO 43 RCL 51 /
LN RCL 37 RCL 44 - *
RCL 43 ENTER↑ X<> 51
- / 365 / ST+ 53
RCL 53 STO 52 RTN

392♦LBL 21
RCL 27 RCL 26 / 1
RCL 37 RCL 40 / - *

402♦LBL 22
1 RCL 21 100 / -
STO 00 RCL 20 100 /
- * STO 56 RCL 00 /
RCL 48 RCL 49 + Y↑X
STO 32 RTN

423♦LBL 23
XROM "CUOd" STO 46
RCL 42 FS? 03
XROM "CBO" FC? 03
RCL 17 FC? 03
XROM "CBOb" STO 27
FS? 08 STO 26 RCL 13
FC? 03 RCL 17 FC? 03
XROM "CRSb" RCL 46
XROM "CUOb" FS? 03
XROM "CUO" STO 46
FS? 08 STO 45 END
```

14. INFCOEF — Calculating the Water Influx Coefficient

The van Everdingen and Hurst unsteady-state method of predicting water influx is discussed briefly in the abstract for the program INFLUX. The INFCOEF and INFLUX programs have been combined due to the substantial overlap of calculations performed. However, to the user they function as separate programs.

The unsteady-state water influx coefficient B is required as an input to predict water influx. An error in B results directly in an error in the calculated water influx. So, it is particularly important to estimate B as carefully as possible. INFCOEF allows the user to compute B either directly from a theoretical equation or from performance.

The theoretical equation is simple, and requires inputs of aquifer properties for porosity and compressibility, as well as reservoir thickness, radius, and angle subtended by the aquifer. Unfortunately, as in all simple equations, this one should only be used when performance estimates are unavailable or too erratic for valid interpretation.

The second method requires reservoir performance data. The water influx is calculated at each historical point using OILMBE for oil reservoirs or GASMBE for gas reservoirs. The time, pressure, and water influx must then be input. The dimensionless time (Td) is calculated by the program, allowing the user to read or interpolate Tables 14-1 or 14-2 and find the corresponding dimensionless influx (QTd). The program calculates the implied value of B for each time-pressure-water influx point. This process can be repeated for different values of Re/RW to give a better estimate of the area of the aquifer.

Equations

B from theory:

$$B = 1.119\ POR\ (CWI + CFR)\ RW^2\ H\ (\measuredangle/360)$$

B from performance:

$$B_i = \frac{We_i}{\Sigma(\Delta P_i\ QTd_i)}$$

$$Td_i = \frac{1.734(10^{-5})\ K\ TIME_i}{POR\ UWI(CWI + CFR)RW^2}$$

QTd_i is read or interpolated from Table 14-1: (Re/RW = ∞) or Table 14-2 (Re/RW = 1.5 to 10) at Td_i.

$$\Delta P_i = \tfrac{1}{2}\ (P_{i-1} - P_i),\ P_{-1} = P_0 = PI$$

$$POR = \frac{\%POR}{100}$$

Nomenclature

Symbol	*Variable Name*	*Input or Output*	*English Units*	*SI Units*
B*	Water influx coefficient	O	BBL/PSI	M3/KPA
B_i*	Water influx coefficient at $TIME_i$, P_i, and We_i	O	BBL/PSI	M3/KPA
CWI	Initial water isothermal compressibility	I	1/PSI	1/KPA
H	Formation thickness†	I	FT	M
K	Permeability	I	MD	MD
P_i*	Pressure	I	PSI	KPA
PI*	Initial pressure	I	PSI	KPA
QTd_i	Dimensionless water influx at Td_i	I	–	–
RW	Internal radius of aquifer	I	FT	M
Re/RW	Radio of external to internal aquifer radii	I	–	–
$TIME_i$*	Time	I	YR	YR
Td_i	Dimensionless time	O	–	–
UWI	Initial water viscosity	I	CP	PA*S
We_i*	Water influx at $TIME_i$ and P_i	I	BBL	M3
$\measuredangle$	Angle subtended by the reservoir	I	–	–

Note: For B_i, P_i, QTd_i, $TIME_i$, Td_i, and We_i, i = 1, 2, 3, . . ., n, where n is the number of TIME values input by the user.

*The units for these variables are saved by the program.

†In the case of deviated wells or slanted beds, use the true vertical thickness instead of the measured thickness of the formation.

Yes/No Questions

CORR? Yes: Use Pac correlations to estimate PVT properties.
No: Input PVT properties.

EDIT? Yes: Allow editing of TIME, P, and We values.
No: No editing necessary.

SKIP? Yes: Skip input of PVT data.
No: Allow input of PVT data.

Table 14-1 QTd versus Td at Re/RW = ∞

Td	QTd	Td	QTd	Td	QTd	Td	QTd	Td	QTd
0.00	0.000	38	20.080	89	39.272	300	105.968	590	187.166
0.01	0.112	39	20.488	90	39.626	305	107.437	600	189.852
0.05	0.278	40	20.894	91	39.979	310	108.904	610	192.533
0.10	0.404	41	21.298	92	40.331	315	110.367	620	195.208
0.15	0.520	42	21.701	93	40.681	320	111.827	625	196.544
0.20	0.606	43	22.101	94	41.034	325	113.284	630	197.878
0.25	0.689	44	22.500	95	41.385	330	114.738	640	200.542
0.30	0.758	45	22.897	96	41.735	335	116.189	650	203.201
0.40	0.898	46	23.291	97	42.084	340	117.638	660	205.854
0.50	1.020	47	23.684	98	42.433	345	119.083	670	208.502
0.60	1.140	48	24.076	99	42.781	350	120.526	675	209.825
0.70	1.251	49	24.466	100	43.129	355	121.966	680	211.145
0.80	1.359	50	24.855	105	44.858	360	123.403	690	213.784
0.90	1.460	51	25.244	110	46.574	365	124.838	700	216.417
1	1.569	52	25.633	115	48.277	370	126.270	710	219.016
2	2.447	53	26.020	120	49.968	375	127.699	720	221.670
3	3.202	54	26.406	125	51.648	380	129.126	725	222.980
4	3.893	55	26.791	130	53.317	385	130.550	730	224.289
5	4.539	56	27.174	135	54.976	390	131.972	740	226.904
6	5.153	57	27.555	140	56.625	395	133.391	750	229.514
7	5.743	58	27.935	145	58.265	400	134.808	760	232.120
8	6.314	59	28.314	150	59.895	405	136.223	770	234.721
9	6.869	60	28.691	155	61.517	410	137.635	775	236.020
10	7.411	61	29.068	160	63.131	415	139.045	780	237.318
11	7.940	62	29.443	165	64.737	420	140.453	790	239.912
12	8.457	63	29.818	170	66.336	425	141.859	800	242.501
13	8.964	64	30.192	175	67.928	430	143.262	810	245.086
14	9.461	65	30.565	180	69.512	435	144.664	820	247.668
15	9.949	66	30.937	185	71.090	440	146.064	825	248.957
16	10.434	67	31.308	190	72.661	445	147.461	830	250.245
17	10.913	68	31.679	195	74.226	450	148.856	840	252.819
18	11.386	69	32.018	200	75.785	455	150.249	850	255.388
19	11.855	70	32.417	205	77.338	460	151.640	860	257.953
20	12.319	71	32.785	210	78.886	465	153.029	870	260.515
21	12.778	72	33.151	215	80.428	470	154.416	875	261.795
22	13.233	73	33.517	220	81.965	475	155.801	880	263.073
23	13.684	74	33.883	225	83.497	480	157.184	890	265.629
24	14.131	75	34.247	230	85.023	485	158.565	900	268.181
25	14.573	76	34.611	235	86.545	490	159.945	910	270.729
26	15.013	77	34.974	240	88.062	495	161.322	920	273.274
27	15.450	78	35.336	245	89.575	500	162.698	925	274.545
28	15.883	79	35.697	250	91.084	510	165.444	930	275.815
29	16.313	80	36.058	255	92.589	520	168.183	940	278.353
30	16.742	81	36.418	260	94.090	525	169.549	950	280.888
31	17.167	82	36.777	265	95.588	530	170.914	960	283.420
32	17.590	83	37.136	270	97.081	540	173.639	970	285.948
33	18.011	84	37.494	275	98.571	550	176.357	975	287.211
34	18.429	85	37.851	280	100.057	560	179.069	980	288.473
35	18.845	86	38.207	285	101.540	570	181.774	990	290.995
36	19.259	87	38.563	290	103.019	575	183.124	1000	293.514
37	19.671	88	38.919	295	104.495	580	184.473	1010	296.030

Table 14–1 cont.

Td	QTd	Td	QTd	Td	QTd	Td	QTd	Td	QTd	Td	QTd
1020	298.543	1440	401.786	2700	693.877	5500	1296.893	30,000	5899.508	$7.0(10)^{8}$	$6.928(10)^{7}$
1025	299.799	1450	404.197	2750	705.090	5600	1317.709	35,000	6780.247	$8.0(10)^{8}$	$7.865(10)^{7}$
1030	301.053	1460	406.606	2800	716.280	5700	1338.486	40,000	7650.096	$9.0(10)^{8}$	$8.797(10)^{7}$
1040	303.560	1470	409.013	2850	727.449	5800	1359.225	50,000	9363.099	$1.0(10)^{9}$	$9.725(10)^{7}$
1050	306.065	1475	410.214	2900	738.598	5900	1379.927	60,000	11,047.299	$1.5(10)^{9}$	$1.429(10)^{8}$
1060	308.567	1480	411.418	2950	749.725	6000	1400.593	70,000	12,708.358	$2.0(10)^{9}$	$1.880(10)^{8}$
1070	311.066	1490	413.820	3000	760.833	6100	1421.224	75,000	13,531.457	$2.5(10)^{9}$	$2.328(10)^{8}$
1075	312.314	1500	416.220	3050	771.922	6200	1441.820	80,000	14,350.121	$3.0(10)^{9}$	$2.771(10)^{8}$
1080	313.562	1525	422.214	3100	782.992	6300	1462.383	90,000	15,975.389	$4.0(10)^{9}$	$3.645(10)^{8}$
1090	316.055	1550	428.196	3150	794.042	6400	1482.912	100,000	17,586.284	$5.0(10)^{9}$	$4.510(10)^{8}$
1100	318.545	1575	434.168	3200	805.075	6500	1503.408	125,000	21,560.732	$6.0(10)^{9}$	$5.368(10)^{8}$
1110	321.032	1600	440.128	3250	816.090	6600	1523.872	$1.5(10)^{5}$	$2.538(10)^{4}$	$7.0(10)^{9}$	$6.220(10)^{8}$
1120	323.517	1625	446.077	3300	827.088	6700	1544.305	$2.0(10)^{5}$	$3.308(10)^{4}$	$8.0(10)^{9}$	$7.066(10)^{8}$
1125	324.760	1650	452.016	3350	838.067	6800	1564.706	$2.5(10)^{5}$	$4.066(10)^{4}$	$9.0(10)^{9}$	$7.909(10)^{8}$
1130	326.000	1675	457.945	3400	849.028	6900	1585.077	$3.0(10)^{5}$	$4.817(10)^{4}$	$1.0(10)^{10}$	$8.747(10)^{8}$
1140	328.480	1700	463.863	3450	859.974	7000	1605.418	$4.0(10)^{5}$	$6.267(10)^{4}$	$1.5(10)^{10}$	$1.288(10)^{9}$
1150	330.958	1725	469.771	3500	870.903	7100	1625.729	$5.0(10)^{5}$	$7.699(10)^{4}$	$2.0(10)^{10}$	$1.697(10)^{9}$
1160	333.433	1750	475.669	3550	881.816	7200	1646.011	$6.0(10)^{5}$	$9.113(10)^{4}$	$2.5(10)^{10}$	$2.103(10)^{9}$
1170	335.906	1775	481.558	3600	892.712	7300	1666.265	$7.0(10)^{5}$	$1.051(10)^{5}$	$3.0(10)^{10}$	$2.505(10)^{9}$
1175	337.142	1800	487.437	3650	903.594	7400	1686.490	$8.0(10)^{5}$	$1.189(10)^{5}$	$4.0(10)^{10}$	$3.299(10)^{9}$
1180	338.376	1825	493.307	3700	914.459	7500	1706.688	$9.0(10)^{5}$	$1.326(10)^{5}$	$5.0(10)^{10}$	$4.087(10)^{9}$
1190	340.843	1850	499.167	3750	925.309	7600	1726.859	$1.0(10)^{6}$	$1.462(10)^{5}$	$6.0(10)^{10}$	$4.808(10)^{9}$
1200	343.308	1875	505.019	3800	936.144	7700	1747.002	$1.5(10)^{6}$	$2.126(10)^{5}$	$7.0(10)^{10}$	$5.643(10)^{9}$
1210	345.770	1900	510.861	3850	946.966	7800	1767.120	$2.0(10)^{6}$	$2.781(10)^{5}$	$8.0(10)^{10}$	$6.414(10)^{9}$
1220	348.230	1925	516.695	3900	957.773	7900	1787.212	$2.5(10)^{6}$	$3.427(10)^{5}$	$9.0(10)^{10}$	$7.183(10)^{9}$
1225	349.460	1950	522.520	3950	968.566	8000	1807.278	$3.0(10)^{6}$	$4.064(10)^{5}$	$1.0(10)^{11}$	$7.948(10)^{9}$
1230	350.688	1975	528.337	4000	979.344	8100	1827.319	$4.0(10)^{6}$	$5.313(10)^{5}$	$1.5(10)^{11}$	$1.17(10)^{10}$
1240	353.144	2000	534.145	4050	990.108	8200	1847.336	$5.0(10)^{6}$	$6.544(10)^{5}$	$2.0(10)^{11}$	$1.55(10)^{10}$
1250	355.597	2025	539.945	4100	1000.858	8300	1867.329	$6.0(10)^{6}$	$7.761(10)^{5}$	$2.5(10)^{11}$	$1.92(10)^{10}$
1260	358.048	2050	545.737	4150	1011.595	8400	1887.298	$7.0(10)^{6}$	$8.965(10)^{5}$	$3.0(10)^{11}$	$2.29(10)^{10}$
1270	360.496	2075	551.522	4200	1022.318	8500	1907.243	$8.0(10)^{6}$	$1.016(10)^{6}$	$4.0(10)^{11}$	$3.02(10)^{10}$
1275	361.720	2100	557.299	4250	1033.028	8600	1927.166	$9.0(10)^{6}$	$1.134(10)^{6}$	$5.0(10)^{11}$	$3.75(10)^{10}$
1280	362.942	2125	563.068	4300	1043.724	8700	1947.065	$1.0(10)^{7}$	$1.252(10)^{6}$	$6.0(10)^{11}$	$4.47(10)^{10}$
1290	365.386	2150	568.830	4350	1054.409	8800	1966.942	$1.5(10)^{7}$	$1.828(10)^{6}$	$7.0(10)^{11}$	$5.19(10)^{10}$
1300	367.828	2175	574.585	4400	1065.082	8900	1986.796	$2.0(10)^{7}$	$2.398(10)^{6}$	$8.0(10)^{11}$	$5.89(10)^{10}$
1310	370.267	2200	580.332	4450	1075.743	9000	2006.628	$2.5(10)^{7}$	$2.961(10)^{6}$	$9.0(10)^{11}$	$6.58(10)^{10}$
1320	372.704	2225	586.072	4500	1086.390	9100	2026.438	$3.0(10)^{7}$	$3.517(10)^{6}$	$1.0(10)^{12}$	$7.28(10)^{10}$
1325	373.922	2250	591.806	4550	1097.024	9200	2046.227	$4.0(10)^{7}$	$4.610(10)^{6}$	$1.5(10)^{12}$	$1.08(10)^{11}$
1330	375.139	2275	597.532	4600	1107.646	9300	2065.996	$5.0(10)^{7}$	$5.689(10)^{6}$	$2.0(10)^{12}$	$1.42(10)^{11}$
1340	377.572	2300	603.252	4650	1118.257	9400	2085.744	$6.0(10)^{7}$	$6.758(10)^{6}$		
1350	380.003	2325	608.965	4700	1128.854	9500	2105.473	$7.0(10)^{7}$	$7.816(10)^{6}$		
1360	382.432	2350	614.672	4750	1139.439	9600	2125.181	$8.0(10)^{7}$	$8.866(10)^{6}$		
1370	384.859	2375	620.372	4800	1150.012	9700	2144.878	$9.0(10)^{7}$	$9.911(10)^{6}$		
1375	386.070	2400	626.066	4850	1160.574	9800	2164.555	$1.0(10)^{8}$	$1.095(10)^{7}$		
1380	387.283	2425	631.755	4900	1171.125	9900	2184.216	$1.5(10)^{8}$	$1.604(10)^{7}$		
1390	389.705	2450	637.437	4950	1181.666	10,000	2203.861	$2.0(10)^{8}$	$2.108(10)^{7}$		
1400	392.125	2475	643.133	5000	1192.198	12,500	2688.967	$2.5(10)^{8}$	$2.607(10)^{7}$		
1410	394.543	2500	648.781	5100	1213.222	15,000	3164.780	$3.0(10)^{8}$	$3.100(10)^{7}$		
1420	396.959	2550	660.093	5200	1234.203	17,500	3633.368	$4.0(10)^{8}$	$4.071(10)^{7}$		
1425	398.167	2600	671.379	5300	1255.141	20,000	4095.800	$5.0(10)^{8}$	$5.032(10)^{7}$		
1430	399.373	2650	682.610	5400	1276.037	25,000	5005.726	$6.0(10)^{8}$	$5.984(10)^{7}$		

Table 14-2 QTd versus Td for Various Values of Re/RW

Re/RW = 1.5		Re/RW = 2.0		Re/RW = 2.5		Re/RW = 3.0		Re/RW = 3.5		Re/RW = 4.0		Re/RW = 4.5	
Td	QTd	Td	QTd	Td	QTd	Td	QTd	Td	QTd	Td	QTd	Td	QTd
$5.0(10)^{-2}$	0.276	$5.0\ (10)^{-3}$	0.278	$1.0(10)^{-1}$	0.408	$3.0\ (10)^{-1}$	0.755	1.00	1.571	2.00	2.442	2.5	2.835
$6.0(10)^{-2}$	0.304	$7.5\ (10)^{-3}$	0.345	$1.5(10)^{-1}$	0.509	$4.0\ (10)^{-1}$	0.895	1.20	1.761	2.20	2.598	3.0	3.196
$7.0(10)^{-2}$	0.330	$1.0\ (10)^{-1}$	0.404	$2.0(10)^{-1}$	0.599	$5.0\ (10)^{-1}$	1.023	1.40	1.910	2.40	2.748	3.5	3.537
$8.0(10)^{-2}$	0.354	$1.25(10)^{-1}$	0.458	$2.5(10)^{-1}$	0.681	$6.0\ (10)^{-1}$	1.143	1.60	2.111	2.60	2.893	4.0	3.859
$9.0(10)^{-2}$	0.375	$1.50(10)^{-1}$	0.507	$3.0(10)^{-1}$	0.758	$7.0\ (10)^{-1}$	1.256	1.80	2.273	2.80	3.031	4.5	4.165
$1.0(10)^{-1}$	0.395	$1.75(10)^{-1}$	0.553	$3.5(10)^{-1}$	0.829	$8.0\ (10)^{-1}$	1.363	2.00	2.427	3.00	3.170	5.0	4.454
$1.1(10)^{-1}$	0.414	$2.00(10)^{-1}$	0.597	$4.0(10)^{-1}$	0.987	$9.0\ (10)^{-1}$	1.465	2.20	2.574	3.25	3.334	5.5	4.727
$1.2(10)^{-1}$	0.431	$2.25(10)^{-1}$	0.638	$4.5(10)^{-1}$	0.962	$1.00(10)^{-1}$	1.563	2.40	2.715	3.50	3.493	6.0	4.986
$1.3(10)^{-1}$	0.446	$2.50(10)^{-1}$	0.678	$5.0(10)^{-1}$	1.024	$1.25(10)^{-1}$	1.791	2.60	2.849	3.75	3.645	6.5	5.231
$1.4(10)^{-1}$	0.461	$2.75(10)^{-1}$	0.715	$5.5(10)^{-1}$	1.083	$1.50(10)^{-1}$	1.997	2.80	2.976	4.00	3.792	7.0	5.464
$1.5(10)^{-1}$	0.474	$3.00(10)^{-1}$	0.751	$6.0(10)^{-1}$	1.140	1.75	2.184	3.00	3.098	4.25	3.932	7.5	5.684
$1.6(10)^{-1}$	0.486	$3.25(10)^{-1}$	0.785	$6.5(10)^{-1}$	1.195	2.00	2.353	3.25	3.242	4.50	4.068	8.0	5.892
$1.7(10)^{-1}$	0.497	$3.50(10)^{-1}$	0.817	$7.0(10)^{-1}$	1.248	2.25	2.507	3.50	3.379	4.75	4.198	8.5	6.089
$1.8(10)^{-1}$	0.507	$3.75(10)^{-1}$	0.848	$7.5(10)^{-1}$	1.299	2.50	2.646	3.75	3.507	5.00	4.323	9.0	6.276
$1.9(10)^{-1}$	0.517	$4.00(10)^{-1}$	0.877	$8.0(10)^{-1}$	1.348	2.75	2.772	4.00	3.628	5.50	4.560	9.5	6.453
$2.0(10)^{-1}$	0.525	$4.25(10)^{-1}$	0.905	$8.5(10)^{-1}$	1.395	3.00	2.886	4.25	3.742	6.00	4.779	10	6.621
$2.1(10)^{-1}$	0.533	$4.50(10)^{-1}$	0.932	$9.0(10)^{-1}$	1.440	3.25	2.990	4.50	3.850	6.50	4.982	11	6.930
$2.2(10)^{-1}$	0.541	$4.75(10)^{-1}$	0.958	$9.5(10)^{-1}$	1.481	3.50	3.081	4.75	3.951	7.00	5.169	12	7.208
$2.3(10)^{-1}$	0.548	$5.00(10)^{-1}$	0.983	1.0	1.526	3.75	3.170	5.00	4.017	7.50	5.313	13	7.457
$2.4(10)^{-1}$	0.551	$5.50(10)^{-1}$	1.028	1.1	1.605	4.00	3.217	5.50	4.222	8.00	5.504	14	7.680
$2.5(10)^{-1}$	0.559	$6.00(10)^{-1}$	1.070	1.2	1.679	4.25	3.317	6.00	4.378	8.50	5.653	15	7.880
$2.6(10)^{-1}$	0.565	$6.50(10)^{-1}$	1.108	1.3	1.747	4.50	3.381	6.50	4.516	9.00	5.790	16	8.060
$2.8(10)^{-1}$	0.574	$7.00(10)^{-1}$	1.143	1.4	1.811	4.75	3.439	7.00	4.639	9.50	5.917	18	8.365
$3.0(10)^{-1}$	0.582	$7.50(10)^{-1}$	1.174	1.5	1.870	5.00	3.491	7.50	4.749	10	6.035	20	8.611
$3.2(10)^{-1}$	0.588	$8.00(10)^{-1}$	1.203	1.6	1.921	5.50	3.581	8.00	4.846	11	6.246	22	8.809
$3.4(10)^{-1}$	0.594	$9.00(10)^{-1}$	1.253	1.7	1.975	6.00	3.656	8.50	4.932	12	6.425	24	8.968
$3.6(10)^{-1}$	0.599	1.00	1.295	1.8	2.022	6.50	3.717	9.00	5.009	13	6.580	26	9.097
$3.8(10)^{-1}$	0.603	1.1	1.330	2.0	2.106	7.00	3.767	9.50	5.078	14	6.712	28	9.200
$4.0(10)^{-1}$	0.606	1.2	1.358	2.2	2.178	7.50	3.809	10.00	5.138	15	6.825	30	9.283
$4.5(10)^{-1}$	0.613	1.3	1.382	2.4	2.241	8.00	3.843	11	5.241	16	6.922	34	9.404
$5.0(10)^{-1}$	0.617	1.4	1.402	2.6	2.294	9.00	3.894	12	5.321	17	7.004	38	9.481
$6.0(10)^{-1}$	0.621	1.6	1.432	2.8	2.310	10.00	3.928	13	5.385	18	7.076	42	9.532
$7.0(10)^{-1}$	0.623	1.7	1.444	3.0	2.380	11.00	3.954	14	5.435	20	7.189	46	9.565
$8.0(10)^{-1}$	0.624	1.8	1.453	3.4	2.444	12.00	3.967	15	5.476	22	7.272	50	9.586
		2.0	1.468	3.8	2.491	14.00	3.985	16	5.506	24	7.332	60	9.612
		2.5	1.487	4.2	2.525	16.00	3.993	17	5.531	26	7.377	70	9.621
		3.0	1.495	4.6	2.551	18.00	3.997	18	5.551	30	7.434	80	9.623
		4.0	1.499	5.0	2.570	20.00	3.999	20	5.579	34	7.461	90	9.624
		5.0	1.500	6.0	2.599	22.00	3.999	25	5.611	38	7.481	100	9.625
				7.0	2.613	24.00	4.000	30	5.621	42	7.490		
				8.0	2.619			35	5.624	46	7.494		
				9.0	2.622			40	5.625	50	7.497		
				10.0	2.624								

Re/RW = 5.0		Re/RW = 6.0		Re/RW = 7.0		Re/RW = 8.0		Re/RW = 9.5		Re/RW = 10.0	
Td	QTd	Td	QTd	Td	QTd	Td	QTd	Td	QTd	Td	QTd
3.0	3.195	6.0	5.148	9.00	6.861	9	6.861	10	7.417	15	9.965
3.5	3.542	6.5	5.440	9.50	7.127	10	7.398	15	9.945	20	12.32
4.0	3.875	7.0	5.724	10	7.389	11	7.920	20	12.26	22	13.22
4.5	4.193	7.5	6.002	11	7.902	12	8.431	22	13.13	24	14.09
5.0	4.499	8.0	6.273	12	8.397	13	8.930	24	13.98	26	14.95
5.5	4.792	8.5	6.537	13	8.876	14	9.418	26	14.79	28	15.78
6.0	5.074	9.0	6.795	14	9.341	15	9.895	28	15.59	30	16.59
6.5	5.345	9.5	7.047	15	9.791	16	10.361	30	16.35	32	17.38
7.0	5.605	10.0	7.293	16	10.23	17	10.82	32	17.10	34	18.16
7.5	5.854	10.5	7.533	17	10.65	18	11.26	34	17.82	36	18.91
8.0	6.094	11	7.767	18	11.06	19	11.70	36	18.52	38	19.65
8.5	6.325	12	8.220	19	11.46	20	12.13	38	19.19	40	20.37
9.0	6.547	13	8.651	20	11.85	22	12.95	40	19.85	42	21.07
9.5	6.760	14	9.063	22	12.58	24	13.74	42	20.48	44	21.76

Re/RW = 5.0		Re/RW = 6.0		Re/RW = 7.0		Re/RW = 8.0		Re/RW = 9.5		Re/RW = 10.0	
Td	QTd	Td	QTd	Td	QTd	Td	QTd	Td	QTd	Td	QTd
10	6.965	15	9.456	24	13.27	26	14.50	44	21.09	46	22.42
11	7.350	16	9.829	26	13.92	28	15.23	46	21.60	48	23.07
12	7.706	17	10.19	28	14.53	30	15.92	48	22.26	50	23.71
13	8.035	18	10.53	30	15.11	34	17.22	50	22.82	52	24.33
14	8.339	19	10.85	35	16.39	38	18.41	52	23.36	54	24.94
15	8.620	20	11.16	40	17.49	40	18.97	54	23.89	56	25.53
16	8.870	22	11.74	45	18.43	45	20.26	56	24.39	58	26.11
18	9.338	24	12.26	50	19.24	50	21.42	58	24.88	60	26.67
20	9.731	25	12.50	60	20.54	55	22.46	60	25.36	65	28.02
22	10.07	31	13.74	70	21.45	60	23.40	65	26.48	70	29.29
24	10.35	35	14.40	80	22.13	70	24.98	70	27.52	75	30.49
26	10.59	39	14.93	90	22.63	80	26.26	75	28.48	80	31.61
28	10.80	51	16.05	100	23.00	90	27.28	80	29.36	85	32.67
30	10.98	60	16.56	120	23.47	100	28.11	85	30.18	90	33.66
34	11.26	70	16.92	140	23.71	120	29.31	90	30.93	95	34.60
38	11.46	80	17.14	160	23.85	140	30.08	95	31.63	100	35.48
42	11.61	90	17.27	180	23.92	160	30.58	100	32.27	120	38.51
46	11.71	100	17.36	200	23.96	180	30.91	120	34.39	140	40.89
50	11.79	110	17.41	500	24.00	200	31.12	140	35.92	160	42.75
60	11.91	120	17.45			240	31.34	160	37.04	180	41.21
70	11.96	130	17.46			280	31.43	180	37.85	200	45.36
80	11.98	140	17.48			320	31.47	200	38.44	240	46.95
90	11.99	150	17.49			360	31.49	240	39.17	280	47.94
100	12.00	160	17.49			400	31.50	280	39.56	320	48.54
120	12.00	180	17.50			500	31.50	320	39.77	360	48.91
		200	17.50					360	39.88	400	49.14
		220	17.50					400	39.94	440	49.28
								440	39.97	480	49.36
								480	39.98		

Example 1

The SW Franklin field has produced for 3.24 YR with production as indicated in Table 14–3. It is suspected that the reservoir has a fairly strong water drive with an aquifer of unknown extent. The reservoir data is almost the same as the NW Franklin field (see the example for OILPVT and Example 2 of OILMBE) but an original oil in place of 54,500,000 BBL and no water production. Other differences are as follows:

Porosity = 15.5%
Water saturation = 38.1%
Water salinity = 2.2%
Outer radius of aquifer = 4,000 FT

Determine the water influx coefficient and aquifer size for this reservoir.

The initial step in solving this problem is to load and run OILMBE to calculate water influx at each point. These calculated water influx values have been included in Table 14–3. The next step is to load and run INFCOEF, entering the times, pressures, and calculated values of water influx. The value of Re/RW is for annotation only. As a first trial, we will use infinite aquifer values.

Table 14–1 is interpolated at each output Td value to determine the corresponding QTd. This is just a simple linear interpolation. For example, Td1 is 5.929. The interpolation calculation is done with (5, 4.539) and (6, 5.153), the two (Td, Qtd) points surrounding the desired QTd:

$$\frac{6 - 5.929}{6 - 5} = \frac{5.153 - \text{QTd}}{5.153 - 4.539}$$

QTd = 5.109

A similar interpolation was performed for each Td value. Only the interpolated QTd values are shown in the keystrokes.

Since the calculated values of B are declining somewhat throughout the history, a smaller aquifer is indicated. Obviously, no aquifers are truly infinite; nonetheless, at early times (and, occasionally, for the life of the field) they may behave as if they were essentially infinite.

The next trial is for Re/RW = 5. Now the QTd values must come from Table 14–2. Whenever a calculated Td value is beyond the range of Table 14–2 at that Re/RW, QTd must be read from Table 14–1 at that Td (again, at early times the aquifer is infinite acting). Note that the calculated values of B are increasing throughout the history. Subsequent trials for Re/RW = 6, 7, and 8 were also performed. Figure 14–1 plots the calculated values of B as a function of time and Re/RW. It appears that Re/RW = 7 is a best fit, with B = 625 BBL/PSI. This is an average of all values except the second, which is excessively low to be considered valid.

Table 14–3

TIME (YR)	P (PSI)	NP (BBL)	GP (MCF)	We (BBL)
0.0	2,850	-	-	-
0.32	2,818	72,757	37,105	51,244
1.25	2,792	220,332	112,325	205,749
2.1	2,767	465,009	237,294	485,646
3.01	2,725	785,448	401,380	839,418
3.24	2,699	1,103,036	562,548	1,210,851

Keystrokes (SIZE> =044)	**Display**	**Comments**
[XEQ] [ALPHA] OILMBE [ALPHA]	*CORR? Y/N:*	Last character is Y or N
Y [R/S]	*SKIP? Y/N:*	Last character is Y or N
N [R/S]	*Tc=?*	
396.944 [R/S]	*Pc=?*	
660.1595 [R/S]	*STD T=?*	
60 [R/S]	*STD P=?*	
14.65 [R/S]	*SEP T=?*	
95 [R/S]	*SEP P=?*	
125 [R/S]	*OIL G=?*	
31.7 [R/S]	*GAS G=?*	
.762 [R/S]	*T=?*	GAS GS printed
155 [R/S]	*RSI=?*	
510 [R/S]	*%NACL=?*	PBP printed
2.2 [R/S]	*%POR=?*	
15.5 [R/S]	*%SWC=?*	
38.2 [R/S]	*PI=?*	
2850 [R/S]	*P=?*	
2818 [R/S]	*NP=?*	
72757 [R/S]	*GP=?*	
37105 [R/S]	*WP=?*	
0 [R/S]	*N We*	
[A]	*N=?*	N known
54.5 [EEX] 6 [R/S]	*N We*	We printed
[R/S]	*P=?*	
2792 [R/S]	*NP=?*	
220332 [R/S]	*GP=?*	
112325 [R/S]	*WP=?*	
[R/S]	*N We*	
[A]	*N=?*	
[R/S]	*N We*	We printed
[R/S]	*P=?*	
2767 [R/S]	*NP=?*	
465009 [R/S]	*GP=?*	
237294 [R/S]	*WP=?*	

Keystrokes (SIZE> =044)	**Display**	**Comments**
[R/S]	*N We*	
[A]	*N=?*	
[R/S]	*N We*	We printed
[R/S]	*P=?*	
2725 [R/S]	*NP=?*	
785448 [R/S]	*GP=?*	
401380 [R/S]	*WP=?*	
[R/S]	*N We*	
[A]	*N=?*	
[R/S]	*N We*	We printed
[R/S]	*P=?*	
2699 [R/S]	*NP=?*	
1103036 [R]	*GP=?*	
562548 [R/S]	*WP=?*	
[R/S]	*N We*	
[A]	*N=?*	
[R/S]	*N We*	We printed
Load INFCOEF		
[/] [FIX] 3	*1,210,851.286*	FIX 3 to match accuracy of Tables 14–1 and 14–2
[XEQ] [ALPHA] INFCOEF [ALPHA]	*SIZE> =130.0000*	
[XEQ] [ALPHA] SIZE [ALPHA] 130 [R/S]	*CORR? Y/N:Y*	
[R/S]	*SKIP? Y/N:N*	
Y [R/S]	*RW=?*	
4000 [R/S]	*K=?*	
62 [R/S]	*PI=?*	
[R/S]	*H, ∡ T,P,We*	
[E]	*TIME1=?*	TIME,P, and We known
.32 [R/S]	*P1=?*	
2818 [R/S]	*We1=?*	
51244 [R/S]	*TIME2=?*	
1.25 [R/S]	*P2=?*	
2792 [R/S]	*We2=?*	
205749 [R/S]	*TIME3=?*	
2.1 [R/S]	*P3=?*	
2767 [R/S]	*We3=?*	
485646 [R/S]	*TIME4=?*	
3.01 [R/S]	*P4=?*	
2725 [R/S]	*We4=?*	
839418 [R/S]	*TIME5=?*	
3.24 [R/S]	*P5=?*	
2699 [R/S]	*We5=?*	
1210851 [R/S]	*TIME6=?*	
[R/S]	*EDIT? Y/N:N*	
[R/S]	*Re/RW=?*	
[EEX] 99 [R/S]	*QTd1=?*	Td1 printed
5.109 [R/S]	*QTd2=?*	B1 and Td2 printed
13.755 [R/S]	*QTd3=?*	B2 and Td3 printed
20.45 [R/S]	*QTd4=?*	B3 and Td4 printed

Keystrokes (SIZE>=044)	Display	Comments
27.084 [R/S]	*QTd5=?*	B4 and Td5 printed
28.701 [R/S]	*Re/RW=?*	B5 printed
5 [R/S]	*QTd1=?*	Td1 printed (printout of B_i and Td_i values proceeds as above)
5.034 [R/S]	*QTd2=?*	
10.232 [R/S]	*QTd3=?*	
11.494 [R/S]	*QTd4=?*	
11.859 [R/S]	*QTd5=?*	
11.91 [R/S]	*Re/RW=?*	
6 [R/S]	*QTd1=?*	
5.109 [R/S]	*QTd2=?*	
12.041 [R/S]	*QTd3=?*	
14.918 [R/S]	*QTd4=?*	
16.32 [R/S]	*QTd5=?*	
16.561 [R/S]	*Re/RW=?*	
7 [R/S]	*QTd1=?*	
5.109 [R/S]	*QTd2=?*	
12.98 [R/S]	*QTd3=?*	
17.249 [R/S]	*QTd4=?*	
19.989 [R/S]	*QTd5=?*	
20.542 [R/S]	*Re/RW=?*	
8 [R/S]	*QTd1=?*	
5.109 [R/S]	*QTd2=?*	
13.407 [R/S]	*QTd3=?*	
18.664 [R/S]	*QTd4=?*	
22.604 [R/S]	*QTd5=?*	
23.404 [R/S]	*Re/RW=?*	

```
OIL  MATL  BAL

CORR: YES
SKIP: NO
Tc=396.9440 R
Pc=660.1595 PSI
STD T=60.0000 F
STD P=14.6500 PSI
SEP T=95.0000 F
SEP P=125.0000 PSI
OIL G=31.7000 API
GAS G=0.7620
GAS GS=0.7671
T=155.0000 F
RSI=510.0000 SCF/BBL
PBP=2,514.4215 PSI
%NACL=2.2000
%POR=15.5000
%SWC=38.2000
PI=2,850.0000 PSI

P=2,818.0000 PSI
NP=72,757.0000 BBL
GP=37,105.0000 MCF
WP=0.0000 BBL
N=54,500,000.00 BBL
We=51,243.8607 BBL

P=2,792.0000 PSI
NP=220,332.0000 BBL
GP=112,325.0000 MCF
We=205,749.2402 BBL

P=2,767.0000 PSI
NP=465,009.0000 BBL
GP=237,294.0000 MCF
We=485,645.8223 BBL

P=2,725.0000 PSI
NP=785,448.0000 BBL
GP=401,380.0000 MCF
We=839,417.9074 BBL

P=2,699.0000 PSI
NP=1,103,036.000 BBL
GP=562,548.0000 MCF
We=1,210,851.286 BBL
```

```
H2O  INF  COEF

SKIP: YES
RW=4,000.000 FT
K=62.000 MD

TIME1=0.320 YR
P1=2,818.000 PSI
We1=51,244.000 BBL

TIME2=1.250 YR
P2=2,792.000 PSI
We2=205,749.000 BBL

TIME3=2.100 YR
P3=2,767.000 PSI
We3=485,646.000 BBL

TIME4=3.010 YR
P4=2,725.000 PSI
We4=839,418.000 BBL

TIME5=3.240 YR
P5=2,699.000 PSI
We5=1,210,851.000 BBL

Re/RW=1.000E99

Td1=5.929
QTd1=5.109
B1=626.884 BBL/PSI

Td2=23.158
QTd2=13.755
B2=558.735 BBL/PSI

Td3=38.906
QTd3=20.450
B3=567.095 BBL/PSI

Td4=55.765
QTd4=27.084
B4=542.155 BBL/PSI

Td5=60.026
QTd5=28.701
B5=504.390 BBL/PSI

Re/RW=5.000

Td1=5.929
QTd1=5.034
B1=636.224 BBL/PSI

Td2=23.158
QTd2=10.232
B2=664.354 BBL/PSI

Td3=38.906
QTd3=11.494
B3=797.450 BBL/PSI

Td4=55.765
QTd4=11.859
B4=881.163 BBL/PSI

Td5=60.026
QTd5=11.910
B5=902.612 BBL/PSI

Re/RW=6.000

Td1=5.929
QTd1=5.109
B1=626.884 BBL/PSI

Td2=23.158
QTd2=12.041
B2=603.693 BBL/PSI

Td3=38.906
QTd3=14.918
B3=676.240 BBL/PSI

Td4=55.765
QTd4=16.320
B4=716.264 BBL/PSI

Td5=60.026
QTd5=16.561
B5=714.053 BBL/PSI

Re/RW=7.000
```

```
Td1=5.929
QTd1=5.109
B1=626.884 BBL/PSI

Td2=23.158
QTd2=12.980
B2=578.205 BBL/PSI

Td3=38.906
QTd3=17.249
B3=620.488 BBL/PSI

Td4=55.765
QTd4=19.989
B4=634.871 BBL/PSI

Td5=60.026
QTd5=20.542
B5=618.811 BBL/PSI

Re/RW=8.000
```

```
Td1=5.929
QTd1=5.109
B1=626.884 BBL/PSI

Td2=23.158
QTd2=13.407
B2=567.313 BBL/PSI

Td3=38.906
QTd3=18.664
B3=593.912 BBL/PSI

Td4=55.765
QTd4=22.604
B4=592.830 BBL/PSI

Td5=60.026
QTd5=23.404
B5=568.808 BBL/PSI
```

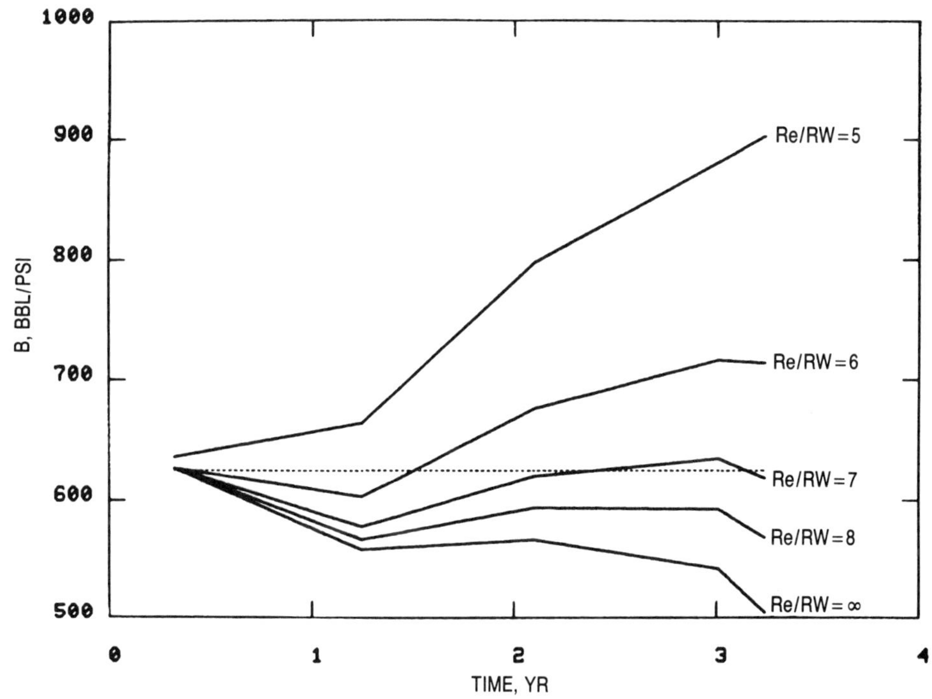

Figure 14–1

Example 2

To calculate B directly (no performance), the thickness (80 FT) and angle subtended by the reservoir (360° in this case) must by input. Recall that the ∡ symbol means angle on the HP-41. With these input, the value of B is calculated to be 1,596 BBL/PSI. This value is over 2.5 times greater than that indicated from performance. Using the first value for prediction (see INFLUX) would result in tremendously overestimated water influx. So, B should always be evaluated from performance.

Keystrokes	Display	Comments
[R/S]	*H, ∡ T,P,We*	
[A]	*H=?*	H and ∡ known
80 [R/S]	*∡=?*	Angle
360 [R/S]	*H, ∡ T,P,We*	B printed
[//] [FIX] 4	*1,595.9691*	Back to FIX 4

```
H=80.000 FT
∡=360.000
B=1,595.969 BBL/PSI
```

User Instructions

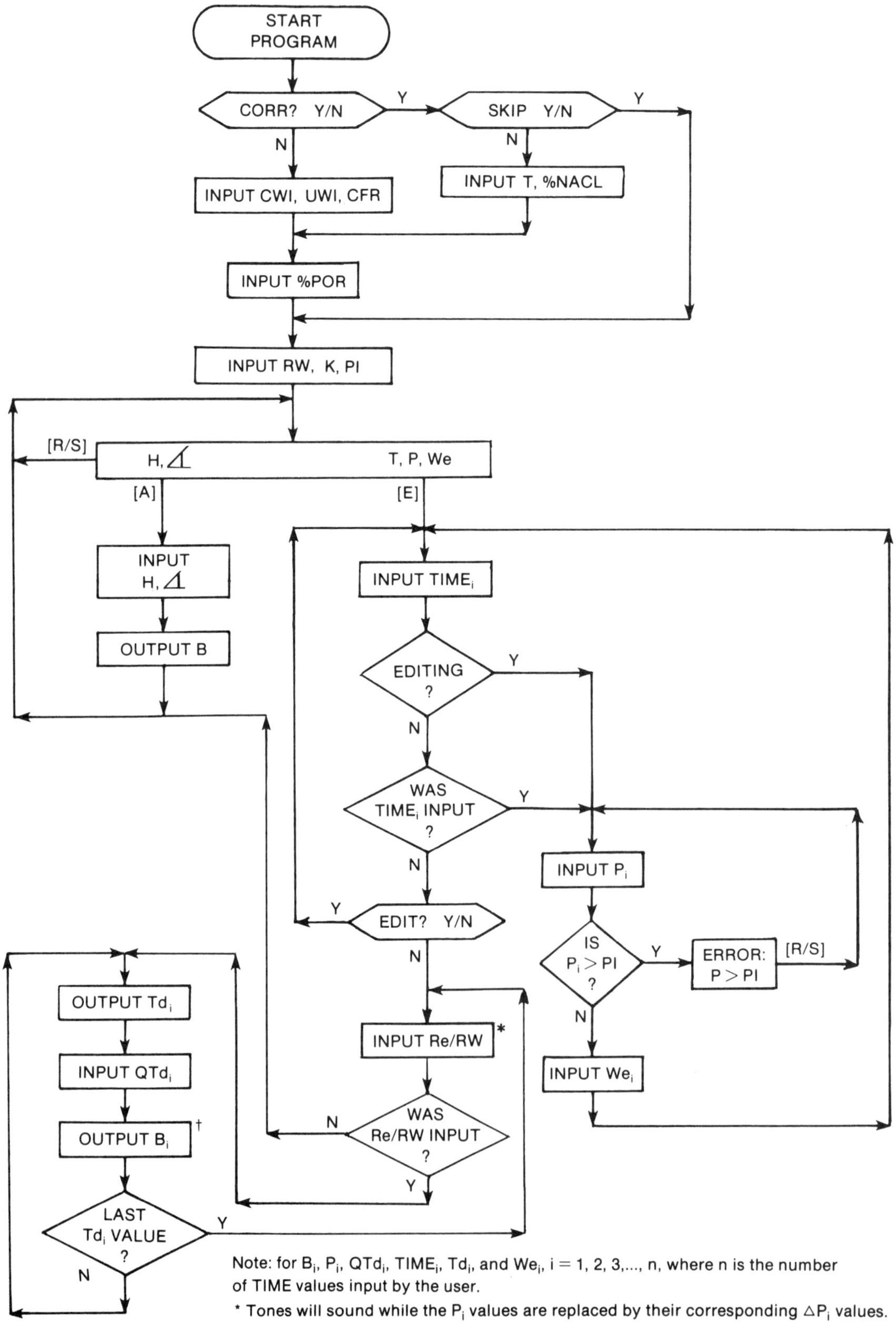

Note: for B_i, P_i, QTd_i, $TIME_i$, Td_i, and We_i, $i = 1, 2, 3, \ldots, n$, where n is the number of TIME values input by the user.

* Tones will sound while the P_i values are replaced by their corresponding ΔP_i values.

† Tones will sound while the $\Sigma(\Delta P_i\ QTd_i)$ term is calculated.

General Information

Memory Requirements

Program length: 1026 bytes (5 cards) (INFCOEF and INFLUX combined)
Minimum size: 130*
Minimum hardware: 41C + quad memory module or 41CV†

*This size will allow up to 20 TIME, P, We values. To accommodate v TIME, P, We values, use size 50 + 4v.
†A 41C + 3 memory modules will allow 1 to 15 TIME, P, We values but will not leave a port available for a printer.

Hidden Options

None

Pac Subroutines Called

TITLE, IN, Y/N?, INU, %POR, %NACL, CCW, CCFR, CUW, CON, INK, OUTK, OUT

Registers

03 Water influx units
04 Water influx units
06 Time units
07 Time units
08 Pressure units
09 Pressure units
26 UWI (CP)
27 ∠ (unused for INFLUX)
28 H (FT) (unused for INFLUX)
29 K (MD)
30 Re/RW
31 RW (FT)
32 CWI (1/PSI)
33 CFR (1/PSI)
34 Scratch
35 PI (PSI)
41 We (BBL)
42 B, B_i (BBL/PSI)
43 Water influx coefficient units
44 Water influx coefficient units
45 Pointer
46 Pointer
47 Pointer
48 Pointer
49 TIME1 (YR)
50 P1, ΔP1 (PSI)
51 We1 (BBL)
52 QTd1
53 TIME2 (YR)
54 P2, ΔP2 (PSI)
55 We2 (BBL)
56 QTd2
57 TIME3 (YR)
etc.

Registers 10–15, 20–25, and 36–40 unused

Flags

00 Set: INFLUX being run.
Clear: INFCOEF being run.

01 Set: Use Pac correlations to estimate PVT properties.
Clear: Input PVT properties.

02 Set: Skip input of PVT data.
Clear: Allow input of PVT data.

03 Set: Allow editing of TIME, P, and We values.
Clear: No editing necessary.

04 Set: Water influx coefficient units not yet input.
Clear: Water influx coefficient units have been input.

07 Set: Time and water influx units not yet input.
Clear: Time and water influx units have been input.

Program Listing

```
 01♦LBL "INFCOEF"
"H2O INF COEF" 129
XROM "TITLE" FC?C 25
PROMPT CF 00 SF 04
SF 07 SF 27 XEQ 15

 12♦LBL 00
"H,∡      T,P,We"
PROMPT GTO 00

 16♦LBL D
 17♦LBL E
XEQ 16 XEQ 17 GTO 00

 21♦LBL A
27 STO 00 "FT"
ASTO 01 CLA ASTO 02
ASTO Z "M" ASTO Y "H"
XROM "INU" 26 STO 00
"∡" XROM "IN" RCL 32
RCL 33 + * RCL 18 *
RCL 31 X↑2 * RCL 28
* 373 * 12 E6 /
STO 42 FS? 08 ADV "B"
XEQ 28 CF 08 GTO 00

 59♦LBL "INFLUX"
"H2O INFLUX" 129
XROM "TITLE" FC?C 25
PROMPT SF 00 CF 04
CF 07 XEQ 15 XEQ 16
ADV SF 08

 72♦LBL 01
41 STO 00 "B" XEQ 26
XEQ 17 GTO 01

 79♦LBL 15
SF 06 "M3" ASTO 03
"⊢/KPA" ASTO 43 "YR"
ASTO 06 "KPA" ASTO 08
CLA ASTO 04 ASTO 07
ASTO 09 ASTO 44 "CORR"
1 XROM "Y/N?" FS? 01
GTO 09 31 STO 00
"1/PSI" ASTO 01 CLA
```

Program Listing (cont.)

```
ASTO 02  ASTO Z  "1/KPA"
ASTO Y  "CW"  XROM "INU"
"1/KPA"  ASTO Y  CLA
ASTO Z  "CFR"
XROM "INU"  25  STO 00
"CP"  ASTO 01  CLA
ASTO 02  ASTO Z  "PA*S"
ASTO Y  "UWI"
XROM "INU"  XROM "%POR"
GTO 10

129♦LBL 09
"SKIP"  2  XROM "Y/N?"
FS? 02  GTO 10  XROM "T"
XROM "%NACL"
XROM "%POR"

138♦LBL 10
30  STO 00  "FT"
ASTO 01  CLA  ASTO 02
ASTO Z  "M"  ASTO Y
"RW"  XROM "INU"  28
STO 00  "MD"  ASTO 01
ASTO Y  CLA  ASTO 02
ASTO Z  "K"  XROM "INU"
34  STO 00  "PI"  XEQ 20
STO 17  FC? 01  RTN
XROM "CCW"  STO 32
XROM "CCFR"  STO 33
XROM "CUW"  STO 26  RTN

174♦LBL 16
CF 03  1.1  STO 45

178♦LBL 02
48  STO 00

181♦LBL 03
FC? 03  ADV  FS? 07
SF 08  XEQ 18  FC? 03
FS? 22  GTO 04  FC? 23
GTO 12

192♦LBL 04
CF 08  XEQ 19  RCL 35
X<>Y  X<=Y?  GTO 11
TONE 3  "P > PI"  PROMPT
1  ST- 00  GTO 04

205♦LBL 11
FS?C 07  SF 08  FC? 00
XEQ 21  CF 08  2  FC? 00
1  ST+ 00  ISG 45
GTO 03

217♦LBL 12
RCL 45  INT  1  -  1 E3
+  LASTX  /  STO 48
STO 45  "EDIT"  3
XROM "Y/N?"  FS? 03
GTO 02  RCL 29  RCL 18
/  RCL 26  /  RCL 32
RCL 33  +  /  RCL 31
X↑2  /
"PSI*MD/CP*FT2-1"
"├/YR"  CON  100  *
STO 34  RCL 35  STO 46
X<> 42  STO 45  RCL 48
FRC  4 E3  *  46.04604
+  STO 47

262♦LBL 05
TONE 5  RCL 47  8  -
RCL IND X  RCL IND 47  -
2  /  STO IND 47  DSE 47
GTO 05  RCL 45  STO 42
TONE 9  RTN

279♦LBL 17
ADV  29  STO 00  "Re/RW"
XROM "IN"  FC? 22  RTN
49  STO 47  RCL 48
STO 45

291♦LBL 06
ADV  RCL 34  RCL IND 47
*  XEQ 23  2  ST+ 47
RCL 47  STO 00  XEQ 24
XEQ 30  FS? 00  GTO 13
/  STO 42  XEQ 27
GTO 14

309♦LBL 13
RCL 42  *  STO 41
XEQ 22

314♦LBL 14
CF 08  2  ST+ 47  ISG 45
GTO 06  GTO 17

321♦LBL 18
"TIME"  XEQ 29  ASTO T
"YR"  ASTO 01  CLA
ASTO 02  ARCL T  RCL 07
RCL 06  RCL Z
XROM "INK"  RDN  STO 06
X<>Y  STO 07  R↑  RTN

340♦LBL 19
"P"  XEQ 29

343♦LBL 20
ASTO T  "PSI"  ASTO 01
CLA  ASTO 02  ARCL T
RCL 09  RCL 08  RCL Z
XROM "INK"  RDN  STO 08
X<>Y  STO 09  R↑  RTN

360♦LBL 21
XEQ 09  XROM "INK"  RDN
STO 03  X<>Y  STO 04  R↑
RTN

369♦LBL 22
XEQ 09  XROM "OUTK"  RDN
STO 03  X<>Y  STO 04  R↑
RTN

378♦LBL 09
"We"  XEQ 29  ASTO T
"BBL"  ASTO 01  CLA
ASTO 02  ARCL T  RCL 04
RCL 03  RCL Z  RTN

391♦LBL 23
"Td"  XEQ 29  XROM "OUT"
RTN

396♦LBL 24
"QTd"  XEQ 29  XROM "IN"
RTN

401♦LBL 25
"B"  XEQ 29

404♦LBL 26
XEQ 10  XROM "INK"  RDN
STO 43  X<>Y  STO 44  R↑
RTN

413♦LBL 27
"B"  XEQ 29

416♦LBL 28
FS?C 04  SF 08  XEQ 10
XROM "OUTK"  RDN  STO 43
X<>Y  STO 44  R↑  RTN

427♦LBL 10
ASTO T  "BBL/PSI"
ASTO 01  ASHF  ASTO 02
CLA  ARCL T  RCL 44
RCL 43  RCL Z  RTN

439♦LBL 29
STO 05  CLST  FS? 41  1
+  FS? 40  2  +  5  /
FS? 39  1  +  FS? 38  2
+  FS? 37  4  +  FS? 36
8  +  FS? 29  CHS
RCL 45  FIX 0  CF 29
ARCL X  X<>Y  X<0?
SF 29  ENTER↑  FRC  5  *
FIX IND Y  X=0?
SCI IND Y  1  X=Y?
ENG IND Z  RCL 05  RTN

483♦LBL 30
RCL 47  1.04804  +
STO 46  50  STO 00  CLX

491♦LBL 07
TONE 5  RCL IND 46
RCL IND 00  *  +  4
ST+ 00  X<>Y  DSE 46
GTO 07  RCL IND 47  X<>Y
TONE 9  END
```

15. INFLUX — Predicting Water Influx

It is more difficult to predict the performance of reservoir behavior from water-drive reservoirs than solution and gas-cap drive reservoirs. Water-drive reservoirs also generally produce with some solution-gas drive and expansion as well. To predict water-drive performance, it is necessary to predict water influx independently of material balance. The material balance is then used to predict oil production. This can become particularly complex.

The best method of predicting water influx is unsteady-state compressible fluid flow theory; this process is often referred to as the van Everdingen and Hurst method. Tables 14–1 and 14–2 present the constant-pressure solution of the radial diffusivity equation for various values of Re/RW where RW is the internal radius of the aquifer that furnished the water encroachment.

The calculation procedure to predict water influx is the same as discussed for INFCOEF. Known times and pressure drops were utilized in the previous program and were quite straightforward. However, there we knew the oil production at each time and were not trying to predict it. Predicting the production involves the following complications.

A future pressure first must be estimated for the average reservoir pressure in the uninvaded zone. This pressure is used for the oil PVT properties. The corresponding pressure at the water-oil contact is used to predict water influx. The value of water influx can then be calculated in a fashion similar to the method presented in the previous program using the water influx constant calculated from performance. The oil production during that period can then be estimated along with the saturations in the uninvaded zone. From these saturations, the average productivity index can be calculated for this pressure drop. At this point, the cumulative production is recalculated using the productivity index. This will require iterating on cumulative production until a value of NP can be obtained, which results in an oil saturation that yields the same value of cumulative production.

At this point, the value of water influx can be calculated from material balance. If it agrees sufficiently with the value estimated by unsteady state calculations, the correct pressure drop has been estimated. If not, another pressure is estimated and the process is repeated until it converges.

The value of oil saturation in the uninvaded zone requires knowing the oil saturation in the bypassed zone, i.e., knowing the displacement efficiency of the reservoir. Slider presents an equation for a certain set of assumptions to calculate saturation in the uninvaded zone. Gas saturation in the water bank, nonpistonlike displacement, and water production will all require modifications to that equation. Furthermore, the prediction procedure is quite sensitive to key values. It is recommended that sensitivities be performed to assure the accuracy of the calculations and to indicate the range of errors associated with estimation errors.

INFLUX calculates water influx as a function of user input times and pressures. It can be performed readily for a variety of Re/RW values. The value of QTd must be input from the Td values calculated by the program. Although this program is combined with INFCOEF, it behaves as a separate program.

Equations

$$We_i = B_i \Sigma(\Delta P_i \, QTd_i)$$

See the equations for INFCOEF.

Nomenclature

Symbol	*Variable Name*	*Input or Output*	*English Units*	*SI Units*
B_i*	Water influx coefficient at $TIME_i$, P_i, and We_i	I	BBL/PSI	M3/KPA
K	Permeability	I	MD	MD
P_i*	Pressure	I	PSI	KPA
PI*	Initial pressure	I	PSI	KPA
QTd_i	Dimensionless water influx at Td_i	I	–	–
RW	Internal radius of aquifer	I	FT	M
Re/RW	Ratio of external to internal aquifer radii	I	–	–
$TIME_i$*	Time	I	YR	YR
Td_i	Dimensionless time	O	–	–
We_i*	Water influx at $TIME_i$ and P_i	O	BBL	M3

Note: For B_i, P_i, QTd_i, $TIME_i$, Td_i, and We_i, i = 1, 2, 3, . . . , n, where n is the number of TIME values input by the user.
*The units for these variables are saved by the program.

Yes/No Questions

See the yes/no questions for INFCOEF.

Example

Using the calculated value of B and Re/RW for the SW Franklin field (see Example 1 of INFCOEF), predict future water influx values. The times and pressures are given in Table 15-1. The QTd values are read from Figure 14-2 for Re/RW = 7. Figure 15-1 graphs predicted water influx and calculated historical water influx.

(Note: The keystrokes assume you have just completed Examples 1 and 2 of INFCOEF.)

Table 15-1

TIME (YR)	P (PSI)
0.00	2,850
0.32	2,818
1.25	2,792
2.1	2,767
3.01	2,725
3.24	2,699
4.0	2,663
5.0	2,618
7.5	2,506
10.0	2,400

Keystrokes	Display	Comments
[/] [FIX] 3		FIX 3 to match accuracy of Tables 14-1 and 14-2
[XEQ] [ALPHA] INFLUX [ALPHA]	*CORR? Y/N:Y*	
[R/S]	*SKIP? Y/N:Y*	
[R/S]	*RW=?*	
[R/S]	*K=?*	
[R/S]	*PI=?*	
[R/S]	*TIME1=?*	
[←]	*0.320*	

Even though the times from the INFCOEF example are still intact, a value for $TIME_i$ must still be entered. If no value is input, the program assumes that the TIME, P data is complete.

Keystrokes	Display	Comments
.32 [R/S]	*P1=?*	
[←]	*16.000*	

The P_i values in the INFCOEF example were replaced by ΔP_i values used to calculate B_i. Therefore, the P values must be entered again.

Keystrokes	Display	Comments
2818 [R/S]	*TIME2=?*	
1.25 [R/S]	*P2=?*	
2792 [R/S]	*TIME3=?*	
2.1 [R/S]	*P3=?*	
2767 [R/S]	*TIME4=?*	
3.01 [R/S]	*P4=?*	
2725 [R/S]	*TIME5=?*	
3.24 [R/S]	*P5=?*	
2699 [R/S]	*TIME6=?*	
4 [R/S]	*P6=?*	
2663 [R/S]	*TIME7=?*	
5 [R/S]	*P7=?*	
2618 [R/S]	*TIME8=?*	
7.5 [R/S]	*P8=?*	
2506 [R/S]	*TIME9=?*	
10 [R/S]	*P9=?*	
2400 [R/S]	*TIME10=?*	
[R/S]	*EDIT? Y/N:N*	
[R/S]	*B=?*	
625 [R/S]	*Re/RW=?*	
7 [R/S]	*QTd1=?*	Td1 printed
5.109 [R/S]	*QTd2=?*	We1 and Td2 printed
12.98 [R/S]	*QTd3=?*	We2 and Td3 printed
17.249 [R/S]	*QTd4=?*	We3 and Td4 printed
19.989 [R/S]	*QTd5=?*	We4 and Td5 printed
20.542 [R/S]	*QTd6=?*	We5 and Td6 printed
21.729 [R/S]	*QTd7=?*	We6 and Td7 printed
22.727 [R/S]	*QTd8=?*	We7 and Td8 printed
23.697 [R/S]	*QTd9=?*	We8 and Td9 printed
23.931 [R/S]	*Re/RW=?*	We9 printed
[/] [FIX] 4	*7.0000*	Back to FIX 4

```
H2O  INFLUX

TIME1=0.320 YR
P1=2,818.000 PSI

TIME2=1.250 YR
P2=2,792.000 PSI

TIME3=2.100 YR
P3=2,767.000 PSI

TIME4=3.010 YR
P4=2,725.000 PSI

TIME5=3.240 YR
P5=2,699.000 PSI
```

```
TIME6=4.000 YR
P6=2,663.000 PSI

TIME7=5.000 YR
P7=2,618.000 PSI

TIME8=7.500 YR
P8=2,506.000 PSI

TIME9=10.000 YR
P9=2,400.000 PSI

B=625.000 BBL/PSI

Re/RW=7.000

Td1=5.929
QTd1=5.109
We1=51,090.002 BBL

Td2=23.158
QTd2=12.980
We2=222,400.631 BBL

Td3=38.906
QTd3=17.249
We3=489,177.197 BBL

Td4=55.765
QTd4=19.989
We4=826,366.575 BBL

Td5=60.026
QTd5=20.542
We5=1,222,961.577 BBL

Td6=74.107
QTd6=21.729
We6=1,644,151.269 BBL

Td7=92.633
QTd7=22.727
We7=2,114,366.268 BBL

Td8=138.950
QTd8=23.697
We8=2,763,483.140 BBL

Td9=185.266
QTd9=23.931
We9=3,731,282.199 BBL
```

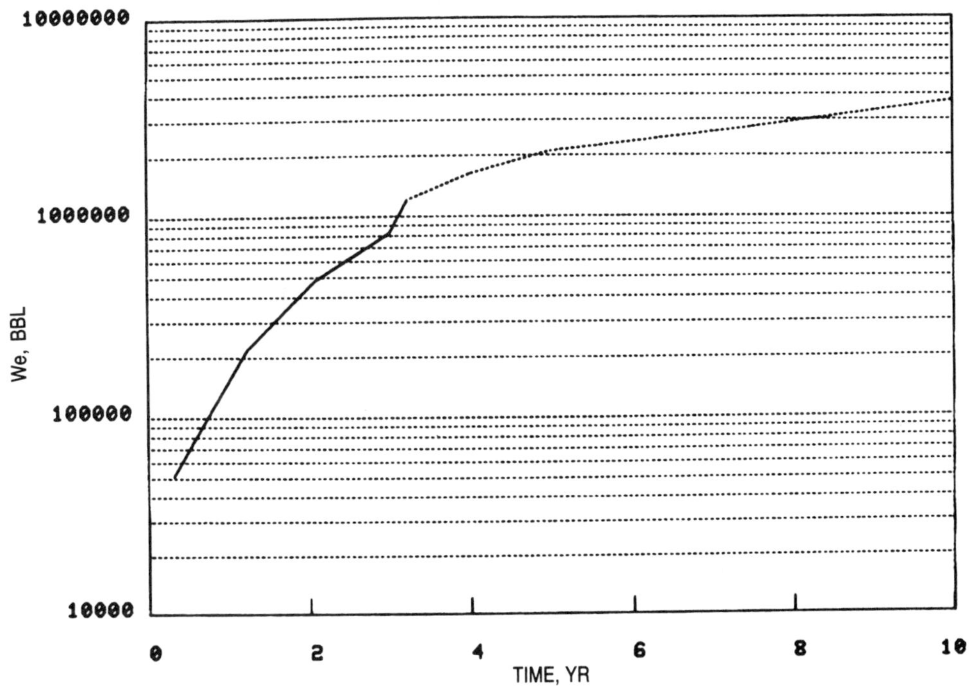

Figure 15–1

General Information

See the general information for INFCOEF.

Registers

See the registers for INFCOEF.

Flags

See the flags for INFCOEF.

Program Listing

See the program listing for INFCOEF.

User Instructions

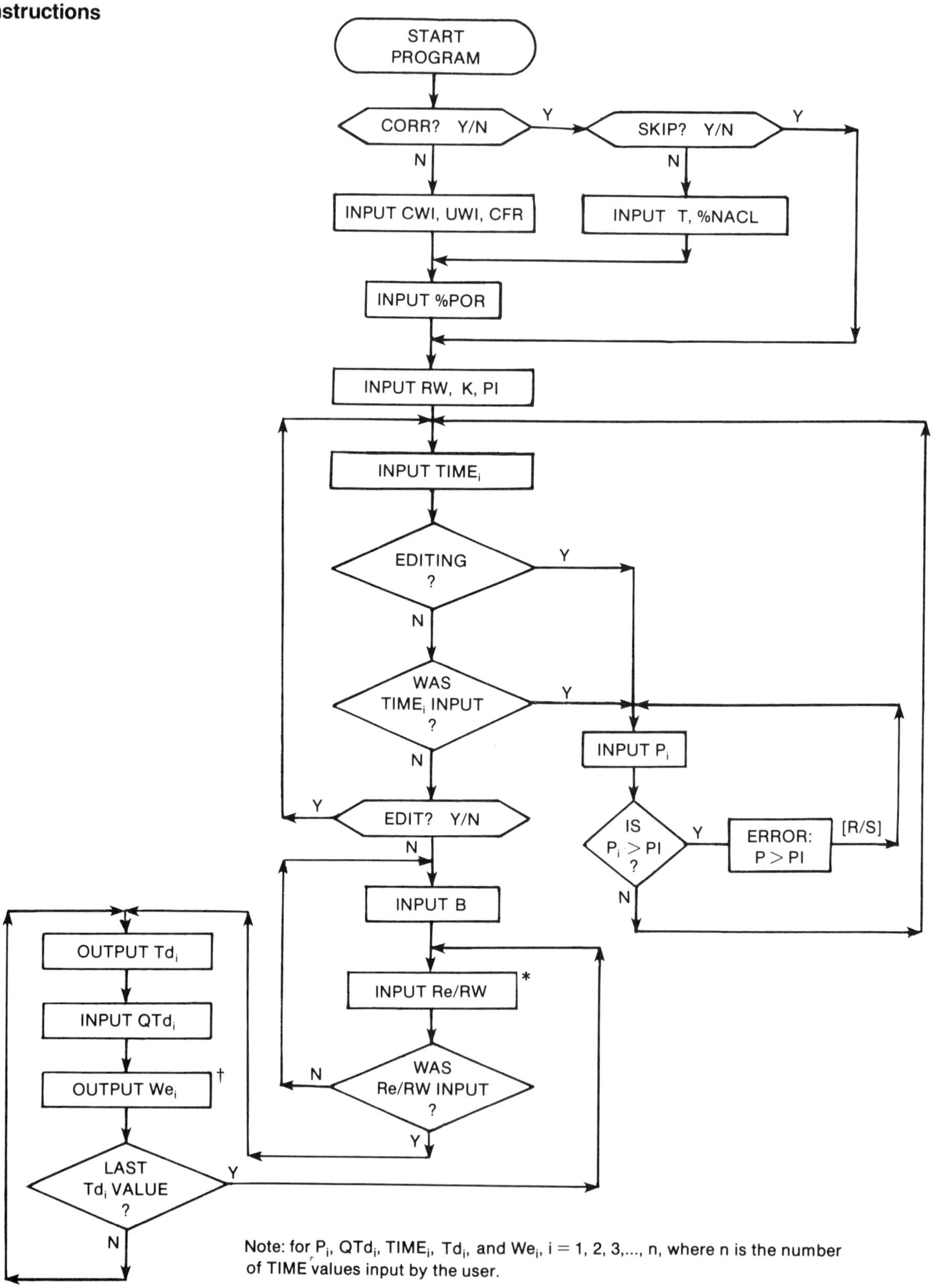

Note: for P_i, QTd_i, $TIME_i$, Td_i, and We_i, i = 1, 2, 3,..., n, where n is the number of TIME values input by the user.

* Tones will sound while the P_i values are replaced by their corresponding ΔP_i values.

† Tones will sound while the $\Sigma(\Delta P_i\ QTd_i)$ term is calculated.

16. GASMBE — Gas Material Balance

For volumetric natural-gas reservoirs with reasonable permeability, the material-balance equation is clearly the preferred analytical method. GASMBE provides a flexible tool to calculate the original gas in place or cumulative water influx. The original gas in place (G) can be calculated at a single point, or the user can perform a regression analysis for a series of pressures and cumulative productions. The user can forecast to a future pressure and calculate cumulative production or to a future production to calculate reservoir pressure. The gas equivalent of condensate production and water content of natural gas in the reservoir are also taken into account (see GASPROD).

Equations

$$G = \frac{GP* \; BG' - (We - BW \; WP*)}{BG' - BGI'}$$

$$GP* = GP + GVCOND + H2OGAS$$

See the equations for GASPROD.

Linear regression:

$$\frac{P}{Z} = A \; GP* + \frac{PI}{ZI}$$

$$\frac{PI}{ZI} = \frac{\Sigma GP*^2 \; \Sigma(P/Z) - \Sigma GP* \; \Sigma GP*(P/Z)}{n\Sigma GP*^2 - (\Sigma GP*)^2}$$

$$A = \frac{\Sigma(P/Z) - n(PI/ZI)}{\Sigma GP*}$$

$$G = \frac{-PI}{ZI} \; \frac{1}{A}$$

$$R{\uparrow}2 = \frac{PI/ZI \; \Sigma(P/Z) + A\Sigma GP*(P/Z) - [\Sigma(P/Z)]^2/n}{\Sigma(P/Z)^2 - [\Sigma(P/Z)]^2/n}$$

where n is the number of P, GP* points input by the user.

GP from average gas-oil ratio:*

$$GP* = GP \left(1 + \frac{GE}{R \; AVG}\right)$$

See the equations for GASPROD.

P from P/Z:
P is calculated iteratively using Newton's method as follows:†

$$P_{i+1} = P_i - \frac{P_i - \frac{P}{Z} z_i}{\frac{P_i}{Pc} CR_i}$$

where Z_i is Z at P_i and CR_i is the pseudoreduced compressibility at P_i. P/Z is also used as the initial guess.

Nomenclature

Symbol	Variable Name	Input or Output	English Units	SI Units
BGI	Initial gas formation volume factor	I	FT3/ SCF	M3/ SCM
COND G	Condensate gravity	I	API	KG/M3
CONDP*	Cumulative condensate production	I	BBL	M3
G*	Initial gas in place	I,O	MCF	SCM
GP*	Cumulative gas production	I,O	MCF	SCM
GP**	Cumulative gas production corrected for gas equivalent of the condensate and for the water content of the gas	O	MCF	SCM
P*	Pressure	I,O	PSI	KPA
PI*	Initial pressure	I,O	PSI	KPA
R AVG*	Average gas-oil ratio	I	MCF/ BBL	SCM/ M3
R↑2	Coefficient of determination	O	–	–
WP*	Cumulative water production	I	BBL	M3
WP**	Cumulative water production corrected for water content of natural gas	O	BBL	M3
We*	Cumulative water influx	I,O	BBL	M3

†Unpublished derivation, E. L. Vogel, Hewlett-Packard Company.
*The units for these variables are saved by the program.

Yes/No Questions

CORR? Yes: Use Pac correlations to estimate PVT properties.
No: Input PVT properties.

	No:	Input PVT properties.
SKIP?	Yes:	Skip input of PVT data.
	No:	Allow input of PVT data.

Example 1

The historical production for the South Texas Vicksburg reservoir is given in Table 16-1. Using GASMBE, fit the best straight line through this data and calculate the initial gas in place (G). The regression is shown in Figure 16-1. Neglect condensate production for this example. The calculated initial pressure (PI) is the least-squares regression result, not usually the value input corresponding to GP = 0. In fact, the initial production (GP = 0) need not necessarily be input.

Calculate the gas recovery to abandonment pressures of 600 and 1,500 PSI. Also, what will the reservoir pressure be when the well has produced 10,000 MMCF?

Table 16-1

TIME (YR)	*P (PSI)*	*GP (MMCF)*	*QG (MCF/DAY)*
0	6,200	0	6,240
0.33	5,911	681.8	5,529
0.80	5,563	1,569.2	4,722
1.21	5,318	2,208.9	4,101
2.20	4,931	3,438.2	3,143
2.94	4,692	4,215.0	2,603

Keystrokes (SIZE> =052)	Display	Comments
[XEQ] [ALPHA] GASMBE [ALPHA]	*SKIP? Y/N:*	Last character is Y or N
N [R/S]	*Tc=?*	
441.8952 [R/S]	*Pc=?*	
661.5656 [R/S]	*T=?*	
180 [R/S]	*COND G=?*	Input a zero value for condensate gravity to neglect condensate production
0 [R/S]	*FIT G,We*	
[A]	*P=?*	Regression option
6200 [R/S]	*GP=?*	
0 [R/S]	*P=?*	
5911 [R/S]	*GP=?*	
681.8 [ALPHA] MMCF [R/S]	*P=?*	
5563 [R/S]	*GP=?*	
1569.2 [R/S]	*P=?*	
5318 [R/S]	*GP=?*	
2208.9 [R/S]	*P=?*	
4931 [R/S]	*GP=?*	
3438.2 [R/S]	*P=?*	
4692 [R/S]	*GP=?*	
4215 [R/S]	*P=?*	

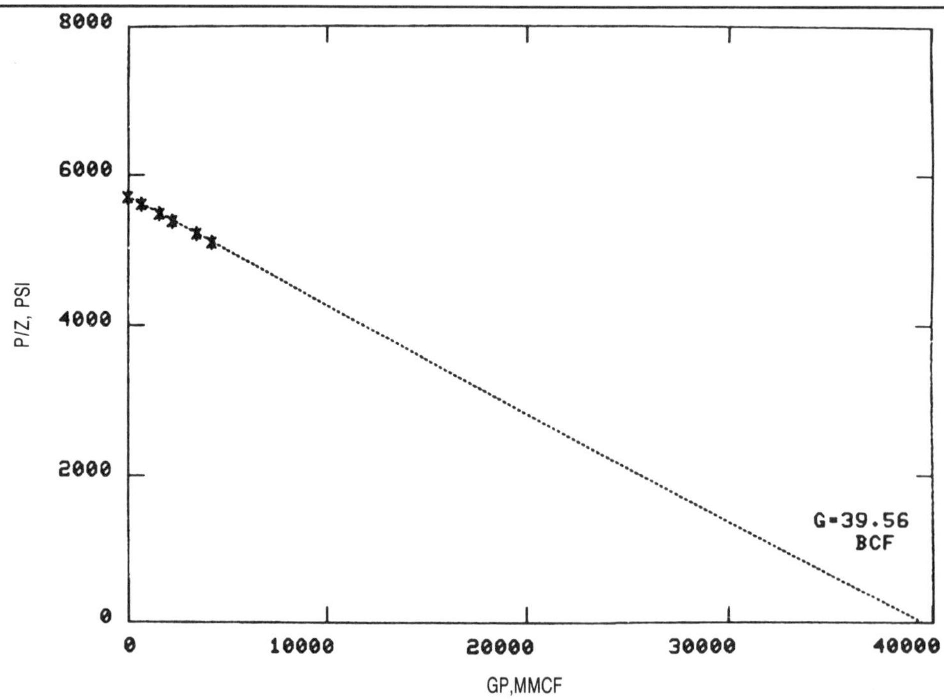

Figure 16–1

Keystrokes (SIZE> =052)	Display	Comments
[R/S]	*P GP*	PI, G, and R↑2 printed
[A]	*P=?*	P known
600 [R/S]	*P GP*	GP printed
[A]	*P=?*	
1500 [R/S]	*P GP*	GP printed
[E]	*GP=?*	GP known
10000 [R/S]	*P GP*	P printed.

Note the program calculates P/Z directly but must use an iterative solution to find P from P/Z

```
GAS MATL BAL

SKIP: NO
Tc=441.8952 R
Pc=661.5656 PSI
T=180.0000 F
COND G=0.0000 API

P=6,200.0000 PSI
GP=0.0000 MCF

P=5,911.0000 PSI
GP=681.8000 MMCF

P=5,563.0000 PSI
GP=1,569.2000 MMCF

P=5,318.0000 PSI
GP=2,208.9000 MMCF

P=4,931.0000 PSI
GP=3,438.2000 MMCF

P=4,692.0000 PSI
GP=4,215.0000 MMCF

PI=6,198.0492 PSI
G=39,563.9294 MMCF
R↑2=0.9999

P=600.0000 PSI
GP=34,947.6976 MMCF

P=1,500.0000 PSI
GP=26,164.1968 MMCF

GP=10,000.0000 MMCF
P=3,397.4062 PSI
```

Example 2

A certain dry gas reservoir has historical production as indicated in Table 16–2. The volumetric estimate of initial gas in place is 337.9 BCF. A plot of P/Z versus GP curves upward, confirming suspicions of water influx. Calculate the cumulative water influx. No water has been produced. The reservoir temperature is 155.4 F, Tc = 349.81 R, Pc = 673.31 PSI, and %NACL = 1.2.

This example is based on an example presented by Ikoku, who describes a method of calculating the water influx coefficient by unsteady-state prediction as well. It is important to note that both INFCOEF and INFLUX can be used for gas reservoirs to calculate the water influx coefficient and predict future water influx, respectively, independent of the material-balance calculations.

Table 16–2

Time (YR)	*P (PSI)*	*GP (MMCF)*
0	2,333	-
2	2,321	2,305
4	2,203	20,257
6	2,028	49,719
8	1,854	80,134
10	1,711	105,930
12	1,531	135,350
14	1,418	157,110
16	1,306	178,300
18	1,227	192,089
20	1,153	205,744

Keystrokes	Display	Comments
[XEQ] [ALPHA] GASMBE [ALPHA]	*SKIP? Y/N:N*	
[R/S]	*Tc=?*	
349.81 [R/S]	*Pc=?*	
673.13 [R/S]	*T=?*	
155.4 [R/S]	*COND G=?*	Input a zero value for condensate gravity to neglect condensate production
0 [R/S]	*FIT G,We*	
[E]	*CORR? Y/N*	G or We known. Last character is Y or N
Y [RS]	*STD T=?*	
60 [R/S]	*STD P=?*	
14.65 [R/S]	*%NACL=?*	
1.2 [R/S]	*PI=?*	
2333 [R/S]	*P=?*	
2321 [R/S]	*GP=?*	
2305 [ALPHA] MMCF [R/S]	*WP=?*	
0 [R/S]	*G We*	
[A]	*G=?*	G known
337.9 [ALPHA] BCF [R/S]	*G We*	We printed
[R/S]	*P=?*	
2203 [R/S]	*GP=?*	
20257 [ALPHA] MMCF [R/S]	*WP=?*	

Keystrokes	Display	Comments
[R/S]	G We	
[A]	G=?	
[R/S]	G We	We printed
[R/S]	P=?	
2028 [R/S]	GP=?	
49719 [R/S]	WP=?	
[R/S]	G We	
[A]	G=?	
[R/S]	G We	We printed
[R/S]	P=?	
1854 [R/S]	GP=?	
80134 [R/S]	WP=?	
[R/S]	G We	
[A]	G=?	
[R/S]	G We	We printed
[R/S]	P=?	
1711 [R/S]	GP=?	
105930 [R/S]	WP=?	
[R/S]	G We	
[A]	G=?	
[R/S]	G We	We printed
[R/S]	P=?	
1531 [R/S]	GP=?	
135350 [R/S]	WP=?	
[R/S]	G We	
[A]	G=?	
[R/S]	G We	We printed
[R/S]	P=?	
1418 [R/S]	GP=?	
157110 [R/S]	WP=?	
[R/S]	G We	
[A]	G=?	
[R/S]	G We	We printed
[R/S]	P=?	
1306 [R/S]	GP=?	
178300 [R/S]	WP=?	
[R/S]	G We	
[A]	G=?	
[R/S]	G We	We printed
[R/S]	P=?	
1227 [R/S]	GP=?	
192089 [R/S]	WP=?	
[R/S]	G We	
[A]	G=?	
[R/S]	G We	We printed
[R/S]	P=?	
1153 [R/S]	GP=?	
205744 [R/S]	WP=?	
[R/S]	G We	
[A]	G=?	
[R/S]	G We	We printed

```
GAS MATL BAL

Tc=349.8100 R
Pc=673.1300 PSI
T=155.4000 F
COND G=0.0000 API
CORR: YES
STD T=60.0000 F
STD P=14.6500 PSI
%NACL=1.2000
PI=2,333.0000 PSI

P=2,321.0000 PSI
GP=2,305.0000 MMCF
WP=0.0000 BBL
G=337.9000 BCF
We=668,625.7429 BBL

P=2,203.0000 PSI
GP=20,257.0000 MMCF
We=1,639,173.640 BBL

P=2,028.0000 PSI
GP=49,719.0000 MMCF
We=6,592,346.979 BBL

P=1,854.0000 PSI
GP=80,134.0000 MMCF
We=13,622,388.80 BBL

P=1,711.0000 PSI
GP=105,930.0000 MMCF
We=21,545,535.70 BBL

P=1,531.0000 PSI
GP=135,350.0000 MMCF
We=28,080,079.59 BBL

P=1,418.0000 PSI
GP=157,110.0000 MMCF
We=39,574,577.49 BBL

P=1,306.0000 PSI
GP=178,300.0000 MMCF
We=52,370,181.80 BBL

P=1,227.0000 PSI
GP=192,089.0000 MMCF
We=60,354,931.09 BBL

P=1,153.0000 PSI
GP=205,744.0000 MMCF
We=70,815,718.80 BBL
```

Example 3

The Plum Nearly (Smackover) field (see Example 1 of GASPVT) has had historical production as indicated in Table 16–3. Since the dew-point pressure is less than 2,000 PSI (see the example for DEW), GOR behavior has remained constant. Additional required information not presented in either of those examples is as follows:

> COND G = 56.2 API
> Water influx is considered negligible
> Water salinity = 15,600 PPM

Calculate the indicated value of G for each of the pressure points. The program corrects the GP values (to GP*) for the gas equivalent of the condensate and for the water content of the gas. The value of WP is corrected (to WP*) for the water content of the gas.

The volumetric estimate for G was 50 BCF. Calculate the indicated value of water influx at the second point using this value. The calculated value of We indicates a very minor water influx to date. The value of GP is 2,098 MMCF, and BG at 12,000 PSI is 0.0027 FT3/SCF. A simple calculation shows that approximately one million reservoir barrels of gas have been produced, with less than 30,000 BBL of water influx. It appears that the assumption of negligible water influx is a good one.

Table 16–3

TIME (YR)	P (PSI)	GP (MMCF)	WP (BBL)	CONDP (BBL)
0	13,600	-	-	-
0.65	12,720	1,054	2,300	16,129
1.28	11,918	2,098	4,800	32,099

Keystrokes	Display	Comments
[XEQ] [ALPHA] GASMBE [ALPHA]	*SKIP? Y/N:N*	
[R/S]	*Tc=?*	
410.9022 [R/S]	*Pc=?*	
673.9813 [R/S]	*T=?*	
255 [R/S]	*COND G=?*	
56.2 [R/S]	*FIT G,We*	
[E]	*CORR? Y/N:Y*	G or We known
[R/S]	*STD T=?*	
[R/S]	*STD P=?*	
15.025 [R/S]	*%NACL=?*	
[R/S]	*PPM=?*	
15600 [R/S]	*PI=?*	
13600 [R/S]	*P=?*	
12720 [R/S]	*GP=?*	
1054 [ALPHA] MMCF [R/S]	*WP=?*	
2300 [R/S]	*CONDP=?*	
16129 [R/S]	*G We*	
[E]	*We=?*	We known
0 [R/S]	*G We*	GP*, WP*, and G printed
[R/S]	*P=?*	
11918 [R/S]	*GP=?*	
2098 [R/S]	*WP=?*	
4800 [R/S]	*CONDP=?*	
32099 [R/S]	*G We*	
[E]	*We=?*	
[R/S]	*G We*	GP*, WP*, and G printed

Keystrokes	Display	Comments
[A]	*G=?*	G known
50 [ALPHA] BCF [R/S]	*G We*	GP*, WP*, and We printed

To calculate gas production in reservoir barrels:

Keystrokes	Display	Comments
2098 [EEX] 6 [ENTER↑].0027 [*]		
[ALPHA] FT3-BBL [ALPHA]	*5,664,600.000*	
[XEQ] [ALPHA] CON [ALPHA]	*1,008,908.349*	Barrels

```
GAS MATL BAL

Tc=410.9022 R
Pc=673.9813 PSI
T=255.0000 F
COND G=56.2000 API
STD P=15.0250 PSI
PPM=15,600.0000
PI=13,600.0000 PSI

P=12,720.0000 PSI
GP=1,054.0000 MMCF
WP=2,300.0000 BBL
CONDP=16,129.0000 BBL
We=0.0000 BBL
GP*=1,070.1679 MMCF
WP*=1,919.5548 BBL
G=51,356.7451 MMCF

P=11,918.0000 PSI
GP=2,098.0000 MMCF
WP=4,800.0000 BBL
CONDP=32,099.0000 BBL
GP*=2,131.8293 MMCF
WP*=3,818.9110 BBL
G=51,413.3431 MMCF

G=50.0000 BCF
GP*=2.1318 BCF
WP*=3,818.9110 BBL
We=28,825.6900 BBL
```

User Instructions

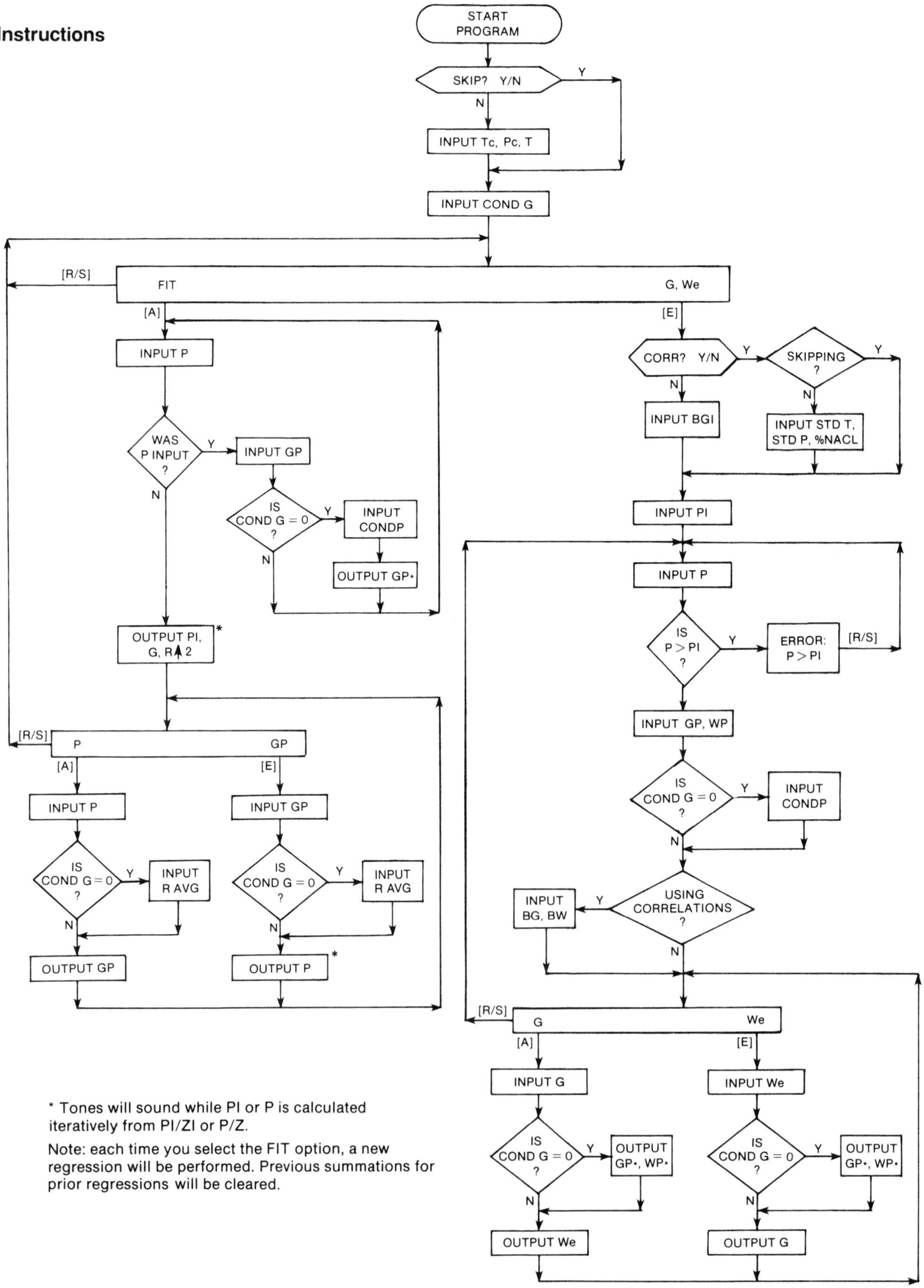

* Tones will sound while PI or P is calculated iteratively from PI/ZI or P/Z.

Note: each time you select the FIT option, a new regression will be performed. Previous summations for prior regressions will be cleared.

General Information

Memory Requirements

Program length: 1235 bytes (6 cards)
Minimum size: 052
Minimum hardware: 41C + quad memory module or 41CV*

*This program can be run in a 41C + 3 memory modules but will not leave a port available for a printer.

Hidden Options

None

Pac Subroutines Called

TITLE, Y/N?, ITcPc, T, INU, CON, STDTP, %NACL, CBG, IN, CBW, CZ, CCR, INK, OUTK

Registers

03	Condensate and water production units
04	Condensate and water production units
06	Gas production units
07	Gas production units
08	Pressure units
09	Pressure units
26	CONDP (BBL)
27	R AVG (MCF/BBL)
28	BGI (FT3/SCF)
29	BG (FT3/SCF)
30	COND G (API)
31	BW
32	Gas equivalent of condensate (MCF/BBL)
33	WP* (BBL)
34	GP* (MCF)
35	PI (PSI)
36	GP (MCF)
38	WP (BBL)
39	G (MCF)
41	We (BBL)
42	ΣGP* (MCF)
43	$\Sigma GP*^2$ (MCF2)
44	Σ(P/Z) (PSI)
45	$\Sigma(P/Z)^2$ (PSI2)
46	ΣGP*(P/Z) (MCF*PSI)
47	n
48	A
49	PI/ZI
50	Gas-oil ratio units
51	Gas-oil ratio units

Registers 12–15, 18, 20, 21, 24, 25, 37, and 40 unused

Note: The summation registers have been moved to start at register 42.

Flags

00	Set:	COND G ≠ 0
	Clear:	COND G = 0
01	Set:	Use Pac correlations to estimate PVT properties.
	Clear:	Input PVT properties.
02	Set:	Skip input of PVT data.
	Clear:	Allow input of PVT data.
03	Set:	Gas-oil ratio units not yet input.
	Clear:	Gas-oil ratio units have been input.
04	Set:	Water production units not yet input.
	Clear:	Water production units have been input.

Program Listing

```
 01♦LBL "GASMBE"
"GAS MATL BAL"  52
XROM "TITLE"  FC?C 25
PROMPT  SF 03  SF 04
SF 06  SF 27  ΣREG 42
CLX  STO 32  "M3"
ASTO 03  "SCM"  ASTO 06
"⊢/M3"  ASTO 50  "KPA"
ASTO 08  CLA  ASTO 04
ASTO 07  ASTO 09
ASTO 51  "SKIP"  2
XROM "Y/N?"  FC? 02
XROM "ITcPc"  FC? 02
XROM "T"  29  STO 00
"API"  ASTO 01  CLA
ASTO 02  ASTO Z  "KG/M3"
ASTO Y  "COND G"
XROM "INU"  ABS  RND
CF 00  X=0?  STO 30
X=0?  GTO 15  SF 00
1.03  RCL 30  "API-SPGR"
CON  -  2.99965  *
STO 32

 61♦LBL 15
"FIT      G,We"  PROMPT
GTO 15

 65♦LBL A
GTO 18

 67♦LBL E
"CORR"  1  XROM "Y/N?"
FC? 01  GTO 00  FC? 02
XROM "STDTP"  FC? 02
XROM "%NACL"  GTO 01

 78♦LBL 00
27  STO 00  "FT3/SCF"
ASTO 01  ASHF  ASTO 02
"M3/SCM"  ASTO Y  CLA
ASTO Z  "BGI"
XROM "INU"

 91♦LBL 01
34  "PI"  XEQ 22  ADV
FC? 01  GTO 02  XEQ 14
CLX  RCL 35  RCL 11  /
XROM "CBG"  STO 28

105♦LBL 16
CF 08  XEQ 21  RCL 35
X<>Y  X<=Y?  GTO 02
TONE 3  "P > PI"  PROMPT
GTO 16

116♦LBL 02
FS?C 04  SF 08  XEQ 24
37  "WP"  XEQ 30  CF 08
FS? 00  XEQ 29  FS? 01
GTO 12  28  STO 00
"FT3/SCF"  ASTO 01  ASHF
ASTO 02  "M3/SCM"
ASTO Y  CLA  ASTO Z
"BG"  XROM "INU"  30
STO 00  "BW"  XROM "IN"

144♦LBL 12
```

Program Listing (cont.)

```
" G        We" PROMPT
GTO 16

148♦LBL A
38 "G" XEQ 25 XEQ 03
X<>Y RCL 39 * 1 E3 *
- STO 41 FS? 00
XEQ 26 FS? 00 XEQ 31
RCL 41 "We" XEQ 32
ADV GTO 12

169♦LBL E
40 "We" XEQ 30 XEQ 03
RCL 41 - X<>Y / 1 E3
/ STO 39 FS? 00
XEQ 26 FS? 00 XEQ 31
XEQ 27 ADV GTO 12

188♦LBL 03
FC? 01 GTO 04 XEQ 14
XROM "CBG" STO 29

194♦LBL 04
FS? 01 XROM "CBW"
FC? 01 RCL 31 STO 31
RCL 36 RCL 38 FC? 00
GTO 06 XEQ 05
2.10777 E-5 RCL Y * 1
+ RCL 32 RCL 26 *
RCL 36 + * STO 34
RCL 38 R↑ LASTX *
"LBM/BBL-SPGR" CON
1 E3 / - STO 33

227♦LBL 06
RCL 31 * X<>Y 1 E3 *
RCL 29 "FT3-BBL" CON
* + RCL 29 RCL 28 -
CON X<>Y RTN

244♦LBL 05
RCL 16 1 % 2657 E-7
RCL 17 227 E-10 * -
RCL 17 * 3.2147 +
Y↑X 60630 RCL 17 /
3.4 + * RCL 19 X≠0?
GTO 06 X<>Y RTN

269♦LBL 06
1757 E-7 * 4893 E-6 +
RCL 19 * 1 X<>Y - *
RTN

281♦LBL 18
CLΣ ADV

284♦LBL 11
XEQ 21 FS? 22 GTO 07
FC? 23 GTO 08

290♦LBL 07
XEQ 24 FS? 00 XEQ 29
ADV CF 04 CF 08
XEQ 14 CZ RCL 17 X<>Y
/ RCL 32 RCL 26 *
RCL 36 + STO 34 Σ+
FS? 00 XEQ 26 GTO 11

312♦LBL 08
RCL 43 RCL 44 *
RCL 42 RCL 46 * -
RCL 47 RCL 43 *
RCL 42 X↑2 - /
STO 49 RCL 44 RCL 47
RCL Z * - RCL 42 /
STO 48 / CHS STO 39
RCL 49 XEQ 20 STO 35
"PI" XEQ 23 RCL 39
XEQ 27 RCL 49 RCL 44
* RCL 48 RCL 46 * +
RCL 44 X↑2 RCL 47 /
- RCL 45 LASTX - /
"R↑2" XROM "OUT"

364♦LBL 19
ADV " P        GP"
PROMPT GTO 15

369♦LBL A
XEQ 21 FS? 00 XEQ 33
XEQ 14 CZ RCL 17 X<>Y
/ RCL 49 - RCL 48 /
STO 34 FC? 00 GTO 08
XEQ 26 RCL 32 RCL 27
/ 1 + /

392♦LBL 08
STO 36 "GP" XEQ 28
GTO 19

397♦LBL E
XEQ 24 FC? 00 GTO 09
XEQ 33 RCL 32 X<>Y /
1 + RCL 36 * STO 34
XEQ 26

411♦LBL 09
RCL 48 * RCL 49 +
XEQ 20 "P" XEQ 23
GTO 19

420♦LBL 20
STO 02 STO 17 XEQ 14
X<>Y STO 01

426♦LBL 10
TONE 5 RCL 01 RCL 17
RCL 11 / CZ STO 00
CLX LASTX CCR LASTX
* RCL 02 RCL 00 *
RCL 17 - X<>Y /
ENTER↑ X<> 17 ST+ 17
/ ABS 1 E-4 X<=Y?
GTO 10 RCL 17 TONE 9
RTN

457♦LBL 14
RCL 16 "F-R" CON
RCL 10 / RCL 17
RCL 11 / RTN

467♦LBL 21
16 "P"

470♦LBL 22
STO 00 XEQ 00
XROM "INK" RDN STO 08
X<>Y STO 09 R↑ RTN

480♦LBL 23
XEQ 00 XROM "OUTK" RDN
STO 08 X<>Y STO 09 R↑
RTN

489♦LBL 00
ASTO T "PSI" ASTO 01
CLA ASTO 02 ARCL T
RCL 09 RCL 08 RCL Z
RTN

500♦LBL 24
35 "GP"

503♦LBL 25
STO 00 XEQ 01
XROM "INK" RDN STO 06
X<>Y STO 07 R↑ RTN

513♦LBL 26
RCL 34 "GP*" GTO 28

517♦LBL 27
RCL 39 "G"

520♦LBL 28
XEQ 01 XROM "OUTK" RDN
STO 06 X<>Y STO 07 R↑
RTN

529♦LBL 01
ASTO T "MCF" ASTO 01
CLA ASTO 02 ARCL T
RCL 07 RCL 06 RCL Z
RTN

540♦LBL 29
25 "CONDP"

543♦LBL 30
STO 00 XEQ 02
XROM "INK" RDN STO 03
X<>Y STO 04 R↑ RTN

553♦LBL 31
RCL 33 "WP*"

556♦LBL 32
XEQ 02 XROM "OUTK" RDN
STO 03 X<>Y STO 04 R↑
RTN

565♦LBL 02
ASTO T "BBL" ASTO 01
CLA ASTO 02 ARCL T
RCL 04 RCL 03 RCL Z
RTN

576♦LBL 33
FS?C 03 SF 08 26
STO 00 "MCF/BBL"
ASTO 01 ASHF ASTO 02
RCL 51 RCL 50 RCL Z
"R AVG" XROM "INK" RDN
STO 50 X<>Y STO 51 R↑
CF 08 END
```

Section 5
Natural Gas Engineering

17. BHPWHP — Bottom-Hole or Surface Pressures for Flowing or Static Gas Wells

Calculates bottom-hole pressure (flowing or static) from surface pressure (flowing or static) and vice versa.

18. GASDEL — Single Well Gas Deliverability

Forecasts gas well deliverability and flowing pressures for volumetric reservoirs. Includes BHPWHP calculation and rate-time conversion.

19. STAB — Stabilized Flow Coefficient

Calculates the stabilized flow coefficient based on the slope of the back pressure deliverability curve and the observed variation in the flow coefficient.

17. BHPWHP — Bottom-Hole or Surface Pressures for Flowing or Static Gas Wells

This program calculates surface pressures from bottom-hole pressures or vice versa for flowing or static gas wells. Naturally, it is preferred to measure bottom-hole pressures with a downhole gauge and surface pressures by gauge or deadweight. For many purposes, a sufficiently accurate value for one value or the other can be obtained by calculating the difference in pressure due to the weight of the gas as well as (in the case of flowing wells) frictional pressure drops.

The mathematical solution to this problem has been evaluated by many authors; most techniques require an iterative procedure to calculate the pressure drop. BHPWHP uses a modified version of the Cullender-Smith technique, which requires iterations for the average Z-factor but converges rapidly. It is incorporated with the program GASDEL but functions independently as far as the user is concerned.

Friction factors are calculated using the equations presented by Cullender and Smith. These equations presume an absolute pipe roughness of 0.00060 inches. Better results may be obtained when measured flowing or static bottom-hole pressures can be compared to flowing or static surface pressures. The user can then vary the values for length, diameter, and effective gas gravity until values calculated by BHPWHP agree with the measurements. Doing so will assure better subsequent calculations for that well.

BHPWHP allows the user to calculate values for deviated wells (see Figure 17-1); no explicit provisions are provided for liquid production. This is best done by altering the effective gas gravity. Practically all of the methods presented are suitable for depths less than about 12,000 FT. For deeper wells, the Cullender-Smith method with shorter integration intervals should be used (see Young).

Major problems affecting these calculations include slugs of unknown amounts of liquid hydrocarbons or water in the well-bore tubing, unusual temperature distributions, variations in gas Z-factor from the correlation, and changes in fluid composition with depth for condensate systems. It is also difficult to select the proper friction factor and specific gravity of the well effluent and the flow rate. A method for evaluating annular flow is described by Ikoku.

Equations

The bottom-hole pressure is solved iteratively from the surface pressure (or vice versa) as follows:

$$BHP^2 = WHP^2\, e^A + \frac{GAS\ G\ T\ AVG\ F\ L(e^A - 1)\ QG^2}{40000\ A\ d^5}$$

where:

	BHP	WHP
QG = 0	PWS	SITP
QG > 0	PWF	FTP

$$A = \frac{GAS\ G\ TVD}{26.67\ T\ AVG'\ Z\ AVG}$$

$$T\ AVG = \frac{T + SURF\ T}{2}$$

$$T\ AVG' = T\ AVG \text{ in R}$$

$$Z\ AVG = \frac{Z_{BHP} + Z_{WHP}}{2}$$

The initial guesses are $WHP + \frac{WHP\ L}{40000}$ if WHP known and $BHP - \frac{BHP\ L}{40000}$ if BHP known.

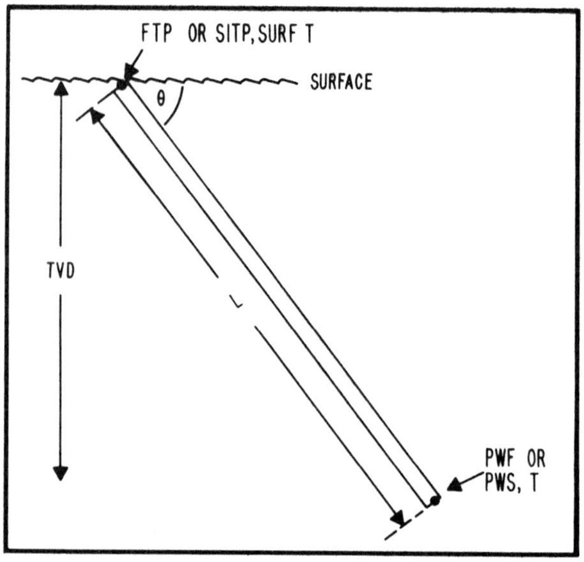

Figure 17–1 Deviated gas well *(after Ikoku)*

Friction factor:

d < 4.277 IN: $F = \frac{0.10797}{d^{2.612}}$

d ≥ 4.277 IN: $F = \frac{0.10337}{d^{2.582}}$

Nomenclature

Symbol	Variable Name	Input or Output	English Units	SI Units
d	Tubing diameter	I	IN	CM
FTP*	Flowing tubing pressure	I,O	PSI	KPA
L	Tubing length (for GASDEL, L = 0 FT implies do not calculate FTP at each pressure step)	I	FT	M
PWF*	Flowing bottom-hole pressure	I,O	PSI	KPA
PWS*	Shutin bottom-hole pressure	I,O	PSI	KPA
QG*	Gas producing rate (for BHPWHP, QG = 0 MCF/DAY implies static conditions)	I	MCF/DAY	SCM/DAY
SITP*	Shutin tubing pressure	I,O	PSI	KPA
SURF T	Surface temperature	I	F	C
TVD	True vertical depth	I	FT	M

*The units for these variables are saved by the program.

Yes/No Questions

SKIP? Yes: Skip input of PVT data.
No: Allow input of PVT data.

Example 1

Calculate the static BHP of a gas well with L = TVD = 5,790 FT, gas gravity = 0.60 (no diluents), SITP = 2,300 PSI, surface temperature = 72 F, bottom-hole temperature = 162 F, and tubing diameter (ID) is 1.996 IN (from Ikoku). Use the TcPc program in the Pac to calculate the pseudocritical properties from the gas gravity.

Keystrokes (SIZE> =044)	Display	Comments
[XEQ] [ALPHA] TcPc [ALPHA]	*COND? Y/N:*	Last character is Y or N
N [R/S]	*GAS G=?*	
.6 [R/S]	*%N2=?*	
0 [R/S]	*%CO2=?*	
0 [R/S]	*%H2S=?*	
0 [R/S]	*GAS G=?*	Tc and Pc printed
[XEQ] [ALPHA] BHPWHP [ALPHA]	*SKIP? Y/N:*	Last character is Y or N
N [R/S]	*Tc=?*	
[R/S]	*Pc=?*	
[R/S]	*T=?*	
162 [R/S]	*L=?*	
5790 [R/S]	*TVD=?*	
5790 [R/S]	*d=?*	
1.996 [R/S]	*GAS G=?*	
[R/S]	*MW=?*	
[R/S]	*SURF T=?*	
72 [R/S]	*QG=?*	QG = 0 for static conditions
0 [R/S]	*PWS SITP*	
[E]	*SITP=?*	SITP known
2300 [R/S]	*PWS SITP*	PWS printed

What value of SITP would be calculated if PWS = 2,640 PSI?

Keystrokes	Display	Comments
[A]	*PWS=?*	PWS known
2640 [R/S]	*PWS SITP*	SITP printed

```
Tc Pc

COND: NO
GAS G=0.6000
%N2=0.0000
%CO2=0.0000
%H2S=0.0000

Tc=358.5000 R
Pc=672.5000 PSI
```

```
BHP-WHP

SKIP: NO
T=162.0000 F
L=5,790.0000 FT
TVD=5,790.0000 FT
d=1.9960 IN
SURF T=72.0000 F

QG=0.0000 MCF/DAY
SITP=2,300.0000 PSI
PWS=2,639.8505 PSI

PWS=2,640.0000 PSI
SITP=2,300.1306 PSI
```

Example 2

Calculate the sandface pressure (PWF) of a flowing gas well indicated by the following surface measurements (from Ikoku):

QG = 5.153 MMCF/DAY
d = 1.996 IN

GAS G = 0.60
L = 5,700 FT
TVD = 5,680 FT
T = 162 F
SURF T = 83 F
FTP = 2,122 PSI

Note: The names for bottom-hole and surface pressures change since QG$\neq$0.

Keystrokes	Display	Comments
[XEQ] [ALPHA] BHPWHP [ALPHA]	*SKIP? Y/N:N*	
Y [R/S]	*L=?*	
5700 [R/S]	*TVD=?*	
5680 [R/S]	*d=?*	
[R/S]	*SURF T=?*	
83 [R/S]	*QG=?*	
5.153 [ALPHA] MMCF/DAY [R/S]	*PWF FTP*	
[E]	*FTP=?*	FTP known
2122 [R/S]	*PWF FTP*	PWF printed

```
BHP-WHP

SKIP: YES
L=5,700.0000 FT
TVD=5,680.0000 FT
SURF T=83.0000 F

QG=5.1530 MMCF/DAY
FTP=2,122.0000 PSI
PWF=2,591.4867 PSI
```

Example 3

This is an example presented by Messer, Raghavan, and Ramey for a deep, hot, sour gas well. Data is as follows:

d = 2 IN
GAS G = 0.75
%METH = 38
%CO2 = 24
%H2S = 38
QG = 8.82 MMCF/DAY
T = 900 R
SURF T = 540 R
FTP = 14,575 PSI
L = TVD = 20,500 FT

Use the PROP program to calculate the pseudo-critical properties from the gas composition.

Keystrokes	Display	Comments
[XEQ] [ALPHA] PROP [ALPHA]	*CLEAR? Y/N:*	Last character is Y or N
Y [R/S]	*%N2=?*	
[R/S]	*%CO2=?*	
24 [R/S]	*%H2S=?*	
38 [R/S]	*%METH=?*	
38 [/] [E]	*SP.HTS? Y/N:*	%TOT, GAS G, Tc, Pc, CWA, Tc*, Pc*, NHV, GHVD, GHVW are printed. Note that the calculated value of GAS G is much higher than the indicated value in the example
[ALPHA] [XEQ] [ALPHA] BHPWHP [ALPHA]	*SKIP? Y/N:Y*	
N [R/S]	*Tc=?*	
[R/S]	*Pc=?*	
[R/S]	*T=?*	
900 [ALPHA] R [R/S]	*L=?*	
20500 [R/S]	*TVD=?*	
20500 [R/S]	*d=?*	
2 [R/S]	*GAS G=?*	
.75 [R/S]	*SURF T=?*	
540 [ALPHA] R [R/S]	*QG=?*	
8.82 [ALPHA] MMCF/DAY [R/S]	*PWF FTP*	
[E]	*FTP=?*	FTP known
14575 [R/S]	*PWF FTP*	PWF printed

The value calculated by BHPWHP for PWF (18,785 PSI) compares to a value of 19,082 PSI in Ikoku's solution of the problem using the extended Sukkar-Cornell integral. Ikoku later recommends the Cullender-Smith method with trapezoidal integration for pressures greater than 10,000 PSI.

Errors in PVT properties, gas gravity, and friction factors generally result in larger errors than do the variations in the solution procedure. As an example, using the value of GAS G calculated by PROP rather than the stated value results in a PWF of 20,380 PSI.

```
GAS PROP

CLEAR: YES
%CO2=24.0000
%H2S=38.0000
%METH=38.0000

%TOT=100.0000
GAS G=1.0222
Tc=517.2912 R
```

```
Pc=1,007.0840 PSI
CWA=31.1288 F
Tc*=486.1624 R
Pc*=933.2500 PSI
NHV=568.8980 BTU/SCF
GHVD=625.7460 BTU/SCF
GHVW=614.8580 BTU/SCF
```

```
BHP-WHP

SKIP: NO
T=900.0000 R
L=20,500.0000 FT
TVD=20,500.0000 FT
d=2.0000 IN
GAS G=0.7500
SURF T=540.0000 R

QG=8.8200 MMCF/DAY
FTP=14,575.0000 PSI
PWF=18,784.8350 PSI
```

General Information

Memory Requirements

Program length:	1340 bytes (6 cards) (BHPWHP and GASDEL combined)
Minimum size:	044
Minimum hardware:	41C + quad memory module or 41CV*

*The program can be run in a 41C + 3 memory modules but will not leave a port available for a printer. If you do not plan to use GASDEL, you can remove it to save space. Simply delete all of the program steps beginning at line 331. This will yield a version of BHPWHP that is 658 bytes long. This shortened version can be run in a 41C + 2 memory modules.

Hidden Options

None

Pac Subroutines Called

TITLE, Y/N?, ITcPc, T, INU, IN, GASG, CON, INK, OUTK, CZ

Registers

03	Gas producing rate units
04	Gas producing rate units
08	Pressure units
09	Pressure units
26	PWS, SITP (PSI)
27	PWF, FTP (PSI)
28	TVD (FT)
29	L (FT)
30	ZI_{BHP}, ZI_{WHP}
31	SURF T (F)
32	T AVG (F)
33	Scratch
34	Scratch
42	d (IN)
43	QG (MCF/DAY)

Registers 12–14, 18–25, and 35–41 unused.

Flags

00	Set:	BHP known.
	Clear:	WHP known.
02	Set:	Skip input of PVT data.
	Clear:	Allow input of PVT data.
06	Set:	Error occurred while calculating FTP.
	Clear:	No error occurred.
07	Set:	Gas producing rate units not yet input.
	Clear:	Gas producing rate units have been input.

User Instructions

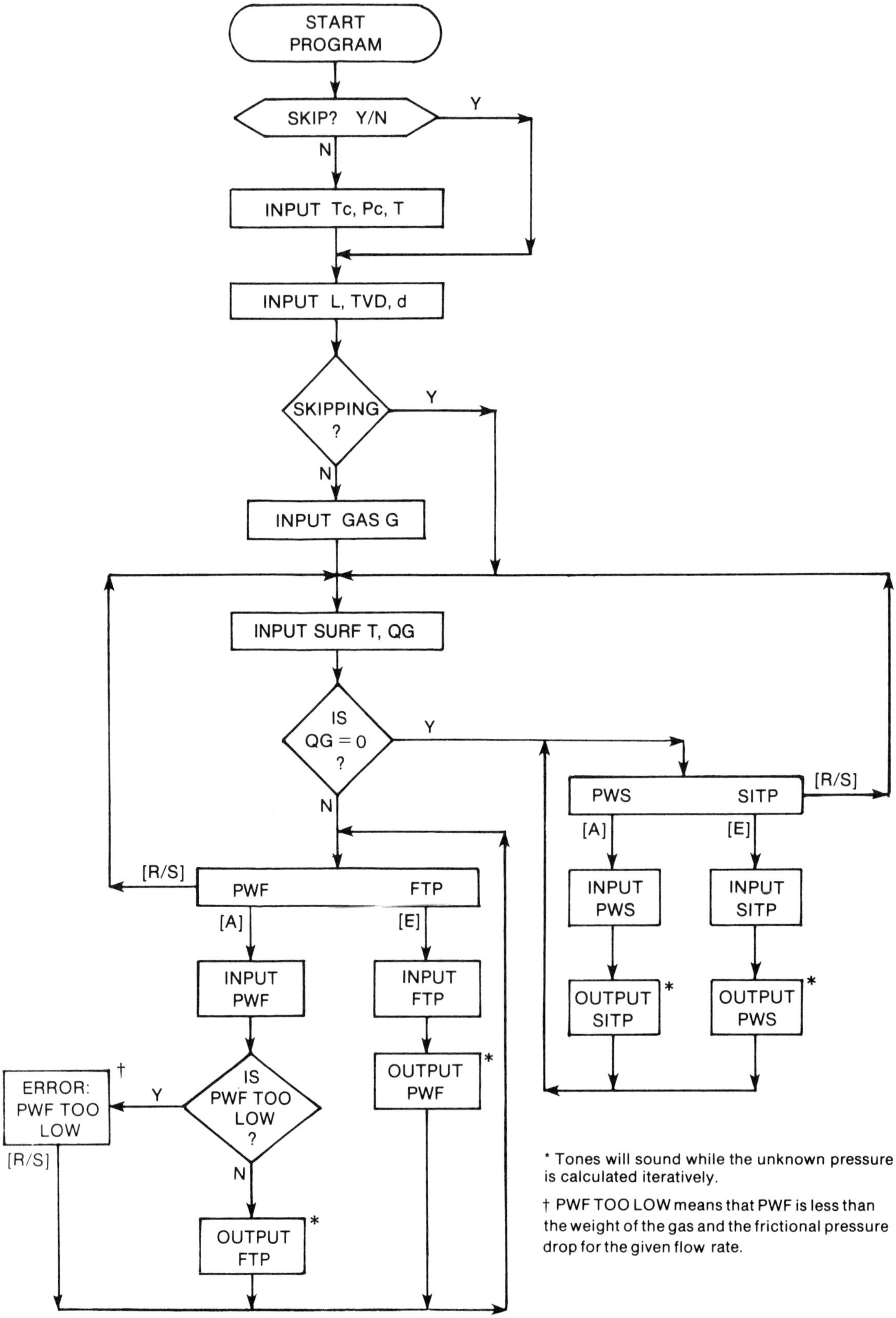

* Tones will sound while the unknown pressure is calculated iteratively.

† PWF TOO LOW means that PWF is less than the weight of the gas and the frictional pressure drop for the given flow rate.

Program Listing

```
 01♦LBL "BHPWHP"
"BHP-WHP" 44
XROM "TITLE" FC?C 25
PROMPT XEQ 00 XEQ 01
XEQ 02 ADV GTO 15

 12♦LBL 00
SF 27 "SCM/DAY"
ASTO 03 ASHF ASTO 04
"KPA" ASTO 08 CLA
ASTO 09 "SKIP" 2
XROM "Y/N?" FS? 02 RTN
XROM "ITcPc" XROM "T"
RTN

 30♦LBL 01
28 STO 00 "FT"
ASTO 01 CLA ASTO 02
ASTO Z "M" ASTO Y "L"
XROM "INU" RTN

 43♦LBL 02
27 STO 00 "M" ASTO Y
CLA ASTO Z "TVD"
XROM "INU" 41 STO 00
"IN" ASTO 01 CLA
ASTO 02 ASTO Z "CM"
ASTO Y "d" XROM "INU"
FC? 02 XROM "GASG"

 65♦LBL 28
30 STO 00 "F" ASTO 01
CLA ASTO 02 ASTO Z
"C" ASTO Y "SURF T"
XROM "INU" RCL 42 X=0?
GTO 05 4.277 RCL 42
X<Y? GTO 03 7.582 Y↑X
.10337 GTO 04

 88♦LBL 03
7.612 Y↑X .10797

 92♦LBL 04
X<>Y /

 95♦LBL 05
RCL 31 RCL 16 + 2 /
STO 32 "F-R" CON *
4 E4 / RCL 15 *
RCL 29 * STO 33
RCL 15 RCL 28 *
RCL 32 CON / 26.67 /
STO 34 RTN

122♦LBL 15
42 "QG" XEQ 18

126♦LBL 29
RCL 43 "PWS     SITP"
X≠0? "PWF       FTP"
PROMPT XEQ 28 GTO 15

134♦LBL A
SF 00 RCL 43 "PWS"
X≠0? "PWF" 25 XEQ 16
CF 08 XEQ 20 FS? 06
GTO 29 STO 27 RCL 43
"SITP" X≠0? "FTP"
X<>Y GTO 05

153♦LBL E
CF 00 RCL 43 "SITP"
X≠0? "FTP" 26 XEQ 16
CF 08 XEQ 20 STO 26
RCL 43 "PWS" X≠0?
"PWF" X<>Y

169♦LBL 05
XEQ 17 ADV GTO 29

173♦LBL 16
STO 00 XEQ 06
XROM "INK" RDN STO 08
X<>Y STO 09 R↑ RTN

183♦LBL 17
XEQ 06 XROM "OUTK" RDN
STO 08 X<>Y STO 09 R↑
RTN

192♦LBL 06
ASTO T "PSI" ASTO 01
CLA ASTO 02 ARCL T
RCL 09 RCL 08 RCL Z
RTN

203♦LBL 18
STO 00 XEQ 07
XROM "INK" RDN STO 03
X<>Y STO 04 R↑ RTN

213♦LBL 19
XEQ 07 XROM "OUTK" RDN
STO 03 X<>Y STO 04 R↑
RTN

222♦LBL 07
ASTO T "MCF/DAY"
ASTO 01 ASHF ASTO 02
CLA ARCL T RCL 04
RCL 03 RCL Z RTN

234♦LBL 20
CF 06 STO 02 RCL 43
X↑2 RCL 33 * STO 00
FS? 00 RCL 16 FC? 00
RCL 31 "F-R" CON
RCL 10 / RCL 02
RCL 11 / CZ STO 30
FS? 00 RCL 31 FC? 00
RCL 16 "F-R" CON
RCL 10 / STO 01
RCL 02 RCL 29 * 4 E4
/ FS? 00 CHS RCL 02
ST* 02 + STO 05

275♦LBL 14
TONE 5 RCL 01 RCL 05
RCL 11 / CZ RCL 30 +
2 / RCL 34 X<>Y /
RCL 00 X<>Y / LASTX
E↑X STO Z 1 - *
FS? 00 GTO 08 X<>Y
RCL 02 * + GTO 09

305♦LBL 08
RCL 02 X<>Y - X<>Y /
X>0? GTO 09 SF 06
TONE 3 "PWF TOO LOW"
PROMPT RTN

318♦LBL 09
SQRT ENTER↑ X<> 05 -
RCL 05 / ABS 1 E-4
X<=Y? GTO 14 RCL 05
TONE 9 RTN

332♦LBL "GASDEL"
"GAS DELIVER" 58
XROM "TITLE" FC?C 25
PROMPT SF 03 SF 07
"SCM" ASTO 06 "YR"
ASTO 54 CLA ASTO 07
ASTO 55 CLX STO 43
STO 44 STO 52 STO 53
XEQ 00 38 "G" XEQ 23
44 STO 00 "C"
XROM "IN" "N"
XROM "IN" XEQ 01 X≠0?
XEQ 02

365♦LBL 13
"PWS     SITP" PROMPT
GTO 13

369♦LBL A
SF 00 25 "PWS" XEQ 16
GTO 10

375♦LBL E
CF 00 26 "SITP"
XEQ 16 STO 26 RCL 29
X≠0? GTO 09 X<>Y
GTO 10

386♦LBL 09
X<>Y XEQ 20 STO 26
CF 08 "PWS" XEQ 17
SF 08

394♦LBL 10
STO 17 CF 00 RCL 29
X≠0? SF 00 XEQ 27
STO 56

402♦LBL 12
" A      %AOF" PROMPT
GTO 12

406♦LBL A
SF 04 48 "A" XEQ 18
ADV CF 07 GTO 21

414♦LBL E
CF 04 48 STO 00
"%AOF" XROM "IN" ADV

421♦LBL 21
CF 08 46 "PWF" XEQ 16
0 RCL 17 XEQ 26
FS?C 07 SF 08 "AOF"
XEQ 19 CF 08 RCL 47
RCL 17 XEQ 26 STO 43
STO 51 "QGI" XEQ 19
```

Program Listing (cont.)

```
441♦LBL 11
49 "QG MIN" XEQ 18
RCL 43 X>Y? GTO 04
TONE 3 "QG MIN > QGI"
PROMPT GTO 11

452♦LBL 04
RCL 50 RCL 45 /
RCL 46 1/X Y↑X RCL 47
X↑2 + SQRT STO 38
"END P" XEQ 17 40
"P DEC" XEQ 16 ADV
RCL 17 STO 57 GTO 05

473♦LBL 22
XEQ 08 FS? 06 GTO 21
RCL 17 "P" XEQ 17
RCL 36 "GP" XEQ 24
ENTER↑ X<> 44 - "DGP"
XEQ 24 RCL 43 STO 5
"QG" FS? 01 "⊢=A"
XEQ 19 RCL 52 ST+ 53
RCL 53 FS?C 03 SF 08
"TIME" XEQ 25 CF 08
RCL 48 "FTP" FS? 00
XEQ 17 ADV

507♦LBL 05
RCL 38 RND RCL 17
RCL 41 - RND X>Y?
GTO 06 X<>Y RCL 17
RND X=Y? GTO 21
RCL 38 RND

523♦LBL 06
LASTX STO 17 GTO 22

527♦LBL 23
STO 00 XEQ 07
XROM "INK" RDN STO 06
X<>Y STO 07 R↑ RTN

537♦LBL 24
XEQ 07 XROM "OUTK" RDN
STO 06 X<>Y STO 07 R↑
RTN

546♦LBL 07
ASTO T "MCF" ASTO 01
CLA ASTO 02 ARCL T
RCL 07 RCL 06 RCL Z
RTN

557♦LBL 25
ASTO T "YR" ASTO 01
CLA ASTO 02 ARCL T
RCL 55 RCL 54 RCL Z
XROM "OUTK" RDN STO 54
X<>Y STO 55 R↑ RTN

574♦LBL 08
XEQ 27 RCL 56 / 1
X<>Y - RCL 39 *
STO 36 RCL 47 RCL 17
XEQ 26 STO 43 RCL 51
+ 2 / RCL 49 X=0?
GTO 10 FS? 04 GTO 09
X<>Y 0 RCL 17 XEQ 26
1 % RCL 49 * R↑
X<>Y

607♦LBL 09
CF 01 X<=Y? SF 01
X<=Y? STO 43 X<=Y?
X<>Y

615♦LBL 10
RCL 36 RCL 44 - R↑ /
365 / STO 52 RCL 47
FC? 00 STO 48 FC? 00
RTN XEQ 20 FC? 06
STO 48 FC? 06 RTN
RCL 57 STO 17 RTN

637♦LBL 26
X↑2 X<>Y X↑2 -
RCL 46 Y↑X RCL 45 *
RTN

647♦LBL 27
RCL 16 "F-R" CON
RCL 10 / RCL 17
RCL 11 / CZ RCL 17
X<>Y / END
```

18. GASDEL — Single Well Gas Deliverability

As the user can tell, the gas material-balance equation is substantially simpler than that for oil material balance (compare GASMBE to OILMBE and OILPRED). Similarly, the prediction of the rate versus time performance is substantially simpler as well. This is due to both the simpler gas material balance and the usual assumption of constant saturations, and therefore constant effective permeability. GASDEL is a fairly flexible program for predicting reservoir performance easily when no water drive is present (Ikoku). Gas composition is assumed to be constant, and the gas equivalent of condensate production is not included in the material balance.

The only two pieces of information required are the P/Z versus cumulative production graph and back-pressure test data. If the P/Z versus cumulative data are unavailable, a value for G (gas in place) must be estimated volumetrically. Figure 18–1 illustrates a typical gas well back-pressure test. Theoretically, these data points form a straight line with a slope N ranging from 0.5 to 1.0. The value of C is best obtained from this test and should represent a stabilized value. This stabilized flow coefficient can be calculated by the method presented by the program STAB. The value of C can be calculated if KH is known; however, if sufficient well tests have been run to evaluate KH with some accuracy, C can be determined from those flow tests.

An allowable value can be input as a constant value or as a percent of the absolute open flow calculated at each pressure. When the well deliverability reaches a minimum rate, the program automatically will allow the user to input a lower value of flowing well pressure (equivalent to adding compression). The BHPWHP calculations to determine flowing tubing pressure from the input flowing bottom-hole pressures are also performed. If these calculations are not desired, the user should specify L = 0.

Equations

$$GP_i = \left(1 - \frac{P_i/Z_i}{PI/ZI}\right) G$$

$$QGI = C(PWS^2 - PWF^2)^N$$

$$QG_i = C(P_i^2 - PWF^2)^N$$

$$QG\,AVG_i = (QG_i + QG_{i-1})/2$$

$$DGP_i = GP_i - GP_{i-1}$$

$$DTIME_i = DGP_i/QG\,AVG_i$$

$$TIME_i = TIME_{i-1} + DTIME_i$$

$$AOF_i = C(P_i^2)^N$$

$$A_i = \frac{\%AOF}{100} AOF_i$$

$$AOF = C(PWS^2)^N$$

$$END\ P = \left[\left(\frac{QG\ MIN}{C}\right)^{1/N} + PWF^2\right]^{1/2}$$

Nomenclature

Symbol	*Variable Name*	*Input or Output*	*English Units*	*SI Units*
A*	Allowable gas producing rate (A = 0 MCF/DAY implies allowable restrictions will be ignored)	I	MCF/ DAY	SCM/ DAY
AOF*	Absolute open flow	O	MCF/ DAY	SCM/ DAY
C	Stabilized flow coefficient	I	-†	-†
END P*	Ending pressure of forecast	O	PSI	KPA
G*	Initial gas in place	I	MCF	SCM

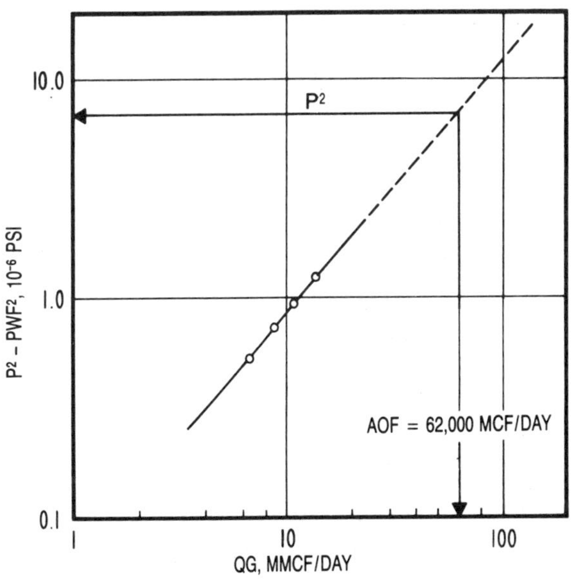

Figure 18–1 Typical gas well back-pressure test *(after Ikoku)*

Symbol	Variable Name	Input or Output	English Units	SI Units
GP*	Cumulative gas production	O	MCF	SCM
DGP*	Delta GP - change in cumulative gas production	O	MCF	SCM
N	Exponent in gas deliverability equation	I	-	-
P*	Pressure	O	PSI	KPA
P DEC*	Pressure decrement of forecast	I	PSI	KPA
QGI*	Initial gas producing rate	O	MCF/DAY	SCM/DAY
QG MIN*	Minimum gas producing rate of forecast	I	MCF/DAY	SCM/DAY
QG=A*	Gas producing rate limited to the allowable rate	O	MCF/DAY	SCM/DAY
TIME*	Cumulative production time	O	YR	YR
%AOF	Allowable percent of absolute open flow at each pressure (%AOF = 0 implies allowable restrictions will be ignored)	I	-	-

See also the nomenclature for BHPWHP.
*The units for these variables are saved by the program.
†Physically, C has units of MCF/DAY*PSI2 or SCM/DAY*KPA2.

Yes/No Questions

SKIP? Yes: Skip input of PVT data.
No: Allow input of PVT data.

Example 1

Forecast the future production for the South Texas Vicksburg reservoir (see Example 2 of GASPVT and Example 1 of GASMBE) at a time of 2.94 YR. Use the remaining gas in place for G. Other required data are:

PWF = 1,200 PSI
N = 0.622
QG = 2,603 MCF/DAY
L = 7,100 FT
TVD = 7,028 FT
d = 1.732 IN
SURF T = 92 F

Use a pressure decrement of 200 PSI for the forecast.

Allowable flow rate (A) = 2.4 MMCF/DAY (recently curtailed)

Add compression to PWF = 500 PSI when the flow rate drops to 1 MMCF/DAY.

The well's economic limit is 100 MCF/DAY.

Keystrokes (SIZE> =058)	Display	Comments
[XEQ] [ALPHA] GASDEL [ALPHA]	*SKIP? Y/N:*	Last character is Y or N
N [R/S]	*Tc=?*	
441.8952 [R/S]	*Pc=?*	
661.5656 [R/S]	*T=?*	
180 [R/S]	*G=?*	
39564 [ENTER↑] 4215 [−]	*35,349.0000*	Calculates remaining gas in place
[ALPHA] MMCF [R/S]	*C=?*	
2603 [ENTER↑] 4692 [//] [x^2]		
1200 [//] [x^2] [−]		
.622 [//] [y^x] [÷]	*0.0735*	C calculated from QG
[R/S]	*N=?*	
.622 [R/S]	*L=?*	
7100 [R/S]	*TVD=?*	
7028 [R/S]	*d=?*	
1.732 [R/S]	*GAS G=?*	
.872 [R/S]	*SURF T=?*	
92 [R/S]	*PWS SITP*	
[A]	*PWS=?*	PWS known. Initial pressure from wellhead or bottom-hole data
4692 [R/S]	*A %AOF*	
[A]	*A=?*	A known
2.4 [ALPHA] MMCF/DAY [R/S]	*PWF=?*	
1200 [R/S]	*QG MIN=?*	AOF and QGI printed
1 [R/S]	*P DEC=?*	END P printed
200 [R/S]	*PWF=?*	Forecast of P,GP, DGP, QG (or QG=A if allowable–restricted), TIME, and FTP printed
500 [R/S]	*QG MIN=?*	AOF and QGI printed
100 [ALPHA] MCF/DAY [R/S]	*P DEC=?*	END P printed
300 [R/S]	*PWF=?*	Remaining forecast printed

GAS DELIVER

```
SKIP: NO
Tc=441.8952 R
Pc=661.5656 PSI
T=180.0000 F
G=35,349.0000 MMCF
C=0.0735
N=0.6220
L=7,100.0000 FT
TVD=7,028.0000 FT
d=1.7320 IN
GAS G=0.8720
SURF T=92.0000 F
PWS=4,692.0000 PSI
A=2.4000 MMCF/DAY

PWF=1,200.0000 PSI
AOF=2.7149 MMCF/DAY
QGI=2.6030 MMCF/DAY
QG MIN=1.0000 MMCF/DAY
END P=2,420.6382 PSI
P DEC=200.0000 PSI

P=4,492.0000 PSI
GP=720.2722 MMCF
DGP=720.2722 MMCF
QG=A=2.4000 MMCF/DAY
TIME=0.8222 YR
FTP=455.1795 PSI

P=4,292.0000 PSI
GP=1,492.9096 MMCF
DGP=772.6375 MMCF
QG=2.3100 MMCF/DAY
TIME=1.7211 YR
FTP=506.2772 PSI

P=4,092.0000 PSI
GP=2,324.4340 MMCF
DGP=831.5244 MMCF
QG=2.1654 MMCF/DAY
TIME=2.7391 YR
FTP=575.6356 PSI

P=3,892.0000 PSI
GP=3,222.2695 MMCF
DGP=897.8355 MMCF
QG=2.0219 MMCF/DAY
TIME=3.9140 YR
FTP=633.1558 PSI

P=3,692.0000 PSI
GP=4,194.7959 MMCF
DGP=972.5264 MMCF
QG=1.8797 MMCF/DAY
TIME=5.2798 YR
FTP=681.9290 PSI

P=3,492.0000 PSI
GP=5,251.3436 MMCF
DGP=1,056.5477 MMCF
QG=1.7387 MMCF/DAY
TIME=6.8798 YR
FTP=723.8229 PSI

P=3,292.0000 PSI
GP=6,402.0873 MMCF
DGP=1,150.7437 MMCF
QG=1.5988 MMCF/DAY
TIME=8.7691 YR
FTP=760.0745 PSI

P=3,092.0000 PSI
GP=7,657.7708 MMCF
DGP=1,255.6834 MMCF
QG=1.4599 MMCF/DAY
TIME=11.0186 YR
FTP=791.5567 PSI

P=2,892.0000 PSI
GP=9,029.1666 MMCF
DGP=1,371.3958 MMCF
QG=1.3221 MMCF/DAY
TIME=13.7196 YR
FTP=818.9146 PSI

P=2,692.0000 PSI
GP=10,526.1333 MMCF
DGP=1,496.9667 MMCF
QG=1.1851 MMCF/DAY
TIME=16.9912 YR
FTP=842.6420 PSI

P=2,492.0000 PSI
GP=12,156.0745 MMCF
DGP=1,629.9412 MMCF
QG=1.0486 MMCF/DAY
TIME=20.9896 YR
FTP=863.1279 PSI

P=2,420.6382 PSI
GP=12,770.5411 MMCF
DGP=614.4667 MMCF
QG=1.0000 MMCF/DAY
TIME=22.6331 YR
FTP=869.7141 PSI

PWF=500.0000 PSI
AOF=1.1918 MMCF/DAY
QGI=1.1599 MMCF/DAY
QG MIN=100.0000 MCF/DAY
END P=599.2140 PSI
P DEC=300.0000 PSI

P=2,120.6382 PSI
GP=15,538.1429 MMCF
DGP=2,767.6018 MMCF
QG=975.5567 MCF/DAY
TIME=29.7347 YR
FTP=215.7042 PSI

P=1,820.6382 PSI
GP=18,562.5458 MMCF
DGP=3,024.4030 MMCF
QG=796.3683 MCF/DAY
TIME=39.0873 YR
FTP=292.2661 PSI

P=1,520.6382 PSI
GP=21,729.4591 MMCF
DGP=3,166.9132 MMCF
QG=622.4589 MCF/DAY
TIME=51.3178 YR
FTP=340.0192 PSI

P=1,220.6382 PSI
GP=24,872.3699 MMCF
DGP=3,142.9108 MMCF
QG=453.6048 MCF/DAY
TIME=67.3219 YR
FTP=371.2449 PSI

P=920.6382 PSI
GP=27,840.9668 MMCF
DGP=2,968.5969 MMCF
QG=288.0677 MCF/DAY
TIME=89.2538 YR
FTP=390.9190 PSI

P=620.6382 PSI
GP=30,555.6694 MMCF
DGP=2,714.7026 MMCF
QG=114.2956 MCF/DAY
TIME=126.2231 YR
FTP=401.6873 PSI

P=599.2140 PSI
GP=30,739.1882 MMCF
DGP=183.5189 MMCF
QG=100.0000 MCF/DAY
TIME=130.9156 YR
FTP=402.1524 PSI
```

Example 2

To produce the reservoir in Example 1 to its economic limit with one well took an unacceptably long time (131 YR). The reservoir engineer would seek to both increase the deliverability from the well and drill additional wells.

In this example, repeat the calculations from Example 1 assuming three additional wells are brought on line at the initial time, each twice as productive as the existing well. In other words, the value for C is 7 times as great as the previous example. Further, input L = 0 to neglect the FTP calculations. Add compression to PWF = 500 PSI when the total flow rate drops to 4000 MCF/DAY.

Assume that the wells will not be allowable restricted. If the user selects A or %AOF = 0, GASDEL will ignore the allowables. Similarly, one could select %AOF = 100; however, this may result in allowable restrictions. The average rate during a pressure decrement cannot exceed the allowable. If the average rate is greater than 100% of the AOF, the average rate will be reduced to the AOF. When the program outputs QGI or QG, it refers to the instantaneous rate. The variable QG = A refers to the average rate.

In practice, the reservoir engineer optimizes field development in much this fashion. Although one well might drain the bulk of the reserves (albeit in centuries), additional wells will accelerate (and usually increase) ultimate recovery. An economic evaluation will indicate which ultimate well spacing is optimal. (Fractured wells present an additional variable when fracture length must be optimized as well.) Reservoir heterogeneities will often require additional wells to drain the reserves in place regardless of the economic considerations.

Keystrokes	Display	Comments
[XEQ] [ALPHA] GASDEL [ALPHA]	*SKIP? Y/N:N*	
Y [R/S]	*G=?*	
[R/S]	*C=?*	
7 [•]	*0.5148*	
[R/S]	*N=?*	
[R/S]	*L=?*	
0 [R/S]	*PWS SITP*	
[A]	*PWS=?*	
[R/S]	*A %AOF*	
[A]	*A=?*	A known
0 [R/S]	*PWF=?*	
1200 [R/S]	*QG MIN=?*	AOF and QGI printed
4 [ALPHA] MMCF/DAY [R/S]	*P DEC=?*	END P printed
[R/S]	*PWF=?*	Forecast printed

The flow rates are increased greatly over those in Example 1. The field reaches a flow rate of 4 MMCF/DAY in 6.3 YR, at which time a new value of PWF and QG MIN will be input. The economic limit for the four wells will be 400 MCF/DAY.

Keystrokes	Display	Comments
500 [R/S]	*QG MIN=?*	AOF and QGI printed
400 [ALPHA] MCF/DAY [R/S]	*P DEC=?*	END P printed
[R/S]	*PWF=?*	Forecast printed

Note that the ultimate recovery from four wells (with three of higher productivity) was 31.2 BCF in 21.6 YR compared with the single well's 30.7 BCF in 131 YR. The engineer will develop estimates of well costs and recoveries for different numbers of wells, couple them with operating expenses, taxes, prices, etc., to determine the optimal number of wells to recover the reserves economically.

```
GAS DELIVER

SKIP: YES
C=0.5148
L=0.0000 FT
A=0.0000 MCF/DAY

PWF=1,200.0000 PSI
AOF=19,004.0414 MCF/DAY
QGI=18,221.0000 MCF/DAY
QG MIN=4.0000 MMCF/DAY
END P=1,799.2672 PSI

P=4,392.0000 PSI
GP=1,099,657.905 MCF
DGP=1,099,657.905 MCF
QG=16.6798 MMCF/DAY
TIME=0.1726 YR

P=4,092.0000 PSI
GP=2,324,434.005 MCF
DGP=1,224,776.100 MCF
QG=15.1577 MMCF/DAY
TIME=0.3834 YR

P=3,792.0000 PSI
GP=3,698,634.850 MCF
DGP=1,374,200.845 MCF
QG=13.6548 MMCF/DAY
TIME=0.6448 YR

P=3,492.0000 PSI
GP=5,251,343.592 MCF
DGP=1,552,708.742 MCF
QG=12.1707 MMCF/DAY
TIME=0.9742 YR

P=3,192.0000 PSI
GP=7,016,130.303 MCF
DGP=1,764,786.711 MCF
QG=10.7046 MMCF/DAY
TIME=1.3970 YR

P=2,892.0000 PSI
GP=9,029,166.569 MCF
DGP=2,013,036.264 MCF
QG=9.2548 MMCF/DAY
TIME=1.9496 YR

P=2,592.0000 PSI
GP=11,324,173.44 MCF
DGP=2,295,006.872 MCF
QG=7.8177 MMCF/DAY
TIME=2.6862 YR

P=2,292.0000 PSI
GP=13,921,565.07 MCF
DGP=2,597,391.630 MCF
QG=6.3862 MMCF/DAY
TIME=3.6882 YR

P=1,992.0000 PSI
GP=16,808,504.90 MCF
DGP=2,886,939.840 MCF
QG=4.9455 MMCF/DAY
TIME=5.0842 YR

P=1,799.2672 PSI
GP=18,785,220.13 MCF
DGP=1,976,715.220 MCF
QG=4.0000 MMCF/DAY
TIME=6.2950 YR

PWF=500.0000 PSI
AOF=5.7679 MMCF/DAY
QGI=5.4866 MMCF/DAY
QG MIN=400.0000 MCF/DAY
END P=542.5427 PSI

P=1,499.2672 PSI
GP=21,956,200.94 MCF
DGP=3,170,980.810 MCF
QG=4,271.8918 MCF/DAY
TIME=8.0755 YR

P=1,199.2672 PSI
GP=25,091,006.89 MCF
DGP=3,134,805.951 MCF
QG=3,092.1988 MCF/DAY
TIME=10.4080 YR

P=899.2672 PSI
GP=28,043,217.00 MCF
DGP=2,952,210.110 MCF
QG=1,933.8055 MCF/DAY
TIME=13.6266 YR

P=599.2672 PSI
GP=30,738,733.95 MCF
DGP=2,695,516.950 MCF
QG=700.2547 MCF/DAY
TIME=19.2339 YR

P=542.5427 PSI
GP=31,218,036.99 MCF
DGP=479,303.0500 MCF
QG=400.0000 MCF/DAY
TIME=21.6209 YR
```

User Instructions

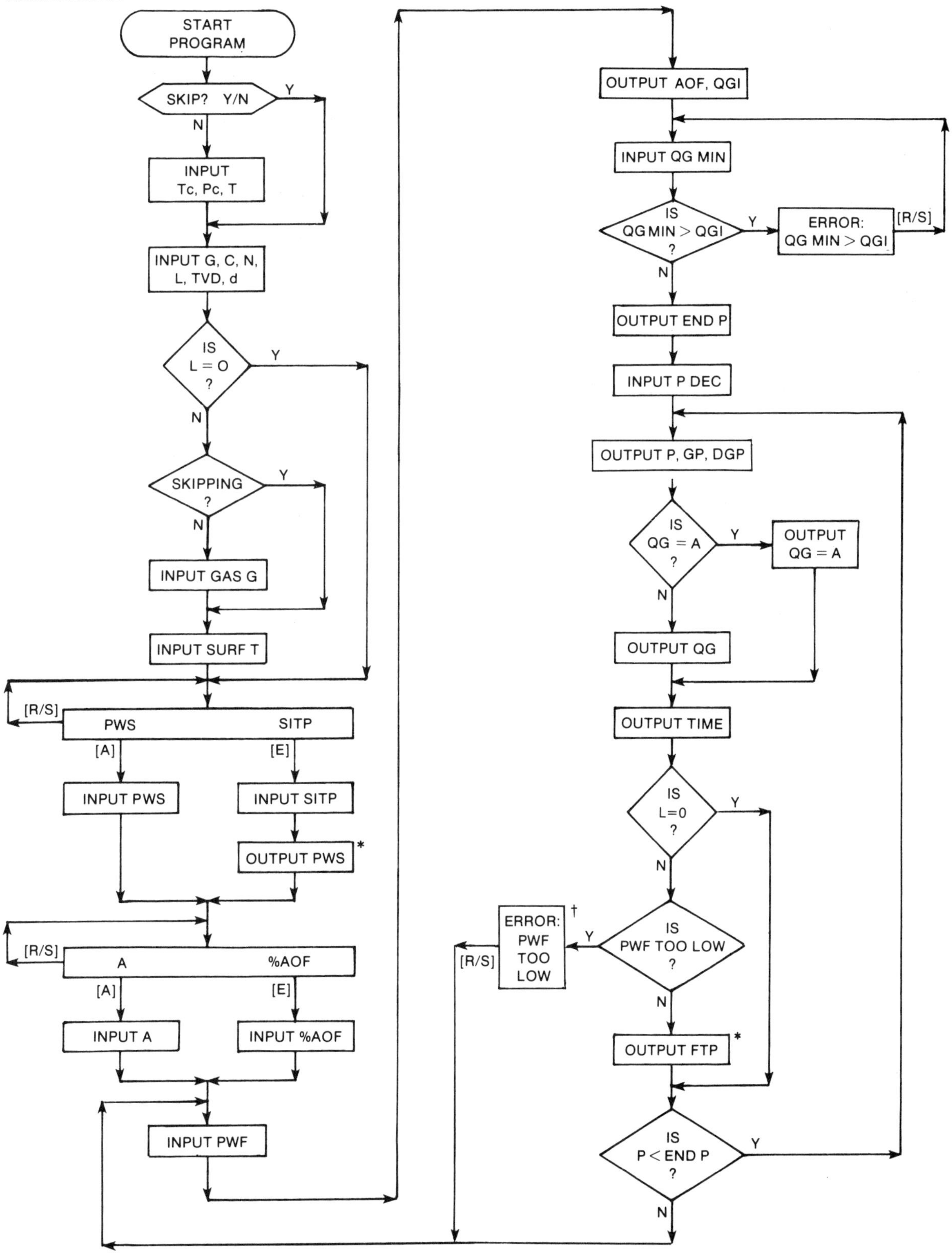

* Tones will sound while the unknown pressure is calculated iteratively.

† PWF TOO LOW means that PWF is less than the weight of the gas and the frictional pressure drop for the given flow rate.

General Information

Memory Requirements

Program length:	1340 bytes (6 cards) (GASDEL and BHPWHP combined)
Minimum size:	058
Minimum hardware:	41C + quad memory module or 41CV*

*The program can be run in a 41C + 3 memory modules but will not leave a port available for a printer.

Hidden Options

None

Pac Subroutines Called

See the Pac subroutines called for BHPWHP.

Registers

06	Gas production units
07	Gas production units
36	GP (MCF)
38	END P (PSI)
39	G (MCF)
41	P DEC (PSI)
44	GP_{i-1} (MCF)
45	C
46	N
47	PWF (PSI)
48	FTP (PSI)
49	A (MCF/DAY), %AOF
50	QG MIN (MCF/DAY)
51	QG_{i-1} (MCF/DAY)
52	$DTIME_i$ (YR)
53	$TIME_{i-1}$ (YR)
54	Time units
55	Time units
56	PWS/Z_{PWS} (PSI)
57	P_{i-1} (PSI)

See also the registers for BHPWHP

Flags

01	Set:	QG AVG ≥ A
	Clear:	QG AVG < A
03	Set:	Time units not yet input.
	Clear:	Time units have been input.
04	Set:	A input.
	Clear:	%AOF input.

See also the flags for BHPWHP.

Program Listing

See the program listing for BHPWHP.

19. STAB — Stabilized Flow Coefficient

Back-pressure testing of gas wells is required by certain regulatory agencies and is useful in some regards for evaluating flow performance. It has been noticed frequently that the back-pressure curve (see Figure 19–1) often keeps a constant slope but shifts to the left with time. This changes the value for C, the flow coefficient. There are more sophisticated methods to predict low permeability and fractured gas well performance. However, the technique presented by Poettman and Schilson can be used to provide a reasonable estimate of the stabilized flow coefficient. The calculations involve functions of time and radius of drainage. For an input value of drainage radius, the time required to reach stabilization is also calculated.

Equations

$$QG = C(SITP^2 - P\ STAB^2)N$$

$$C = C1\left[\frac{\ln(A\sqrt{T1})}{\ln(A\sqrt{T})}\right]^N$$

$$A = T2^X/T1^Y$$

$$X = C2^{1/N}/\ 2(C1^{1/N} - C2^{1/N})$$

$$Y = C1^{1/N}/2(C1^{1/N} - C2^{1/N})$$

$$RAD_i = B\sqrt{T_i}$$

$$END\ T = (Re/B)^2$$

$$B = 0.69688\left|\frac{KG}{\%POR\ UG\ CG\ (1 - SW)}\right|^{1/2}$$

If UG and CG are unknown, they are evaluated at

$$P = \sqrt{\frac{SITP^2 + P\ STAB^2}{2}}$$

Nomenclature

Symbol	Variable Name	Input or Output	English Units	SI Units
C	Stabilized flow coefficient	O	-†	-†

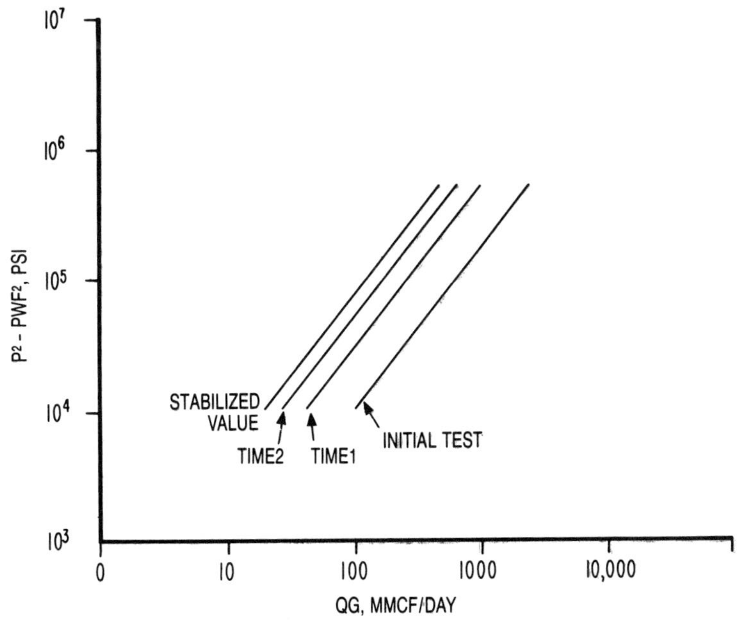

Figure 19–1 Back-pressure test stabilization

Symbol	Variable Name	Input or Output	English Units	SI Units
C1	Stabilized flow coefficient at TIME1	O	-†	-†
C2	Stabilized flow coefficient at TIME2	O	-†	-†
END T*	Ending time of forecast	I,O	HR	HR
KG	Effective permeability to gas	I	MD	MD
N	Slope of back pressure deliverability curve	I	-	-
P STAB*	Stabilized line pressure	I	PSI	KPA
PWF1*	Flowing bottom hole pressure at TIME1	I	PSI	KPA
PWF2*	Flowing bottom hole pressure at TIME2	I	PSI	KPA
QG*	Gas producing rate	O	MCF/DAY	SCM/DAY
QG1*	Gas producing rate at TIME1	I	MCF/DAY	SCM/DAY
QG2*	Gas producing rate at TIME2	I	MCF/DAY	SCM/DAY
RAD	Radius of investigation	O	FT	M
Re	Radius of drainage	I	FT	M
SITP*	Shutin tubing pressure	I	PSI	KPA
TIME*	Time	O	HR	HR
TIME1*	Time at first time-pressure-rate point	I	HR	HR
TIME2*	Time at second time-pressure-rate point	I	HR	HR
T INC*	Time increment of forecast	I	HR	HR

*The units for these variables are saved by the program.
†Physically, C, C1, and C2 have units of MCF/DAY*PSI2 or SCM/DAY*KPA2.

Yes/No Questions

CORR? Yes: Use Pac correlations to estimate PVT properties.
No: Input PVT properties.

SKIP? Yes: Skip input of PVT data.
No: Allow input of PVT data.

Example

Gas well A is completed in the Midcox formation and is spaced on 320 acres. The initial well test information is as follows:

SITP = 2,900 PSI
TIME1 = 3 HR
PWF1 = 2,400 PSI
QG1 = 1,750 MCF/DAY
TIME2 = 14 HR
PWF2 = 2,025 PSI
QG2 = 1,220 MCF/DAY

Other pertinent data are:

Stabilized line pressure = 600 PSI
Permeability = 0.09 MD
Gas gravity = 0.642 (no diluents)
Porosity = 16.2%
Temperature = 140 F
Water saturation = 36.1%
Slope of back pressure curve = 0.82

Calculate the test flow coefficients and the flow coefficients monthly until stabilization occurs. The radius of drainage corresponding to 320 acres is 2,106 FT.

Keystrokes (SIZE> =049)	Display	Comments
[XEQ] [ALPHA] TcPc [ALPHA]	*COND? Y/N:*	Last character is Y or N
N [R/S]	*GAS G=?*	
.642 [R/S]	*%N2=?*	
0 [R/S]	*%CO2=?*	
0 [R/S]	*%H2S=?*	
0 [R/S]	*GAS G=?*	Tc and Pc printed
[XEQ] [ALPHA] STAB [ALPHA]	*CORR? Y/N:*	Last character is Y or N
Y [R/S]	*SKIP? Y/N:*	Last character is Y or N
N [R/S]	*Tc=?*	
[R/S]	*Pc=?*	
[R/S]	*GAS G=?*	
[R/S]	*MW=?*	
[R/S]	*T=?*	
140 [R/S]	*%POR=?*	
16.2 [R/S]	*%SW=?*	
36.1 [R/S]	*KG=?*	
.09 [R/S]	*N=?*	
.82 [R/S]	*TIME1=?*	
3 [R/S]	*PWF1=?*	
2400 [R/S]	*QG1=?*	
1750 [R/S]	*TIME2=?*	
14 [R/S]	*PWF2=?*	

Keystrokes (SIZE> =049)	Display	Comments
2025 [R/S]	*QG2=?*	
1220 [R/S]	*SITP=?*	
2900 [R/S]	*P STAB=?*	
600 [R/S]	*Re ENDT*	C1 and C2 printed

Keystrokes (SIZE> =049)	Display	Comments
[A]	*Re=?*	Re known
2106 [R/S]	*T INC=?*	END T printed
1 [ALPHA] MO [R/S]	*Re ENDT*	Forecast printed

```
Tc  Pc

COND: NO
GAS G=0.6420
%N2=0.0000
%CO2=0.0000
%H2S=0.0000

Tc=371.4980 R
Pc=671.1739 PSI


STAB  FLOW  C

CORR: YES
SKIP: NO
T=140.0000 F
%POR=16.2000
%SW=36.1000
KG=0.0900 MD
N=0.8200
TIME1=3.0000 HR
PWF1=2,400.0000 PSI
QG1=1,750.0000 MCF/DAY
TIME2=14.0000 HR
PWF2=2,025.0000 PSI
QG2=1,220.0000 MCF/DAY
SITP=2,900.0000 PSI
P STAB=600.0000 PSI

C1=0.0095
C2=0.0044

Re=2,106.0000 FT

END T=8,774.3437 HR
T INC=1.0000 MO

TIME=1.0000 MO
C=0.0021
QG=945.1575 MCF/DAY
RAD=607.4527 FT

TIME=2.0000 MO
C=0.0019
QG=869.8494 MCF/DAY
RAD=859.0679 FT

TIME=3.0000 MO
C=0.0018
QG=831.6255 MCF/DAY
RAD=1,052.1390 FT

TIME=4.0000 MO
C=0.0018
QG=806.6813 MCF/DAY
RAD=1,214.9054 FT

TIME=5.0000 MO
C=0.0017
QG=788.4446 MCF/DAY
RAD=1,358.3056 FT

TIME=6.0000 MO
C=0.0017
QG=774.2082 MCF/DAY
RAD=1,487.9492 FT

TIME=7.0000 MO
C=0.0017
QG=762.6085 MCF/DAY
RAD=1,607.1688 FT

TIME=8.0000 MO
C=0.0016
QG=752.8672 MCF/DAY
RAD=1,718.1357 FT

TIME=9.0000 MO
C=0.0016
QG=744.5009 MCF/DAY
RAD=1,822.3581 FT

TIME=10.0000 MO
C=0.0016
QG=737.1897 MCF/DAY
RAD=1,920.9341 FT

TIME=11.0000 MO
C=0.0016
QG=730.7117 MCF/DAY
RAD=2,014.6927 FT

TIME=12.0000 MO
C=0.0016
QG=724.9071 MCF/DAY
RAD=2,104.2779 FT

TIME=12.0196 MO
C=0.0016
QG=724.7989 MCF/DAY
RAD=2,106.0000 FT
```

User Instructions

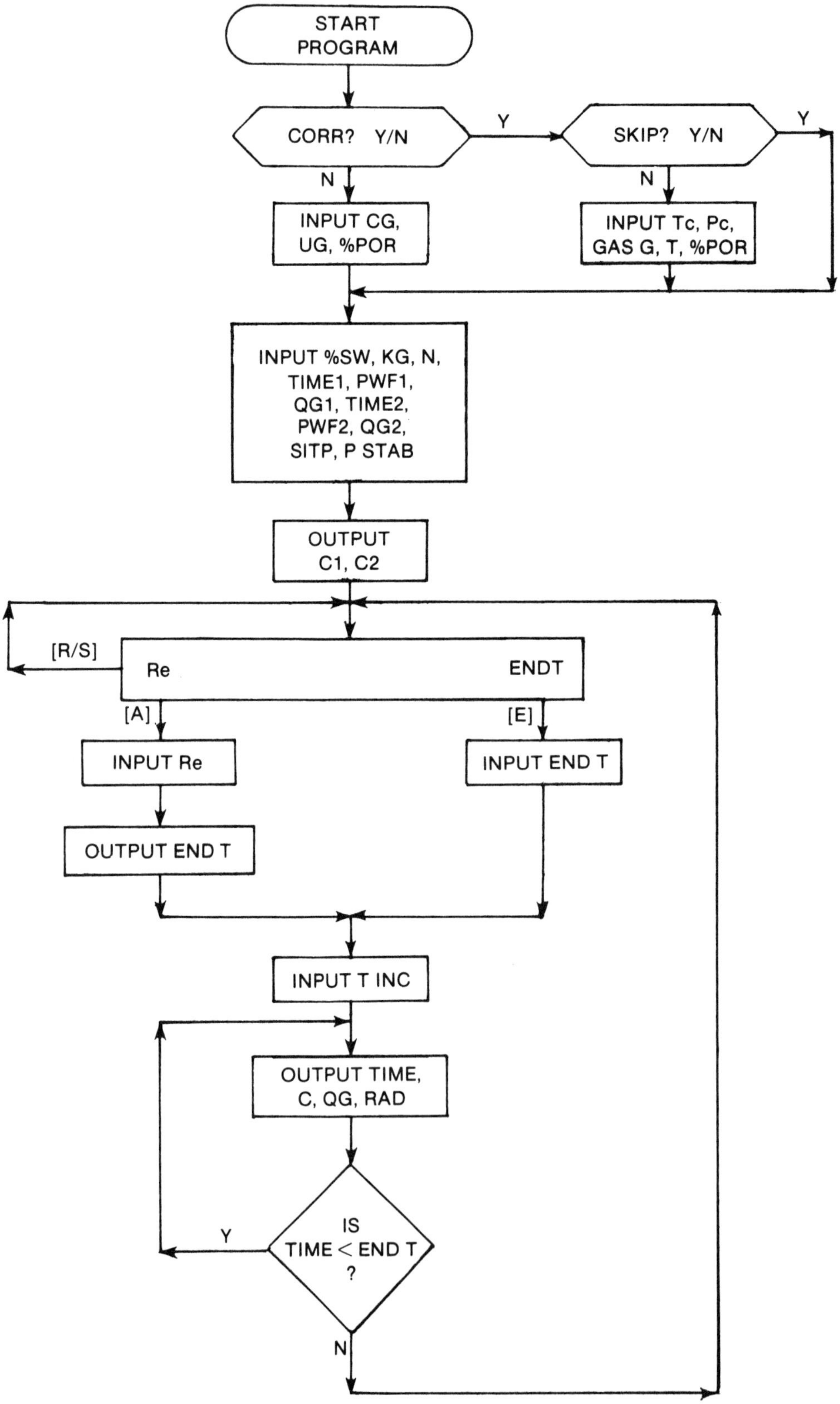

General Information

Memory Requirements

Program length:	784 bytes (4 cards)
Minimum size:	049
Minimum hardware:	41C + 2 memory modules

Hidden Options

None

Pac Subroutines Called

TITLE, Y/N?, ITcPc, GASG, T, %POR, INU, IN, CON, CCG, CUG, OUT, INK, OUTK, OUTU

Registers

03	Time units
04	Time units
06	Gas producing rate units
07	Gas producing rate units
08	Pressure units
09	Pressure units
17	P STAB (PSI)
26	SITP (PSI)
27	Scratch
28	Scratch
29	KG (MD)
30	Re (FT)
31	Scratch
32	C1
33	TIME (HR)
34	T INC (HR)
35	END T (HR)
36	TIME1 (HR)
37	PWF1 (PSI)
38	QG1 (MCF/DAY)
39	TIME2 (HR)
40	PWF2 (PSI)
41	QG2 (MCF/DAY)
42	CG (1/PSI)
43	QG (MCF/DAY)
44	P STAB (PSI)
45	C
46	N
47	PWF2 (PSI)
48	UG (CP)

Registers 12–14, 19, 20, and 22–25 unused

Flags

01	Set:	Use Pac correlations to estimate PVT properties.
	Clear:	Input PVT properties.
02	Set:	Skip input of PVT data.
	Clear:	Allow input of PVT data.

Program Listing

```
 01♦LBL "STAB"
"STAB FLOW C" 49
XROM "TITLE" FC?C 25
PROMPT SF 27 "HR"
ASTO 03 "SCM/DAY"
ASTO 06 ASHF ASTO 07
"KPA" ASTO 08 CLA
ASTO 04 ASTO 09 "CORR"
1 XROM "Y/N?" FC? 01
GTO 00 "SKIP" 2
XROM "Y/N?" FS? 02
GTO 01 XROM "ITcPc"
XROM "GASG" XROM "T"
XROM "%POR" GTO 01

 34♦LBL 00
41 STO 00 "1/PSI"
ASTO 01 CLA ASTO 02
ASTO Z "1/KPA" ASTO Y
"CG" XROM "INU" 47
STO 00 "CP" ASTO 01
CLA ASTO 02 ASTO Z
"PA*S" ASTO Y "UG"
XROM "INU" XROM "%POR"

 58♦LBL 01
20 STO 00 "%SW"
XROM "IN" 28 STO 00
"MD" ASTO 01 ASTO Y
CLA ASTO 02 ASTO Z
"KG" XROM "INU" 45
STO 00 "N" XROM "IN"
35 "TIME1" XEQ 18 36
"PWF1" XEQ 20 37
"QG1" XEQ 16 CF 08 38
"TIME2" XEQ 18 39
"PWF2" XEQ 20 STO 47
40 "QG2" XEQ 16 25
"SITP" XEQ 20 43
"P STAB" XEQ 20 FC? 01
GTO 02 RCL 16 "F-R"
CON RCL 10 / STO 01
X<>Y X↑2 RCL 26 X↑2
+ 2 / SQRT STO 17
RCL 11 / STO 00
XROM "CCG" STO 42
RCL 01 RCL 00
XROM "CUG" STO 48

129♦LBL 02
RCL 29 RCL 42 /
```

Program Listing (cont.)

```
RCL 48  /  1  RCL 21
100  /  -  RCL 18  *  /
SQRT  .69688  *  STO 27
RCL 38  RCL 37  XEQ 06
/  STO 32  ADV  "C1"
XROM "OUT"  RCL 41
RCL 40  XEQ 06  /  "C2"
XROM "OUT"  ADV  RCL 32
/  RCL 46  1/X  Y↑X
ENTER↑  1/X  1  ST- Z  -
2  ST* Z  *  RCL 39
"HR"  CON  X<>Y  1/X
Y↑X  RCL 36  CON  RCL Z
1/X  Y↑X  *  STO 31
RCL 36  CON  SQRT  *  LN
STO 28

194♦LBL 15
"Re      ENDT"  PROMPT
GTO 15

198♦LBL A
29  "Re"  XEQ 22  RCL 27
/  X↑2  STO 35  "END T"
XEQ 19  GTO 03

209♦LBL D
210♦LBL E
34  "END T"  XEQ 18

214♦LBL 03
33  "T INC"  XEQ 18  CLX
STO 33  GTO 04

221♦LBL 14
RCL 33  "TIME"  XEQ 19
"HR"  CON  SQRT  RCL 31
*  LN  RCL 28  X<>Y  /
RCL 46  Y↑X  RCL 32  *
STO 45  "C"  XROM "OUT"
RCL 44  XEQ 06  *
STO 43  "QG"  XEQ 17
RCL 33  SQRT  RCL 27  *
"RAD"  XEQ 23

253♦LBL 04
ADV  RCL 35  RND  RCL 33
RCL 34  +  RND  X<Y?
GTO 05  X<>Y  RCL 33
RND  X=Y?  GTO 15
RCL 35  RND

270♦LBL 05
LASTX  STO 33  GTO 14

274♦LBL 06
X↑2  RCL 26  X↑2  -  CHS
RCL 46  Y↑X  RTN

283♦LBL 16
STO 00  XEQ 07
XROM "INK"  RDN  STO 06
X<>Y  STO 07  R↑  RTN

293♦LBL 17
XEQ 07  XROM "OUTK"  RDN
STO 06  X<>Y  STO 07  R↑
RTN

302♦LBL 07
ASTO T  "MCF/DAY"
ASTO 01  ASHF  ASTO 02
CLA  ARCL T  RCL 07
RCL 06  RCL Z  RTN

314♦LBL 18
STO 00  XEQ 08
XROM "INK"  RDN  STO 03
X<>Y  STO 04  R↑  RTN

324♦LBL 19
XEQ 08  XROM "OUTK"  RDN
STO 03  X<>Y  STO 04  R↑
RTN

333♦LBL 08
ASTO T  "HR"  ASTO 01
CLA  ASTO 02  ARCL T
RCL 04  RCL 03  RCL Z
RTN

344♦LBL 20
STO 00  XEQ 09
XROM "INK"  RDN  STO 08
X<>Y  STO 09  R↑  RTN

354♦LBL 21
XEQ 09  XROM "OUTK"  RDN
STO 08  X<>Y  STO 09  R↑
RTN

363♦LBL 09
ASTO T  "PSI"  ASTO 01
CLA  ASTO 02  ARCL T
RCL 09  RCL 08  RCL Z
RTN

374♦LBL 22
STO 00  XEQ 10
XROM "INU"  RTN

379♦LBL 23
XEQ 10  XROM "OUTU"  RTN

383♦LBL 10
ASTO T  "FT"  ASTO 01
"M"  ASTO Y  CLA
ASTO 02  ASTO Z  ARCL T
END
```

Section 6

Pressure Transient Analysis

20. BLDUTIL — Utilities for Pressure Buildup Analysis

Utility program that calculates Horner time, coefficients for multiple producing rates prior to shutin, and P/UG*Z as a function of pressure.

21. BUILDUP — Analyzing Pressure Buildups

Calculates permeability and skin for oil or gas well buildups. Pressure, pressure squared, and real gas potential equations are allowed for gas buildups.

22. DRAW — Multiple Rate Pressure Drawdown Analysis

Calculates permeability and skin for multiple-rate oil or gas drawdowns. Includes utilities to calculate X and Y axes of drawdown plot.

20. BLDUTIL — Utilities for Pressure Buildup Analysis

Pressure transient testing is a very broad field that involves myriad analytical techniques. It is completely beyond the scope of this book to even reference these methods. However, a few methods for assisting the analysis of Horner plots for buildups and for drawdowns are presented in this chapter. BLDUTIL provides three utilities for assisting in the analysis of buildup tests with multiple or constant rate histories.

The first utility is nearly trivial (by itself) and calculates Horner time from the shutin time. This type of Horner plot is used for buildups preceded by a constant producing rate. For the case of multiple rates prior to shutin, a different plot is in order. The second utility performs the superposition calculations required for this technique (Earlougher). The slope of either of these graphs is utilized in the pressure buildup analysis program, BUILDUP.

The third utility provides a table of P/UG*Z as a function of pressure for gas wells. This is helpful in selecting the type of analysis that should be performed. At low pressures (usually less than 2,000 PSI), this function is nearly linear. In these cases, the pressure squared equations should be used for analysis. At high pressures (usually greater than 4,500 PSI) the pressure equations can be used. At intermediate values, the P/UG*Z term exhibits some curvature. For this range, the real gas potential term MP should be used. It is valid to use MP for the analysis of any pressure range.

Equations

Horner time utility:

$$\text{TIME} = 24\,\text{NP/QO}, \quad \text{TIME} = 24\,\text{GP/QG}$$

$$\text{HORN T} = \frac{\text{TIME} + \text{DTIME}}{\text{DTIME}}$$

Summation utility (shown for oil only):

$$\Sigma = \sum_{i=1}^{n} \frac{\text{QO}_i}{\text{QO}_n} \log\left[\frac{\text{TIME}_n - \text{TIME}_{i-1} + \text{DTIME}}{\text{TIME}_n - \text{TIME}_i + \text{DTIME}}\right]$$

where n is the number of TIME_i values input by the user.

$\text{TIME}_0 = 0$ HR

Nomenclature

Symbol	*Variable Name*	*Input or Output*	*English Units*	*SI Units*
BEG P*	Beginning pressure of table	I	PSI	KPA
DTIME*	Shutin time	I	HR	HR
END P*	Ending pressure of table	I	PSI	KPA
GP	Cumulative gas production	I	MCF	SCM
HORN T	Horner time	O	-	-
NP	Cumulative oil production	I	BBL	M3
P*	Pressure	O	PSI	KPA
P INC*	Pressure increment	I	PSI	KPA
P/UG*Z	Pressure-viscosity-Z factor quotient	O	-†	-†
QG*	Gas producing rate	I	MCF/DAY	SCM/DAY
QG_i*	Gas producing rate at TIME_i	I	MCF/DAY	SCM/DAY
QO*	Oil producing rate	I	BBL/DAY	M3/DAY
QO_i*	Oil producing rate at TIME_i	I	BBL/DAY	M3/DAY
TIME*	Producing time at QG or QO	I	HR	HR
TIME_i*	Producing time at QG_i or QO_i	I	HR	HR
Σ	Superposition term	O	-	-

*The units for these variables are saved by the program.
†Physically, P/UG*Z has units of PSI/CP or KPA/PA*S.

Yes/No Questions

EDIT? Yes: Allow editing of TIME and QO or TIME and QG values.
No: No editing necessary.

OIL? Yes: Oil reservoir.
No: Gas reservoir.

SKIP? Yes: Skip input of PVT data.
No: Allow input of PVT data.

Example 1: Variable Rate Buildup

This example follows Earlougher's Example 5.5. Table 20-1 gives the rate and pressure data for this example. Figure 20-1 illustrates the rate variation

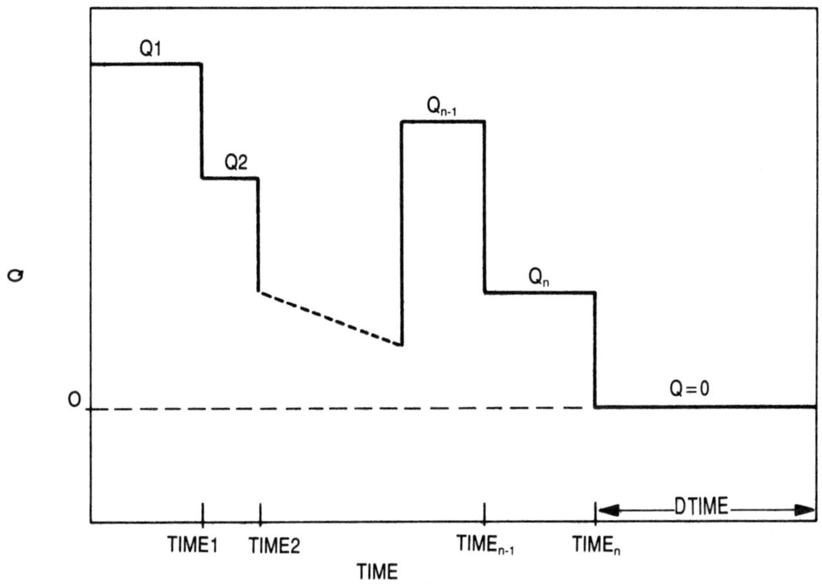

Figure 20–1 *(after Earlougher)*

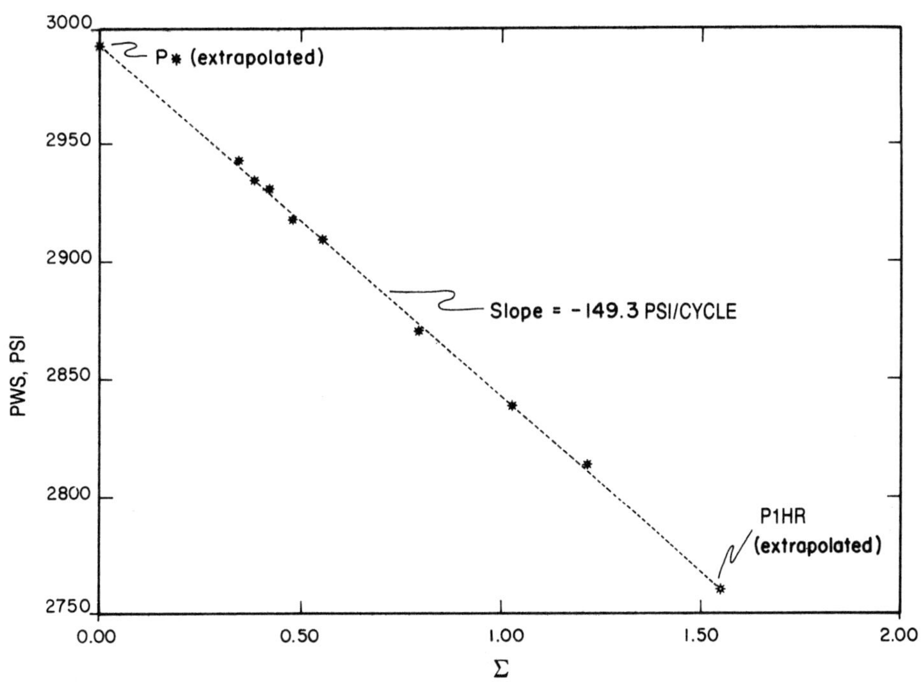

Figure 20–2

preceding a pressure buildup test. Figure 20-2 is a graph of the buildup test shutin pressure (PWS) versus the Σ term calculated in the example, with an indicated slope of 153 PSI/CYCLE. This buildup data is shown in Table 20-2. The permeability for this well will be calculated as an example in BUILDUP. Note that a value for P1HR (2,761.1 PSI) and P* (2,992.6 PSI) can be obtained from this graph by reading the PWS values at a Σ corresponding to DTIME = 1 HR and Σ = 0, respectively.

Table 20-1

TIME (HR)	QO (BBL/DAY)
0	0
3	478.5
6	319.0
9	159.5

Table 20-2

DTIME (HR)	PWS (PSI)	Σ (calculated)
2	2,813	1.2211
3	2,838	1.0280
5	2,872	0.7949
7	2,895	0.6533
9	2,910	0.5563
11	2,919	0.4851
13	2,930	0.4305
15	2,935	0.3871
17	2,942	0.3517
1	—*	1.5528

*Σ at DTIME = 1 HR used to calculate P1HR.

Keystrokes (SIZE> =063)	Display	Comments
[XEQ] [ALPHA] BLDUTIL [ALPHA]	*OIL? Y/N:*	Last character is Y or N
Y [R/S]	*HORN Σ*	
[C]	*TIME1=?*	Summation option
3 [R/S]	*QO1=?*	
478.5 [R/S]	*TIME2=?*	
6 [R/S]	*QO2=?*	
319 [R/S]	*TIME3=?*	
9 [R/S]	*QO3=?*	
159.5 [R/S]	*TIME4=?*	
[R/S]	*EDIT? Y/N:N*	
[R/S]	*DTIME=?*	Σ printed
2 [R/S]	*DTIME=?*	Σ printed
3 [R/S]	*DTIME=?*	Σ printed
5 [R/S]	*DTIME=?*	Σ printed
7 [R/S]	*DTIME=?*	Σ printed
9 [R/S]	*DTIME=?*	Σ printed
11 [R/S]	*DTIME=?*	Σ printed
13 [R/S]	*DTIME=?*	Σ printed
15 [R/S]	*DTIME=?*	Σ printed
17 [R/S]	*DTIME=?*	Σ printed
1 [R/S]	*DTIME=?*	Σ at DTIME = 1 HR used to calculate P1HR

```
BUILDUP UTIL

OIL: YES

TIME1=3.0000 HR
QO1=478.5000 BBL/DAY

TIME2=6.0000 HR
QO2=319.0000 BBL/DAY

TIME3=9.0000 HR
QO3=159.5000 BBL/DAY

DTIME=2.0000 HR
Σ=1.2211

DTIME=3.0000 HR
Σ=1.0280

DTIME=5.0000 HR
Σ=0.7949

DTIME=7.0000 HR
Σ=0.6533

DTIME=9.0000 HR
Σ=0.5563

DTIME=11.0000 HR
Σ=0.4851

DTIME=13.0000 HR
Σ=0.4305

DTIME=15.0000 HR
Σ=0.3871

DTIME=17.0000 HR
Σ=0.3517

DTIME=1.0000 HR
Σ=1.5528
```

Example 2

A well in the Plum Nearly (Smackover) reservoir had produced 66.17 MMCF with a current rate of 2.210 MMCF/DAY, which has been more or less constant for the well's 1-month producing history. At the time of this buildup, a second well was being drilled. Table 20–3 shows the shutin times for this buildup. Calculate the pseudo-producing time (TIME) and Horner times (HORN T) for this buildup.

Table 20–3

DTIME (HR)
0.5
1.0
2.0
3.0
6.0
12.0
18.0
24.0
36.0
48.0
60.0
74.64 (bomb changed)
96.0
156.0
216.0

Keystrokes	Display	Comments
[XEQ] [ALPHA]BLDUTIL [ALPHA]	*OIL? Y/N:Y*	
N [R/S]	*HORN Σ P/UZ*	
[A]	*GP=?*	Horner time option
66.17 [ALPHA]MMCF[R/S]	*QG=?*	TIME printed
2.21 [ALPHA]MMCF/DAY [R/S]	*DTIME=?*	HORN T printed
.5 [R/S]	*DTIME=?*	HORN T printed
1 [R/S]	*DTIME=?*	HORN T printed
2 [R/S]	*DTIME=?*	HORN T printed
3 [R/S]	*DTIME=?*	HORN T printed
6 [R/S]	*DTIME=?*	HORN T printed
12 [R/S]	*DTIME=?*	HORN T printed
18 [R/S]	*DTIME=?*	HORN T printed
24 [R/S]	*DTIME=?*	HORN T printed
36 [R/S]	*DTIME=?*	HORN T printed
48 [R/S]	*DTIME=?*	HORN T printed
60 [R/S]	*DTIME=?*	HORN T printed
74.64 [R/S]	*DTIME=?*	HORN T printed
96 [R/S]	*DTIME=?*	HORN T printed
156 [R/S]	*DTIME=?*	HORN T printed
216 [R/S]	*DTIME=?*	HORN T printed

```
BUILDUP UTIL

OIL: NO
GP=66.1700 MMCF
QG=2.2100 MMCF/DAY
TIME=718.5882 HR

DTIME=0.5000 HR
HORN T=1,438.1765

DTIME=1.0000 HR
HORN T=719.5882

DTIME=2.0000 HR
HORN T=360.2941

DTIME=3.0000 HR
HORN T=240.5294

DTIME=6.0000 HR
HORN T=120.7647

DTIME=12.0000 HR
HORN T=60.8824

DTIME=18.0000 HR
HORN T=40.9216

DTIME=24.0000 HR
HORN T=30.9412

DTIME=36.0000 HR
HORN T=20.9608

DTIME=48.0000 HR
HORN T=15.9706

DTIME=60.0000 HR
HORN T=12.9765

DTIME=74.6400 HR
HORN T=10.6274

DTIME=96.0000 HR
HORN T=8.4853

DTIME=156.0000 HR
HORN T=5.6063

DTIME=216.0000 HR
HORN T=4.3268
```

Example 3

Calculate P/UG*Z for the well in Example 2 and determine where each of the buildup analysis types should be applicable. Use a 500 PSI range up to the initial reservoir pressure of 13,600 PSI, starting at 100 PSI. Figure 20–3 is a graph of P/UZ as a function of pressure. For the buildup data in this test (pressure varies from 10,000 to 13,000 PSI), either the pressure or real-gas potential equations should yield valid results. From Example 1 of GASPVT, Tc = 410.9022 R, Pc = 673.9813 PSI, T = 255 F, and GAS G = 0.8588.

Keystrokes	Display	Comments
[R/S]	*HORN Σ P/UZ*	
[E]	*SKIP? Y/N:*	P/UZ option. Last character is Y or N
N [R/S]	*Tc=?*	
410.9022 [R/S]	*Pc=?*	
673.9813 [R/S]	*T=?*	
255 [R/S]	*GAS G=?*	
.8588 [R/S]	*BEG P=?*	
100 [R/S]	*END P=?*	
13600 [R/S]	*P INC=?*	
500 [R/S]	*HORN Σ P/UZ*	Table printed

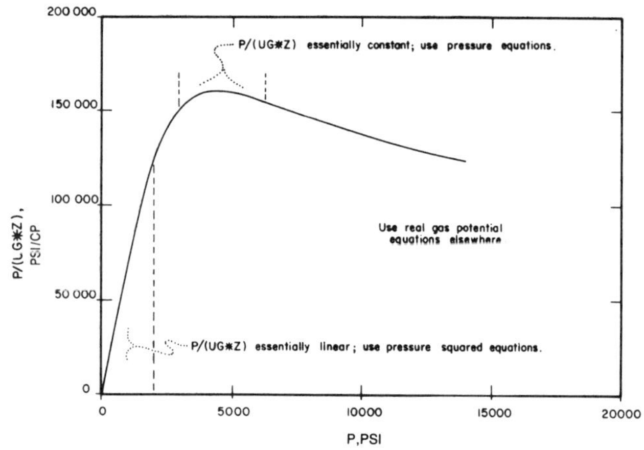

Figure 20 - 3

```
SKIP: NO
Tc=410.9022 R
Pc=673.9813 PSI
T=255.0000 F
GAS G=0.8588
BEG P=100.0000 PSI
END P=13,600.0000 PSI
P INC=500.0000 PSI

P=100.0000 PSI
P/UG*Z=7,423.3797

P=600.0000 PSI
P/UG*Z=44,550.7752

P=1,100.0000 PSI
P/UG*Z=79,309.6592

P=1,600.0000 PSI
P/UG*Z=109,009.2852

P=2,100.0000 PSI
P/UG*Z=131,921.9207

P=2,600.0000 PSI
P/UG*Z=147,801.8827

P=3,100.0000 PSI
P/UG*Z=157,670.7450

P=3,600.0000 PSI
P/UG*Z=163,055.4373

P=4,100.0000 PSI
P/UG*Z=165,374.8054

P=4,600.0000 PSI
P/UG*Z=165,710.0992

P=5,100.0000 PSI
P/UG*Z=164,809.5979

P=5,600.0000 PSI
P/UG*Z=163,166.6078

P=6,100.0000 PSI
P/UG*Z=161,098.1447

P=6,600.0000 PSI
P/UG*Z=158,804.9346

P=7,100.0000 PSI
P/UG*Z=156,412.6948

P=7,600.0000 PSI
P/UG*Z=153,999.2665

P=8,100.0000 PSI
P/UG*Z=151,612.0631

P=8,600.0000 PSI
P/UG*Z=149,279.1858

P=9,100.0000 PSI
P/UG*Z=147,016.4917

P=9,600.0000 PSI
P/UG*Z=144,832.0919

P=10,100.0000 PSI
P/UG*Z=142,729.2197

P=10,600.0000 PSI
P/UG*Z=140,708.0772

P=11,100.0000 PSI
P/UG*Z=138,767.0079

P=11,600.0000 PSI
P/UG*Z=136,903.2676

P=12,100.0000 PSI
P/UG*Z=135,113.5118

P=12,600.0000 PSI
P/UG*Z=133,394.1091

P=13,100.0000 PSI
P/UG*Z=131,741.3452

P=13,600.0000 PSI
P/UG*Z=130,151.5422
```

User Instructions

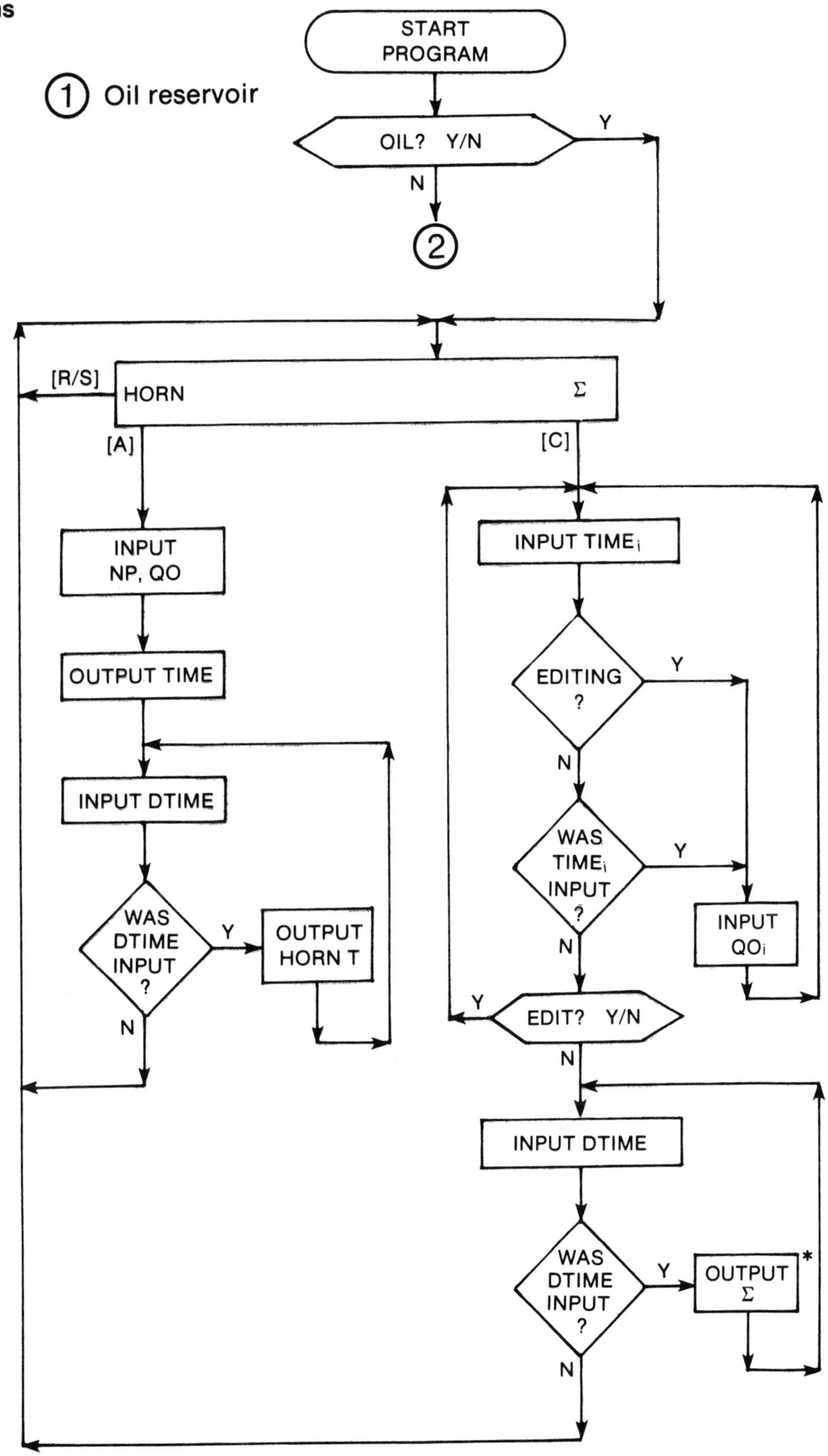

Note: for $TIME_i$ and QO_i, i = 1, 2, 3,..., n, where n is the number of TIME values input by the user.

* Tones will sound while Σ is calculated.

User Instructions

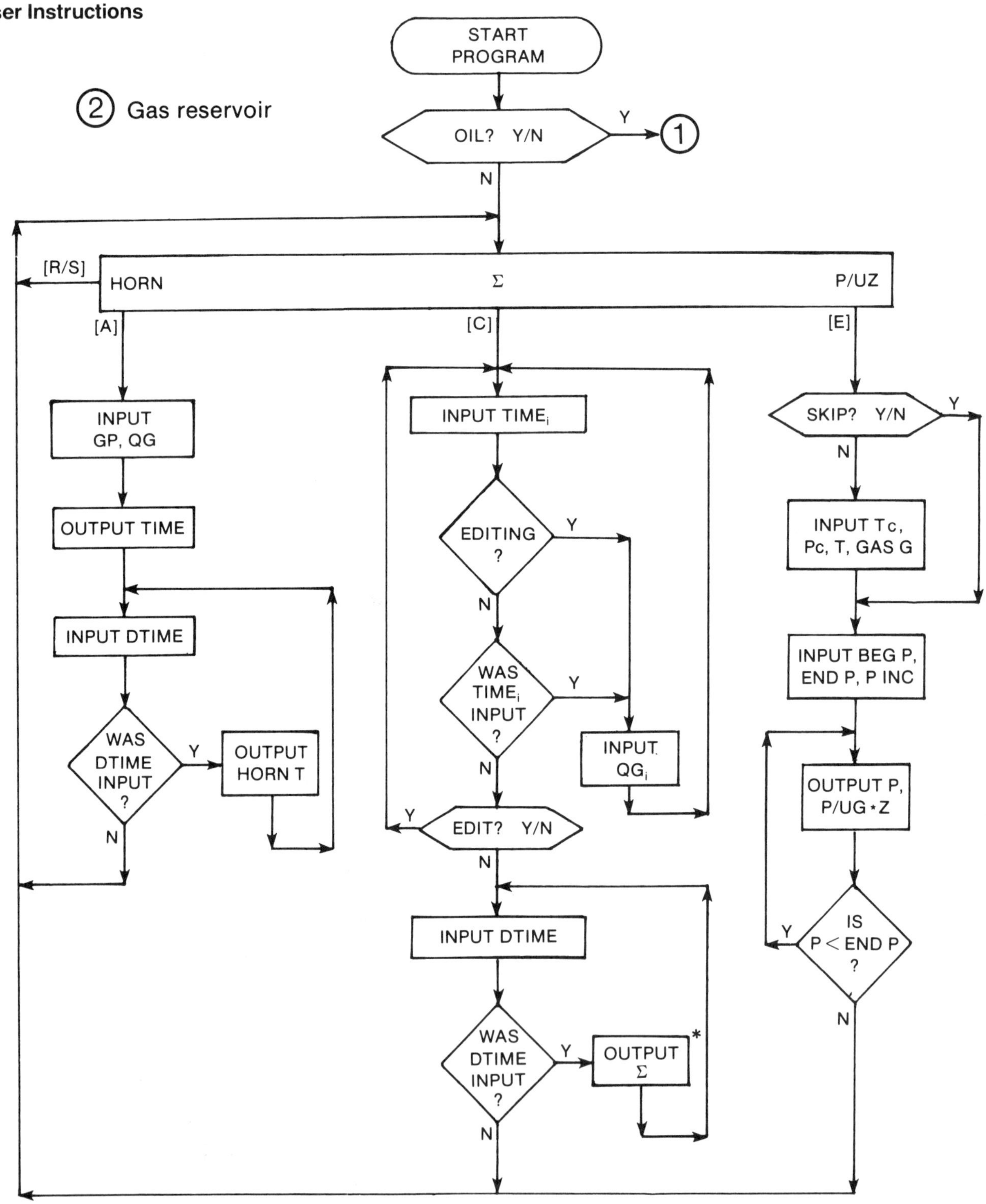

Note: for TIME$_i$ and QG$_i$, i = 1, 2, 3,..., n, where n is the number of TIME values input by the user.

* Tones will sound while Σ is calculated.

General Information

Memory Requirements

Program length:	780 bytes (4 cards)
Minimum size:	063*
Minimum hardware:	41C + 2 memory modules

*This size will allow up to 10 TIME, QO or TIME, QG values. To accommodate v TIME, QO or TIME, QG values, use size 43 + 2v.

Hidden Options

None

Pac Subroutines Called

TITLE, Y/N?, INU, OUT, INK, OUTK, ITcPc, T, GASG, CZ, CUG, CON

Registers

03	Time units
04	Time units
06	Oil or gas producing rate units
07	Oil or gas producing rate units
08	Pressure units
09	Pressure units
17	BEG P, P (PSI)
26	TIME (HR)
27	DTIME (HR)
28	QG (MCF/DAY), QO (BBL/DAY)
31	Pointer
32	$TIME_n$ (HR)
33	QG_n (MCF/DAY), QO_n (BBL/DAY)
34	Pointer
36	GP (MCF), NP (BBL)
38	END P (PSI)
41	P INC (PSI)
42	TIME1 (HR)
43	QG1 (MCF/DAY), QO1 (BBL/DAY)
44	TIME2 (HR)
45	QG2 (MCF/DAY), QO2 (BBL/DAY)
46	TIME3 (HR)
	etc.

Registers 12–14, 18–25, 29, 30, 35, 37, 39, and 40 unused

Flags

01	Set:	Oil or gas producing rate and time units not yet input.
	Clear:	Oil or gas producing rate and time units have been input.
02	Set:	Skip input of PVT data.
	Clear:	Allow input of PVT data.
03	Set:	Allow editing of TIME and QO or TIME and QG values.
	Clear:	No editing necessary.
04	Set:	Pressure units not yet input.
	Clear:	Pressure units have been input.
07	Set:	Oil reservoir.
	Clear:	Gas reservoir.

Program Listing

```
 01♦LBL "BLDUTIL"
"BUILDUP UTIL" 63
XROM "TITLE" FC?C 25
PROMPT SF 01 SF 04
SF 27 "OIL" 7
XROM "Y/N?" "M3"
FC? 07 "SCM" "├/DAY"
ASTO 06 ASHF ASTO 07
"HR" ASTO 03 "KPA"
ASTO 08 CLA ASTO 04
ASTO 09

 27♦LBL 15
"HORN Σ" FC? 07
"├ P/UZ" PROMPT GTO 15

 33♦LBL D
 34♦LBL E
FS? 07 GTO 15 GTO 21

 38♦LBL A
 39♦LBL B
35 STO 00 "BBL"
FC? 07 "MCF" ASTO 01
CLA ASTO 02 ASTO Z
"M3" FC? 07 "SCM"
ASTO Y "NP" FC? 07
"GP" XROM "INU"
FS?C 01 SF 08 27
STO 00 "QO" FC? 07
"QG" XEQ 17 RCL 36
X<>Y / 24 * STO 26
"TIME" XEQ 20 CF 08

 74♦LBL 14
ADV XEQ 18 FS? 22
GTO 00 FC? 23 GTO 15

 81♦LBL 00
RCL 26 X<>Y / 1 +
"HORN T" XROM "OUT"
GTO 14
```

Program Listing (cont.)

```
 90♦LBL C
CF 03  1.1  STO 31

 94♦LBL 13
41  STO 00

 97♦LBL 12
FC? 03  ADV  FS?C 01
SF 08  XEQ 19  FC? 03
FS? 22  GTO 01  FC? 23
GTO 02

108♦LBL 01
"QO"  FC? 07  "QG"
XEQ 16  CF 08  ISG 31
GTO 12

116♦LBL 02
RCL 31  INT  1  -  STO Y
1 E3  +  LASTX  /
STO 31  X<>Y  2  *  40
+  RCL IND X  STO 32
CLX  1  +  RCL IND X
STO 33  X<>Y  1 E3  /
42  +  STO 34  "EDIT"  3
XROM "Y/N?"  FS? 03
GTO 13  ADV

151♦LBL 11
XEQ 18  FS? 22  GTO 03
FC? 23  GTO 15

157♦LBL 03
RCL 40  STO 35  RCL 27
ST+ 32  CLX  STO 01
STO 40  RCL 34  STO 00

167♦LBL 10
TONE 5  RCL 00  2  -
RCL IND X  RCL 32  -
RCL IND 00  LASTX  -  /
LOG  ISG 00  RCL IND 00
*  RCL 33  /  ST+ 01
ISG 00  GTO 10  RCL 35
STO 40  RCL 27  ST- 32
RCL 01  TONE 9  "Σ"
XROM "OUT"  ADV  GTO 11

198♦LBL 16
XEQ 06

200♦LBL 17
ASTO T  "BBL"  FC? 07
"MCF"  "F/DAY"  ASTO 01
ASHF  ASTO 02  CLA
ARCL T  RCL 07  RCL 06
RCL Z  XROM "INK"  RDN
STO 06  X<>Y  STO 07  R↑
RTN

221♦LBL 18
26  STO 00  "DTIME"
GTO 04

226♦LBL 19
"TIME"  XEQ 06

229♦LBL 04
XEQ 05  XROM "INK"  RDN
STO 03  X<>Y  STO 04  R↑
RTN

238♦LBL 20
XEQ 05  XROM "OUTK"  RDN
STO 03  X<>Y  STO 04  R↑
RTN

247♦LBL 05
ASTO T  "HR"  ASTO 01
CLA  ASTO 02  ARCL T
RCL 04  RCL 03  RCL Z
RTN

258♦LBL 06
STO 05  CLST  FS? 41  1
+  FS? 40  2  +  5  /
FS? 39  1  +  FS? 38  2
+  FS? 37  4  +  FS? 36
8  +  FS? 29  CHS
RCL 31  FIX 0  CF 29
ARCL X  X<>Y  X<0?
SF 29  ENTER↑  FRC  5  *
FIX IND Y  X=0?
SCI IND Y  1  X=Y?
ENG IND Z  RCL 05  RTN

302♦LBL 21
"SKIP"  2  XROM "Y/N?"
FS? 02  GTO 07
XROM "ITcPc"  XROM "T"
XROM "GASG"

311♦LBL 07
FS?C 04  SF 08  16
"BEG P"  XEQ 22  CF 08
37  "END P"  XEQ 22  40
"P INC"  XEQ 22  ADV

325♦LBL 09
RCL 17  "P"  XEQ 23
XEQ 02  CZ  STO 01  CLX
LASTX  XROM "CUG"
RCL 17  X<>Y  /  RCL 01
/  "P/UG*Z"  XROM "OUT"
ADV  RCL 38  RND  RCL 17
RCL 41  +  RND  X<Y?
GTO 08  X<>Y  RCL 17
RND  X=Y?  GTO 01
RCL 38  RND

358♦LBL 08
LASTX  STO 17  GTO 09

362♦LBL 01
RCL 41  ST+ 17  GTO 15

366♦LBL 02
RCL 16  "F-R"  CON
RCL 10  /  RCL 17
RCL 11  /  RTN

376♦LBL 22
STO 00  XEQ 03
XROM "INK"  RDN  STO 08
X<>Y  STO 09  R↑  RTN

386♦LBL 23
XEQ 03  XROM "OUTK"  RDN
STO 08  X<>Y  STO 09  R↑
RTN

395♦LBL 03
ASTO T  "PSI"  ASTO 01
CLA  ASTO 02  ARCL T
RCL 09  RCL 08  RCL Z
END
```

21. BUILDUP — Analyzing Pressure Buildups

Is there a petroleum engineer that has never seen a Horner plot for a pressure buildup? Theis and Horner suggested this plot, which is constructed for essentially every pressure buildup, regardless of whether the test even approached meeting the assumptions implicit in the analysis. Nonetheless, the Horner plot is a very valuable tool for analyzing pressure buildups. Matthews & Russell and Earlougher are mandatory references for anyone performing these calculations. Quite often, the constant rate prior to shutin assumption is not met; an analysis method presented by Odeh and Selig should yield nearly equivalent results for multiple rate histories.

Even a brief discussion of each of the factors influencing pressure buildup analysis would require a lengthy discourse. A mind-boggling number of factors affect the shape of pressure buildup curves. These various factors may complicate but often prohibit legitimate analysis. Wellbore storage effects, phase separation, skin damage or improvement, hydraulic fractures, interference, faults, reservoir boundaries, stratified layers, fissured reservoirs, and gas or water contacts are some of the physical culprits that must be considered. Nonstabilized rates prior to shutin and physical problems (bad gauges, leaking lubricators, poor calibration, etc.) are items that can also wreak havoc on analysis.

We plot the Horner style plot for practically every buildup test as well. It is, in our opinion, absolutely preferable to the Miller-Dyes-Hutchison technique because:

1. It is valid for shorter production times
2. The straight line lasts longer
3. It is more useful in bounded reservoirs.

However, the method presented by Earlougher (multiple rates prior to shutin) is preferred when production rates prior to shutin are varied widely within a relatively short time. The program BLDUTIL performs the tedious calculation of the summation term required for this method.

We heartily recommend additional analysis for all pressure buildups. Another essential plot is the shutin minus the flowing pressure (PWS-PWF) versus shutin time on a log-log graph. Identification of the correct straight line is much easier and more certain when confirmed by this method. There is an engineering adage that comes to mind: "Combine an engineer, a straight edge, and some data, and a straight line will inevitably result."

Perhaps it is just a string of bad luck, but it seems that a disproportionate number of buildups that come our way do not have the correct straight line. Low permeability and hydraulically fractured wells almost never have it. One thing to check is the implication of P* from the Horner plot. Recall that it should never be larger than the initial pressure and usually is greater than the average pressure. What appears to be an absolutely straight line (for a week's worth of data or more) may point to a value of P* many hundreds of PSI below the initial pressure after very little production. The implied gas in place may be absurdly low, which brings up questions about the validity of the straight line. Again, the log-log plot and type curves should always be used to confirm this buildup analysis.

BUILDUP allows analyses with pressure, pressure squared, and real gas potential equations, calculating permeability, skin factor, skin pressure drop, and flow efficiency. Gas well buildups should be analyzed with the appropriate equations (see BLDUTIL for P/UG*Z utility). The use of the real-gas potential equations is always valid. Analysis with the wrong equations for the pressure range of the buildup can result in significant errors for nearly perfect data.

Equations

For oil reservoirs:

Above bubble point:

$$KH = 162.6 \frac{QO\ UO\ BO}{b}$$

$$A = \log \left(\frac{K}{POR\ UO\ CT\ RW^2} \right)$$

Below bubble point:

$$KH = 162.6 \frac{QO\ UOb\ BOb}{b}$$

$$A = \log \left(\frac{K}{POR\ UOb\ CTb\ RW^2} \right)$$

$$K/U = \frac{162.6}{b\,H} [BOb\ QO + BG(QG - QO\ RSb) + BW\ QW]$$

PVT properties evaluated at $P\ AVG = \frac{P* + PWF}{2}$

$$SKIN = 1.1513 \left(\frac{P1HR - PWF}{b} - A + 3.2275 \right)$$

$$DPSKIN = 0.87\ b\ SKIN$$

For gas reservoirs:

P equations:

$$KH = 28960 \frac{QG\ UG\ Z\ T'\ STD\ P}{bP\ P\ STD\ T'}$$

$$SKIN^{*} = 1.1513 \left(\frac{P1HR - PWF}{bP} - A + 3.2275 \right)$$

$$DPSKIN = 0.87\ bP\ SKIN$$

$$\text{PVT properties evaluated at P AVG} = \frac{P{*} + PWF}{2}$$

P↑2 equations:

$$KH = 57920 \frac{QG\ UG\ Z\ T'\ STD\ P}{bP{\uparrow}2\ STD\ T'}$$

$$SKIN\dagger = 1.1513 \left(\frac{P1HR{\uparrow}2 - PWF{\uparrow}2}{bP{\uparrow}2} - A + 3.2275 \right)$$

PVT properties evaluated at P AVG =

$$\sqrt{\frac{P{*}^2 + PWF^2}{2}}$$

MP (real gas potential) equations:

$$KH = 1638 \frac{QG\ T'}{bMP}$$

$$SKIN\dagger = 1.1513 \left(\frac{MP1HR - MPWF}{bMP} - A + 3.2275 \right)$$

*At high flow rates, the skin factor could include additional effects due to turbulent flow.

† At high flow rates, the skin factor could include additional effects due to turbulent flow.

**The calculations for DPSKIN require a slope in PSI/CYCLE. Consequently, DPSKIN and %EFF cannot be calculated from pressure squared or real gas potential equations for gas reservoirs, since their slopes are PSI2/CYCLE and PSI2/CP*CYCLE, respectively.

PVT properties evaluated at P AVG =

$$\sqrt{\frac{P{*}^2 + PWF^2}{2}}$$

$$A = \log \left(\frac{K}{POR\ UG\ CT\ RW^2} \right)$$

$$\%EFF^{**} = 100 \frac{P{*} - PWF - DPSKIN}{P{*} - PWF}$$

$$T' = T \text{ in } R$$

$$STD\ T' = STD\ T \text{ in } R$$

$$POR = \frac{\%POR}{100}$$

Nomenclature

Symbol	Variable Name	Input or Output	English Units	SI Units
b*	Slope of Horner straight line for oil reservoir	I	PSI†	KPA†
bMP*	Slope of Horner straight line for gas reservoirs using real gas potential equations	I	PSI2/ CP†	KPA2/ PA*S†
bP*	Slope of Horner straight line for gas reservoirs using pressure equations	I	PSI†	KPA†
bP↑2*	Slope of Horner straight line for gas reservoirs using pressure squared equations	I	PSI2†	KPA2†
DPSKIN*	Pressure drop across skin	O	PSI	KPA
H	Formation thickness**	I	FT	M
K	Permeability	I	MD	MD
KH	Permeability thickness	O	MD*FT	MD*M
K/U	Total mobility	O	MD/ CP	MD/ PA*S
MP1HR*	Real gas potential at P1HR	I	PSI2/ CP	KPA2/ PA*S
MPWF*	Real gas potential at PWF	I	PSI2/ CP	KPA2/ PA*S
PWF*	Flowing bottom hole pressure	I	PSI	KPA

Symbol	*Variable Name*	*Input or Output*	*English Units*	*SI Units*
PWF↑2*	PWF squared	I	PSI2	KPA2
P1HR*	Pressure at one hour from Horner plot	I	PSI	KPA
P1HR↑2*	P1HR squared	I	PSI2	KPA2
P**	False pressure from Horner plot	I	PSI	KPA
P*↑2*	P* squared	I	PSI2	KPA2
QG	Stabilized gas flow rate prior to shutin	I	MCF/DAY	SCM/DAY
QO*	Stabilized oil flow rate prior to shutin	I	BBL/DAY	M3/DAY
QW*	Stabilized water flow rate prior to shutin	I	BBL/DAY	M3/DAY
RW	Effective wellbore radius	I	FT	M
SKIN	Skin factor	O	-	-
%EFF	Flow efficiency	O	-	-

*The units for these variables are saved by the program.

†The units for the slopes are per cycle.

**In the case of deviated wells or slanted beds, use the true vertical thickness instead of the measured thickness of the formation.

Yes/No Questions

CORR? Yes: Use Pac correlations to estimate PVT properties.
No: Input PVT properties.

OIL? Yes: Oil reservoir.
No: Gas reservoir.

SKIP? Yes: Skip input of PVT data.
No: Allow input of PVT data.

Example 1

Calculate the permeability for the oil-well buildup with a multiple rate flow history (see Example 1 of BLDUTIL). Use the following values:

BO = 1.0
UO = 0.6 CP
CT = $4.0*10^{-4}$ 1/PSI
PWF = 2,510 PSI
P* = 2,992.9 PSI
P1HR = 2,761.1 PSI
b = 149.3 PSI/CYCLE
H = 20 FT
RW = 0.3 FT
%POR = 10
QO = 159.5 BBL/DAY
PBP = 2,100 PSI

The calculated value for KH is 104 MD-FT, compared to a theoretical value of 106 MD*FT for this simulated data (Odeh and Selig). A conventional Horner plot (neglecting rate variation prior to shutin) yielded a permeability estimate of 77 MD*FT.

Keystrokes (SIZE> =054)	**Display**	**Comments**
[XEQ] [ALPHA]BUILDUP [ALPHA]	*OIL? Y/N:*	Last character is Y or N
Y [R/S]	*CORR? Y/N:*	Last character is Y or N
N [R/S]	*PBP=?*	
2100 [R/S]	*%POR=?*	
10 [R/S]	*H=?*	
20 [R/S]	*RW=?*	
.3 [R/S]	*QO=?*	
159.5 [R/S]	*P*=?*	
2992.9 [R/S]	*b=?*	
149.3 [R/S]	*P1HR=?*	
2761.1 [R/S]	*PWF=?*	
2510 [R/S]	*BO=?*	
1 [R/S]	*UO=?*	
.6 [R/S]	*CT=?*	
4 [EEX] [CHS]4 [R/S]	*P*=?*	KH, K, SKIN, DPSKIN, and %EFF printed

```
BUILDUP

OIL: YES
CORR: NO
PBP=2,100.0000 PSI
%POR=10.0000
H=20.0000 FT
RW=0.3000 FT
QO=159.5000 BBL/DAY

P*=2,992.9000 PSI
b=149.3000 PSI

P1HR=2,761.1000 PSI
PWF=2,510.0000 PSI
BO=1.0000
UO=0.6000 CP
CT=0.0004 1/PSI

KH=104.2252 MD*FT
K=5.2113 MD
SKIN=-1.6960
DPSKIN=-220.2986 PSI
%EFF=145.6199
```

Example 2

Analyze the gas-well buildups for the second example of BLDUTIL using the pressure, pressure squared, and real-gas potential equations. Let BUILDUP calculate the PVT properties. This well is the Plum Nearly (Smackover) field used in the

GASPVT, DEW, and GASMBE examples. Data required are:

T = 255 F
H = 62 FT
Tc = 410.9022 R
Pc = 673.9813 PSI
PWF = 5,075 PSI
%POR = 9.1
Water salinity = 15,600 PPM
RW = 3.5 IN
%SW = 34.0
GAS G = 0.8588
STD T = 60 F
STD P = 15.025 PSI
QG = 2,210 MCF/DAY

Table 21-1 shows the values for shutin bottom-hole pressure (PWS), shutin bottom-hole pressure squared (PWS↑2), and shutin real-gas potential (MPWS) along with the tabulated values of HORN T calculated in Example 2 of BLDUTIL. Figures 21-1 through 21-3 are the Horner plots. The values calculated or extrapolated from these plots are shown in Table 21-2.

Because of the magnitude of the numbers, especially for the pressure squared and real gas potential equations, the display format was changed to [ENG] 4.

Table 21-1

DTIME (HR)	*HORN T*	*PWS (PSI)*	*PWS↑2 (PSI2)×10⁶*	*MPWS (PSI2/CP)×10⁶*
0.5	1,438.1765	10,085	101.7	2,698
1.0	719.5882	10,427	108.7	2,793
2.0	360.2941	10,765	115.9	2,886
3.0	240.5294	11,043	121.9	2,961
6.0	120.7647	11,392	129.8	3,055
12.0	60.8824	11,701	136.9	3,138
18.0	40.9216	11,879	141.1	3,185
24.0	30.9412	12,004	144.1	3,218
36.0	20.9608	12,179	148.3	3,264
48.0	15.9706	12,347	152.4	3,308
60.0	12.9765	12,397	153.7	3,321
74.64	10.6274	12,487	155.9	3,344
96.0	8.4853	12,589	158.4	3,371
156.0	5.6063	12,779	163.3	3,420
216.0	4.4218	12,898	166.3	3,451

Table 21-2

Figure	*Data*
21-1	P* = 13,560 PSI bP = 1,050 PSI/CYCLE P1HR = 10,560 PSI
21-2	P*↑2 = 183.3 (10⁶) PSI2 bP↑2 = 26.1 (10⁶) PSI2/CYCLE P1HR↑2 = 109.2 (10⁶) PSI2
21-3	P* = 13,574 PSI bMP = 270 (10⁶) PSI2/CP*CYCLE MP1HR = 2,833 (10⁶) PSI2/CP MPWF = 1,207 (10⁶) PSI2/CP

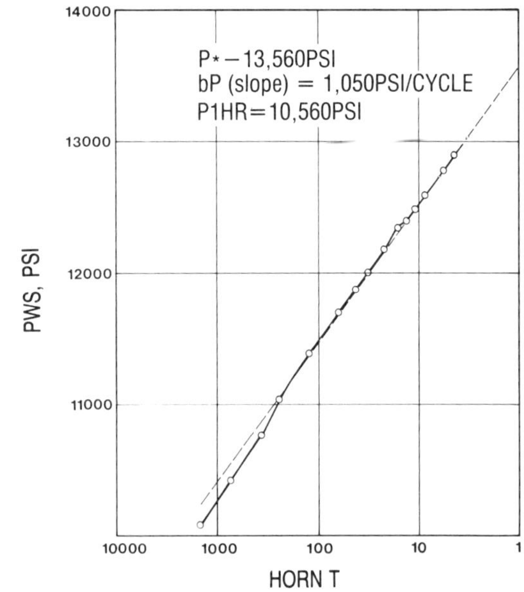

Figure 21–1

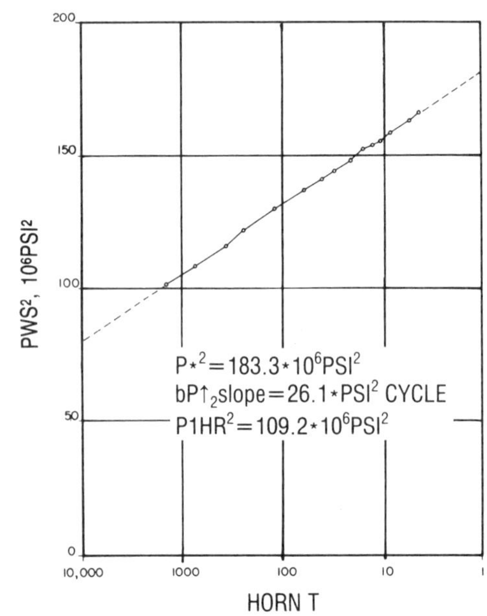

Figure 21–2

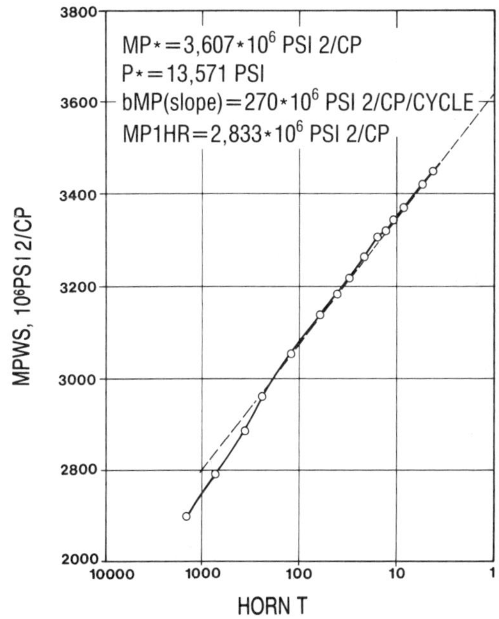

Figure 21–3

Keystrokes	Display	Comments
[//] [ENG] 4		
[XEQ] [ALPHA] BUILDUP [ALPHA]	*OIL? Y/N:Y*	
N [R/S]	*CORR? Y/N:N*	
Y [R/S]	*SKIP? Y/N:*	Last character is Y or N
N [R/S]	*Tc=?*	
410.9022 [R/S]	*Pc=?*	
673.9813 [R/S]	*STD T=?*	
60 [R/S]	*STD P=?*	
15.025 [R/S]	*GAS G=?*	
.8588 [R/S]	*T=?*	
255 [R/S]	*%NACL=?*	
[R/S]	*PPM=?*	
15600 [R/S]	*%SW=?*	
34 [R/S]	*%POR=?*	%SG printed
9.1 [R/S]	*H=?*	
62 [R/S]	*RW=?*	
3.5 [ALPHA] IN [R/S]	*QG=?*	
2210 [R/S]	*P P↑2 MP*	
[A]	*P*=?*	Pressure equations
13560 [R/S]	*bP=?*	
1050 [R/S]	*P1HR=?*	
10560 [R/S]	*PWF=?*	
5075 [R/S]	*P P↑2 MP*	KH, K, SKIN, DPSKIN, and %EFF printed
[C]	*P*↑2=?*	Pressure squared equations
183.3 [EEX] 6 [R/S]	*bP↑2=?*	
26.1 [EEX] 6 [R/S]	*P1HR↑2=?*	
109.2 [EEX] 6 [R/S]	*PWF↑2=?*	
5075 [//] [x^2] [R/S]	*P P↑2 MP*	KH, K, and SKIN printed
[E]	*P*=?*	Real gas potential equations
13574 [R/S]	*bMP=?*	
270 [EEX] 6 [R/S]	*MP1HR=?*	
2833 [EEX] 6 [R/S]	*MPWF=?*	
1207 [EEX] 6 [R/S]	*PWF=?*	
5075 [R/S]	*P P↑2 MP*	KH, K, and SKIN printed
[//] [FIX] 4	*2.4458*	Back to FIX 4

```
BUILDUP

OIL: NO
CORR: YES
SKIP: NO
Tc=410.90E0 R
Pc=673.98E0 PSI
STD T=60.000E0 F
STD P=15.025E0 PSI
GAS G=858.80E-3
T=255.00E0 F
PPM=15.600E3
%SW=34.000E0
%SG=66.000E0
%POR=9.1000E0
H=62.000E0 FT
RW=3.5000E0 IN
QG=2.2100E3 MCF/DAY

P*=13.560E3 PSI
bP=1.0500E3 PSI
P1HR=10.560E3 PSI
PWF=5.0750E3 PSI

KH=8.6234E0 MD*FT
K=139.09E-3 MD
SKIN=1.6132E0
DPSKIN=1.4736E3 PSI
%EFF=82.633E0

P*↑2=183.30E6 PSI2
bP↑2=26.100E6 PSI2
P1HR↑2=109.20E6 PSI2
PWF↑2=25.756E6 PSI2

KH=7.2850E0 MD*FT
K=117.50E-3 MD
SKIN=-668.82E-3

P*=13.574E3 PSI
bMP=270.00E6 PSI2/CP
MP1HR=2.8330E9 PSI2/CP
MPWF=1.2070E9 PSI2/CP
PWF=5.0750E3 PSI

KH=9.5842E0 MD*FT
K=154.58E-3 MD
SKIN=2.4458E0
```

User Instructions

(1) Oil reservoir

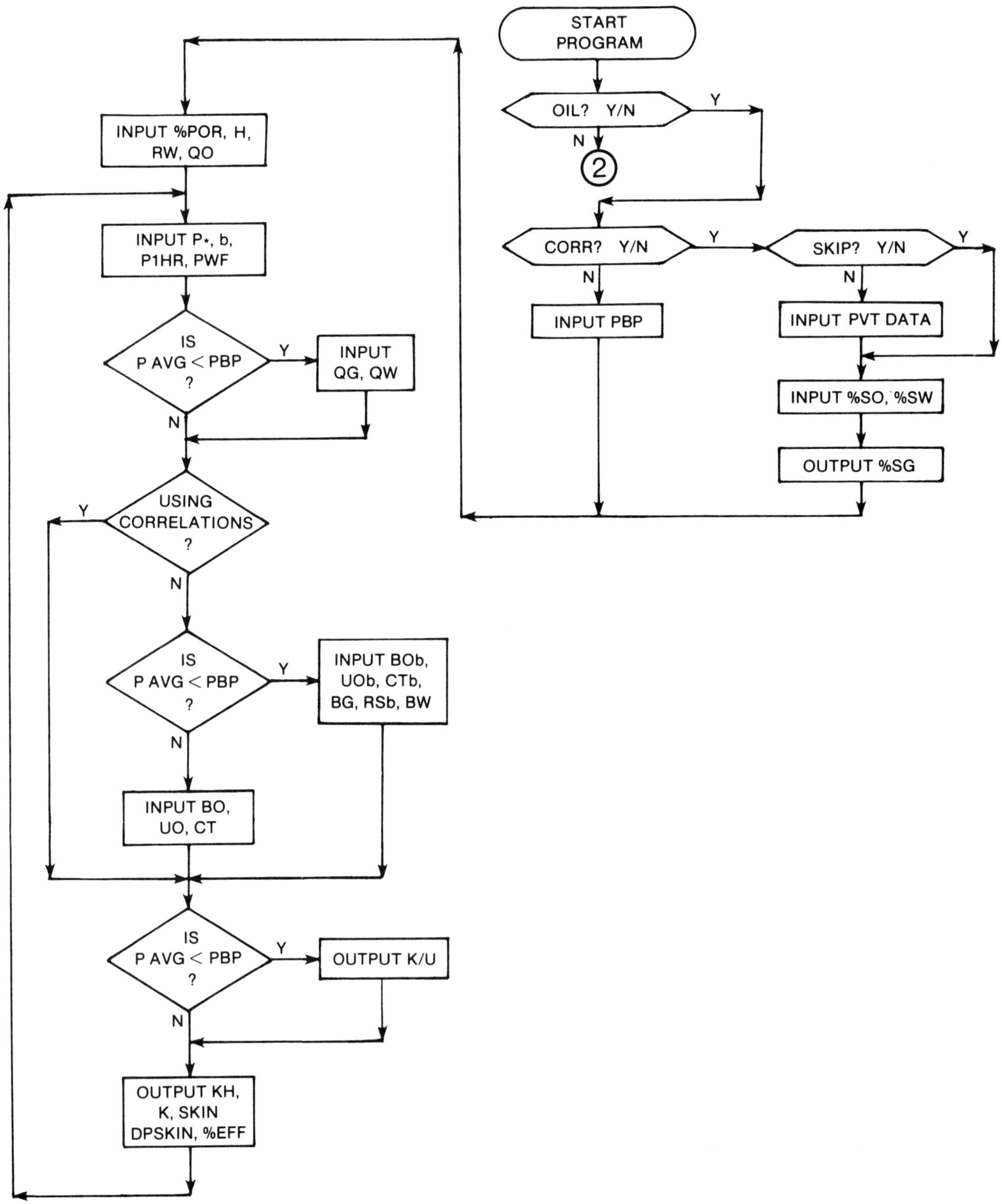

User Instructions

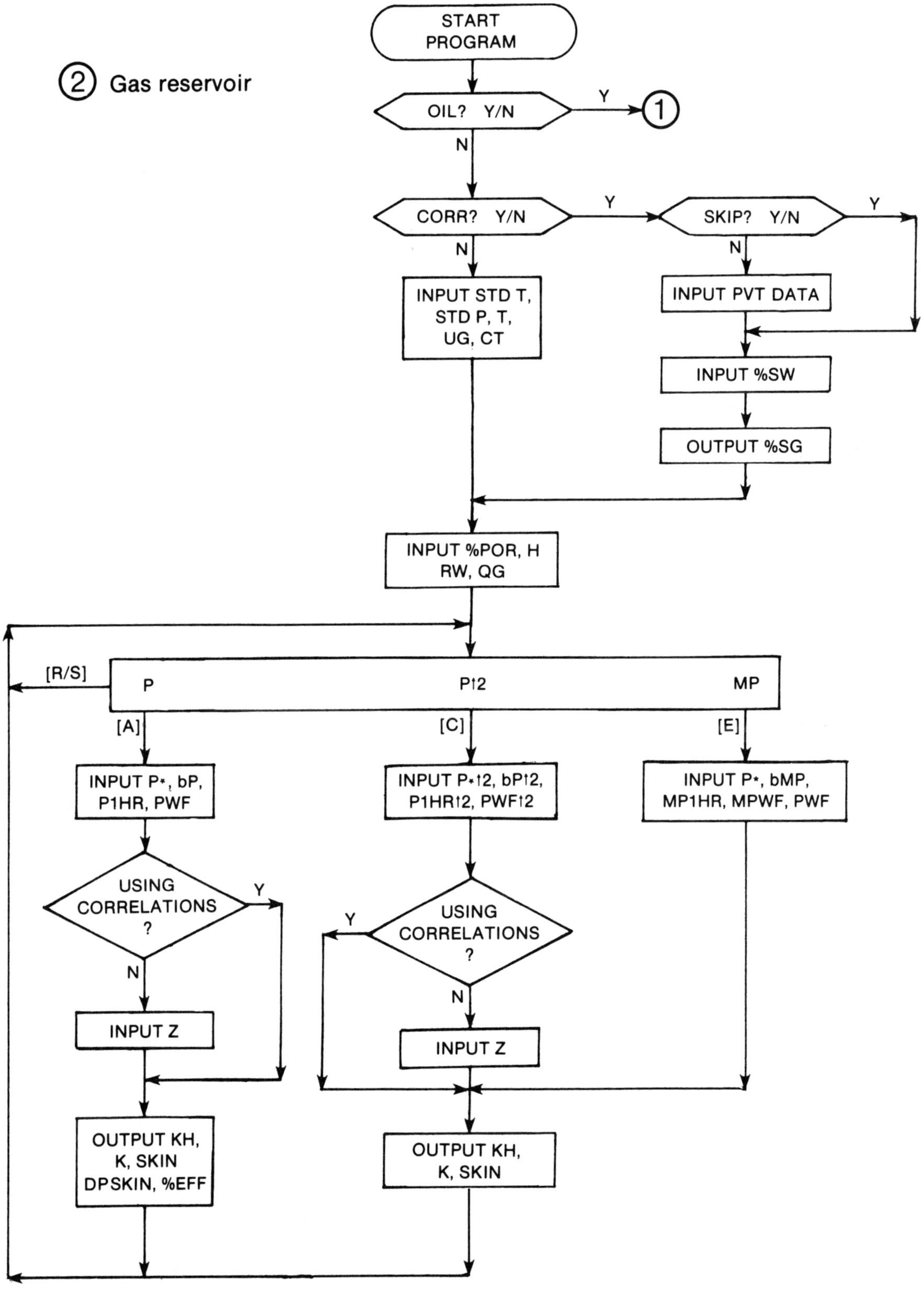

General Information

Memory Requirements

Program length:	1383 bytes (7 cards)
Minimum size:	054
Minimum hardware:	41C + quad memory module or 41CV*

*The program can be run in a 41C + 3 memory modules but will not leave a port available for a printer.

Hidden Options

None

Pac Subroutines Called

TITLE, Y/N?, ITcPc, STDTP, W8, GASG, T, %NACL, IN, OUT, T, INU, CUOd, CBG, CBOb, CRSb, CUOb, CBW, CON, OUTU, CBO, CUO, CUG, CCT, CCTb, INK, OUTK'

Registers

03	Oil or water flow rate or gas potential units
04	Oil or water flow rate or gas potential units
06	Scratch
07	Scratch
08	Pressure units
09	Pressure units
17	P AVG (PSI)
26	Z
27	BG (FT3/SCF)
28	H (FT)
29	K (MD)
30	RSb (SCF/BBL)
31	RW (FT)
32	BO, BOb
33	UG, UOd, UO, UOb (CP)
34	CT, CTb (1/PSI)
42	QG (MCF/DAY), QO (BBL/DAY)
43	P* (PSI), P*↑2 (PSI2)
44	QG (MCF/DAY)
45	QW (BBL/DAY)
46	b (PSI)*, bP (PSI2)*, bMP (PSI2/CP)*
47	P1HR (PSI), P1HR↑2 (PSI2), MP1HR (PSI2/CP)
48	PWF (PSI), PWF↑2 (PSI2), MPWF (PSI2/CP)
49	PWF (PSI)
50	Pressure squared units
51	Pressure squared units
52	Scratch
53	BW

Registers 35–41 used

*The units for the slopes are per cycle.

Flags

00	Set:	Gas potential units not yet input.
	Clear:	Gas potential units have been input.
01	Set:	Use Pac correlations to estimate PVT properties.
	Clear:	Input PVT properties.
02	Set:	Skip input of PVT data.
	Clear:	Allow input of PVT data.
03	Set:	P AVG ≥ PBP.
	Clear:	P AVG < PBP.
04	Set:	Pressure squared units not yet input.
	Clear:	Pressure squared units have been input.
05	Set:	Pressure units not yet input.
	Clear:	Pressure units have been input.
07	Set:	Oil reservoir.
	Clear:	Gas reservoir.

Program Listing

```
 01♦LBL "BUILDUP"
"BUILDUP" 54
XROM "TITLE" FC?C 25
PROMPT SF 00 SF 04
SF 05 SF 06 SF 27
"OIL" 7 XROM "Y/N?"
"KPA" ASTO 08 "⊦2"
ASTO 50 "⊦/PA*S"
FS? 07 "M3/DAY"
ASTO 03 ASHF ASTO 04
CLA ASTO 09 ASTO 51
"CORR" 1 XROM "Y/N?"
FC? 01 GTO 01 "SKIP"
2 XROM "Y/N?" FS? 02
GTO 00 XROM "ITcPc"
XROM "STDTP" FS? 07
XROM "W8" FC? 07
XROM "GASG" FC? 07
XROM "T" XROM "%NACL"

 47♦LBL 00
19 FC? 07 20 STO 00
"%SO" FS? 07 XROM "IN"
CLX FC? 07 STO 20
"%SW" XROM "IN" RCL 20
+ 100 X<>Y - "%SG"
XROM "OUT" GTO 03

 68♦LBL 01
FS? 07 GTO 02
XROM "STDTP" XROM "T"
32 STO 00 "CP"
ASTO 01 CLA ASTO 02
ASTO Z "PA*S" ASTO Y
"UG" XROM "INU"
"1/PSI" ASTO 01 CLA
ASTO 02 ASTO Z "1/KPA"
ASTO Y "CT" XROM "INU"
GTO 03

 94♦LBL 02
13 STO 00 "PSI"
ASTO 01 CLA ASTO 02
ASTO Z "KPA" ASTO Y
"PBP" XROM "INU"

106♦LBL 03
XROM "%POR" 27 STO 00
"FT" ASTO 01 CLA
ASTO 02 ASTO Z "M"
ASTO Y "H" XROM "INU"
30 STO 00 "M" ASTO Y
CLA ASTO Z "RW"
XROM "INU" 41 STO 00
"QO" FS? 07 XEQ 26
FC? 07 XEQ 27 FC? 07
GTO 19

136♦LBL 15
ADV XEQ 21 CF 08
XEQ 20 RCL 17 RND
RCL 14 RND CF 03
```

Program Listing (cont.)

```
X<=Y?  SF 03  43  STO 00
FC? 03  XEQ 27  "QW"
FC? 03  XEQ 26  FS? 01
GTO 16  31  STO 00  "BO"
FC? 03  "⊦b"  XROM "IN"
"CP"  ASTO 01  CLA
ASTO 02  ASTO Z  "PA*S"
ASTO Y  "UO"  FC? 03
"⊦b"  XROM "INU"
"1/PSI"  ASTO 01  CLA
ASTO 02  ASTO Z  "1/KPA"
ASTO Y  "CT"  FC? 03
"⊦b"  XROM "INU"  162.6
RCL 42  *  RCL 46  /
STO 52  FS? 03  GTO 18
26  STO 00  "FT3/SCF"
ASTO 01  ASHF  ASTO 02
"M3/SCM"  ASTO Y  CLA
ASTO Z  "BG"  XROM "INU"
29  STO 00  "SCF/BBL"
ASTO 01  ASHF  ASTO 02
"SCM/M3"  ASTO Y  CLA
ASTO Z  "RSb"
XROM "INU"  52  STO 00
"BW"  XROM "IN"  GTO 04

222♦LBL 16
XROM "CUOd"  STO 33
162.6  XEQ 07  RCL 05
ST/ 52  FS? 03  GTO 17
RDN  XROM "CBG"  STO 27
RCL 17  XROM "CBOb"
STO 32  RCL 17
XROM "CRSb"  STO 30
RCL 33  XROM "CUOb"
STO 33  XROM "CBW"
STO 53

245♦LBL 04
RCL 44  1 E3  *  RCL 42
RCL 30  *  -  "FT3-BBL"
CON  RCL 27  *  RCL 32
RCL 42  *  +  RCL 53
RCL 45  *  +  RCL 52  *
RCL 42  /  RCL 28  /
STO 29  "MD/CP"  ASTO 01
CLA  ASTO 02  "MD/PA*S"
ASTO Y  ASHF  ASTO Z
"K/U"  XROM "OUTU"
GTO 18

283♦LBL 17
RCL 14  XROM "CBOb"
XROM "CBO"  STO 32
RCL 13  RCL 33
XROM "CUOb"  XROM "CUO"
STO 33

293♦LBL 18
RCL 33  RCL 32  *
ST* 52  XEQ 08  XEQ 12
GTO 15

301♦LBL 19
" P  P↑2  MP"  PROMPT
GTO 19

305♦LBL E
ADV  FS?C 05  SF 08
XEQ 21  CF 08  FS?C 00
SF 08  45  STO 00  "bMP"
XEQ 25  CF 08  "MP1HR"
XEQ 25  "MPWF"  XEQ 25
"PWF"  XEQ 22  RCL 43
R-P  2  SQRT  /  STO 17
.0565743645  XEQ 06
XEQ 08  GTO 19

334♦LBL A
ADV  FS?C 05  SF 08
XEQ 21  CF 08  XEQ 20
25  STO 00  "Z"  FC? 01
XROM "IN"  RCL 17  1/X
XEQ 05  XEQ 12  GTO 19

351♦LBL C
ADV  FS?C 04  SF 08  42
STO 00  "P*↑2"  XEQ 24
CF 08  45  STO 00
"bP↑2"  XEQ 24  "P1HR↑2"
XEQ 24  "PWF↑2"  XEQ 24
SQRT  STO 49  LASTX
RCL 43  +  2  /  SQRT
STO 17  25  STO 00  "Z"
FC? 01  XROM "IN"  2
XEQ 05  GTO 19

385♦LBL 05
XEQ 06  RCL 06  RCL 07
FS? 01  CZ  FC? 01
RCL 26  RCL 33  *
RCL 23  *  RCL 22  "F-R"
CON  /  ST* 52  XEQ 08
RTN

404♦LBL 06
162600  "FT3-BBL"  CON
*

409♦LBL 07
RCL 42  *  RCL 46  /
RCL 16  "F-R"  CON
STO 05  *  STO 52
FC? 01  RTN  RCL 05
RCL 10  /  STO 06
RCL 17  RCL 11  /
STO 07  FS? 07  RTN
XROM "CUG"  STO 33  RTN

435♦LBL 08
RCL 34  FC? 01  GTO 09
RCL 06  RCL 07  RCL 17
FC? 07  XROM "CCT"
FC? 07  GTO 09  FS? 03
XROM "CCT"  FC? 03
XROM "CCTb"

450♦LBL 09
RCL 18  100  /  *
RCL 31  X↑2  *  RCL 52
RCL 28  /  FS? 07
FS? 03  GTO 10  X<> 29
GTO 11

466♦LBL 10
STO 29  RCL 33  /

470♦LBL 11
X<>Y  /  LOG  RCL 47
RCL 48  -  RCL 46  /
X<>Y  -  3.2275  +
1.1513  *  X<> 52  ADV
"MD*FT"  ASTO 01  CLA
ASTO 02  ASTO Z  "MD*M"
ASTO Y  "KH"
XROM "OUTU"  RCL 29
"MD"  ASTO 01  ASTO Y
CLA  ASTO 02  ASTO Z
"K"  XROM "OUTU"  RCL 52
"SKIN"  XROM "OUT"  RTN

509♦LBL 12
87  %  RCL 46  *
"DPSKIN"  XEQ 23  RCL 49
RCL 43  -  /  1  +  100
*  "%EFF"  XROM "OUT"
RTN

527♦LBL 20
45  STO 00  "b"  FC? 07
"bP"  XEQ 22  "P1HR"
XEQ 22  "PWF"  XEQ 22
STO 49  RCL 43  +  2  /
STO 17  RTN

545♦LBL 21
42  STO 00  "P*"

549♦LBL 22
XEQ 13  XROM "INK"  RDN
STO 08  X<>Y  STO 09  R↑
RTN

558♦LBL 23
XEQ 13  XROM "OUTK"  RDN
STO 08  X<>Y  STO 09  R↑
RTN

567♦LBL 13
ASTO T  "PSI"  ASTO 01
CLA  ASTO 02  ARCL T
RCL 09  RCL 08  RCL Z
RTN

578♦LBL 24
ASTO T  "PSI2"  ASTO 01
CLA  ASTO 02  ARCL T
RCL 51  RCL 50  RCL Z
XROM "INK"  RDN  STO 50
X<>Y  STO 51  R↑  RTN

595♦LBL 25
ASTO T  "PSI2/CP"
GTO 14

599♦LBL 26
ASTO T  "BBL/DAY"

602♦LBL 14
ASTO 01  ASHF  ASTO 02
CLA  ARCL T  RCL 04
RCL 03  RCL Z
XROM "INK"  RDN  STO 03
X<>Y  STO 04  R↑  RTN

618♦LBL 27
"MCF/DAY"  ASTO 01  ASHF
ASTO 02  "SCM/DAY"
ASTO Y  ASHF  ASTO Z
"QG"  XROM "INU"  END
```

22. DRAW — Multiple Rate Pressure Drawdown Analysis

A properly conducted drawdown test theoretically can determine as much information as a buildup test. In fact, the variety of type curves valid for drawdowns makes them applicable in many situations where buildups are not applicable. And for some obvious reasons, managers prefer to see positive (rather than zero) flow rates. When a constant flow rate is obtained, the flowing pressure can be plotted against the log of the flowing time to calculate the applicable slope. The permeability and skin can then be calculated using the same equations as presented in BUILDUP.

However, drawdown tests are flowing tests and are often difficult to control. Jockeying the choke around to maintain a constant rate may result in complex transients invalidating the analysis. DRAW presents a method of analyzing multiple rate drawdowns for both oil and gas wells. A less complex method is available for the two-rate case; however, this method is equally valid. It is important to note that the magnitude of error possible in analyzing a multiple rate test as a constant rate test is very high. Example 4.1 of Earlougher indicates that a constant rate test evaluation for multiple rate data resulted in permeability estimate over 100 percent too high.

DRAW also provides two drawdown utilities. The first calculates the Y axis of the drawdown plot, and the second calculates the X axis (superposition term) of the plot.

Equations

DP/Q utility:

$$DP/QO = \frac{PI - PWF}{QO}, \quad DP/QG = \frac{PI - PWF}{QG}$$

Summation utility (shown for oil only):

$$\Sigma = \sum_{i=1}^{n} \left[\frac{QO_i - QO_{i-1}}{QO_n} \log (TIME - TIME_{i-1}) \right]$$

where n is the number of TIME values input by the user.

$$TIME_0 = 0 \text{ HR}, \quad QO_0 = 0 \text{ BBL}$$

Drawdown analysis:

For oil reservoirs:

Above bubble point:

$$KH = 162.6 \frac{UO\ BO}{b}$$

$$A = \log \left(\frac{K}{POR\ UO\ CT\ RW^2} \right)$$

Below bubble point:

$$KH = 162.6 \frac{UOb\ BOb}{b}$$

$$A = \log \left(\frac{K}{POR\ UOb\ CTb\ RW^2} \right)$$

For gas reservoirs:

$$KH = 28960 \frac{UG\ Z\ T'\ STD\ P}{b\ P\ STD\ T'}$$

$$A = \log \left(\frac{K}{POR\ UG\ CT\ RW^2} \right)$$

$$SKIN^* = 1.1513 \left(\frac{a}{b} - A + 3.2275 \right)$$

$$DPSKIN = 0.87\ b\ QO\ SKIN$$

$$DPSKIN = 0.87\ b\ QG\ SKIN$$

$$\%EFF = 100 \frac{P_* - PWF - DPSKIN}{P_* - PWF}$$

PVT properties evaluated at PI

T′ = T in R

STD T′ = STD T in R

$$POR = \frac{\%POR}{100}$$

*At high flow rates for gas reservoirs, the skin factor could include additional effects due to turbulent flow.

Nomenclature

Symbol	*Variable Name*	*Input or Output*	*English Units*	*SI Units*
a	Intercept of multiple rate drawdown plot at Σ = 0	I	–*	–*
b	Slope of multiple rate drawdown plot	I	–*	–*
DPSKIN†	Pressure drop across skin	O	PSI	KPA
DP/QG	Delta P (PI − PWF) divided by QG	O	–*	–*
DP/QO	Delta P (PI − PWF) divided by QO	O	–*	–*
H	Formation thickness**	I	FT	M
K	Permeability	O	MD	MD
KH	Permeability thickness	O	MD∗FT	MD∗M
PI†	Initial pressure	I	PSI	KPA
PWF†	Flowing bottom-hole pressure	I	PSI	KPA
QG†	Gas flow rate	I	MCF/DAY	SCM/DAY
QG_i†	Gas flow rate at $TIME_i$	I	MCF/DAY	SCM/DAY
QO†	Oil flow rate	I	BBL/DAY	M3/DAY
QO_i†	Oil flow rate at $TIME_i$	I	BBL/DAY	M3/DAY
RW	Effective wellbore radius	I	FT	M
SKIN	Skin factor	O	–	–
TIME†	Producing time at QG or QO	I	HR	HR
$TIME_i$†	Producing time at QG_i or QO_i	I	HR	HR
%EFF	Flow efficiency	O	–	–
Σ	Superposition term used as X axis of multiple rate drawdown plot	O	–	–

*Physically, the intercept, a, and DP/QG have units of PSI∗DAY/MCF or KPA∗DAY/SCM for gas reservoirs, and the intercept and DP/QO have units of PSI∗DAY/BBL or KPA∗DAY/M3 for oil reservoirs. The units of the slope, b, are the same units per cycle.

†The units for these variables are saved by the program.

**In the case of deviated wells or slanted beds, use the true vertical thickness instead of the measured thickness of the formation.

Yes/No Questions

CORR?	Yes: Use Pac correlations to estimate PVT properties.
	No: Input PVT properties.
EDIT?	Yes: Allow editing of TIME and QO or TIME and QG values.
	No: No editing necessary.
OIL?	Yes: Oil reservoir.
	No: Gas reservoir.
SKIP?	Yes: Skip input of PVT data.
	No: Allow input of PVT data.

Example

This example is after Earlougher, pp. 32–33. The producing rate during a 48-hour drawdown test declined from 1,580 to 983 BBL/DAY with PWF measured as indicated in Table 22–1. Reservoir data are:

PI = 2,906 PSI
BO = 1.27
UO = 0.6 CP
H = 40 FT
CT = $2\ (10^{-5})$ 1/PSI
%POR = 12
PBP = 2,100 PSI
RW = 0.3 FT

Table 22–1

TIME (HR)	*QO (BBL/DAY)*	*PWF (PSI)*
1.00	1,580	2,023
1.50	1,580	1,968
1.89	1,580	1,941
2.40	1,580	–
3.00	1,490	1,892
3.45	1,490	1,882
3.98	1,490	1,873
4.50	1,490	1,867
4.80	1,490	–
5.50	1,440	1,853
6.05	1,440	1,843
6.55	1,440	1,834
7.00	1,440	1,830
7.20	1,440	–
7.50	1,370	1,827
8.95	1,370	1,821
9.60	1,370	–
10.0	1,300	1,815
12.0	1,300	1,797
14.4	1,260	–
15.0	1,190	1,775
18.0	1,190	1,771
19.2	1,190	–
20.0	1,160	1,772
21.6	1,160	–
24.0	1,137	1,756
28.8	1,106	–
30.0	1,080	1,751
33.6	1,000	–
36.0	1,000	–
36.2	983	1,756
48.0	983	1,743

Using DRAW, calculate the DP/QO term (Y axis) and the summation term (X axis). Plot the drawdown test and determine the correct slope (b) and intercept (a). This plot is shown in Figure 22–1. Finally, calculate the permeability of the well.

Although the table and the plot show all the times, rates, and PWF values, the keystrokes will only show the first straight line on the plot (through TIME = 8.95 HR).

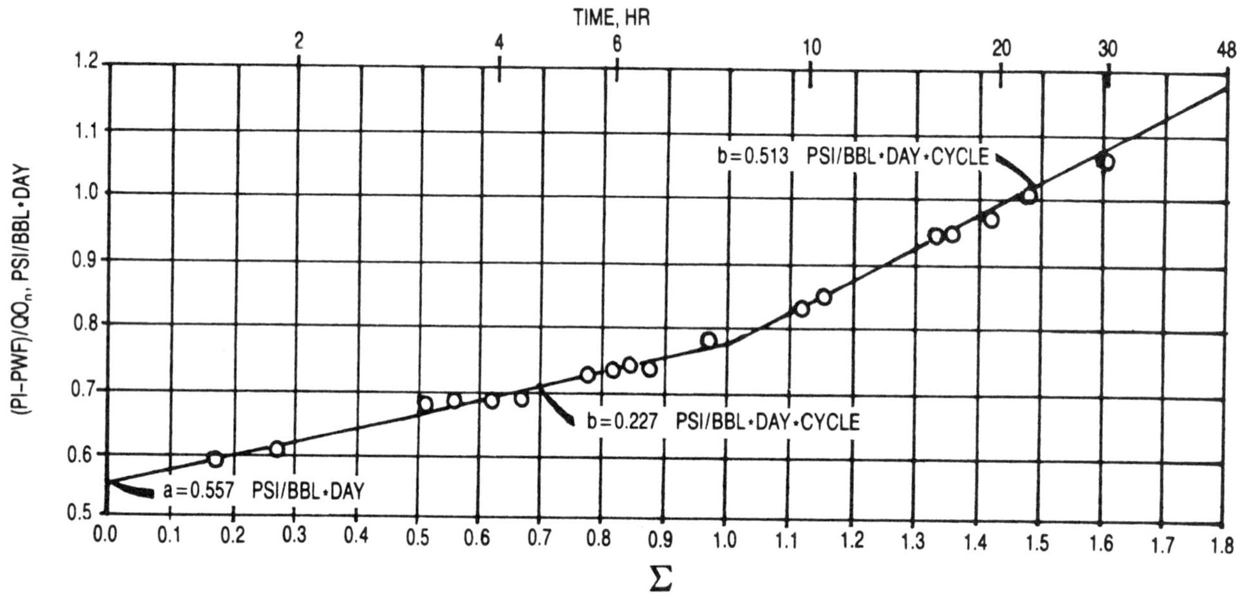

Figure 22-1 *(after Earlougher)*

Keystrokes (SIZE> =089)	Display	Comments
[XEQ] [ALPHA]DRAW [ALPHA]	*OIL? Y/N:*	Last character is Y or N
Y [R/S]	*DP/Q Σ KH*	
[A]	*PI=?*	Y axis option
2906 [R/S]	*PWF=?*	
2023 [R/S]	*QO=?*	
1580 [R/S]	*PWF=?*	DP/QO printed
1968 [R/S]	*QO=?*	
[R/S]	*PWF=?*	DP/QO printed
1941 [R/S]	*QO=?*	
[R/S]	*PWF=?*	DP/QO printed
1892 [R/S]	*QO=?*	
1490 [R/S]	*PWF=?*	
1882 [R/S]	*QO=?*	
[R/S]	*PWF=?*	
1873 [R/S]	*QO=?*	
[R/S]	*PWF=?*	
1867 [R/S]	*QO=?*	
[R/S]	*PWF=?*	
1853 [R/S]	*QO=?*	
1440 [R/S]	*PWF=?*	
1843 [R/S]	*QO=?*	
[R/S]	*PWF=?*	
1834 [R/S]	*QO=?*	
[R/S]	*PWF=?*	
1830 [R/S]	*QO=?*	
[R/S]	*PWF=?*	
1827 [R/S]	*QO=?*	
1370 [R/S]	*PWF=?*	
1821 [R/S]	*QO=?*	
[R/S]	*PWF=?*	
[R/S]	*DP/Q Σ KH*	
[C]	*TIME1=?*	X axis option
1 [R/S]	*QO1=?*	
1580 [R/S]	*TIME2=?*	Σ printed
1.5 [R/S]	*QO2=?*	
[R/S]	*TIME3=?*	Σ printed
1.89 [R/S]	*QO3=?*	
[R/S]	*TIME4=?*	Σ printed
2.4 [R/S]	*QO4=?*	
[R/S]	*TIME5=?*	
3 [R/S]	*QO5=?*	
1490 [R/S]	*TIME6=?*	
3.45 [R/S]	*QO6=?*	
[R/S]	*TIME7=?*	
3.98 [R/S]	*QO7=?*	
[R/S]	*TIME8=?*	
4.5 [R/S]	*QO8=?*	
[R/S]	*TIME9=?*	
4.8 [R/S]	*QO9=?*	
[R/S]	*TIME10=?*	
5.5 [R/S]	*QO10=?*	
1440 [R/S]	*TIME11=?*	
6.05 [R/S]	*QO11=?*	

Keystrokes (SIZE> =089)	Display	Comments
[R/S]	*TIME12=?*	
6.55 [R/S]	*QO12=?*	
[R/S]	*TIME13=?*	
7 [R/S]	*QO13=?*	
[R/S]	*TIME14=?*	
7.2 [R/S]	*QO14=?*	
[R/S]	*TIME15=?*	
7.5 [R/S]	*QO15=?*	
1370 [R/S]	*TIME16=?*	
8.95 [R/S]	*QO16=?*	
[R/S]	*TIME17=?*	
[R/S]	*DP/Q Σ KH*	
[E]	*CORR? Y/N:*	KH option. Last character is Y or N
N [R/S]	*PBP=?*	
2100 [R/S]	*%POR=?*	
12 [R/S]	*H=?*	
40 [R/S]	*RW=?*	
.3 [R/S]	*QO=?*	
[←]	*1,370.0000*	This is the last QO for the first straight line in Figure 22–1
[R/S]	*PI=?*	
2906 [R/S]	*b=?*	Slope of first straight line in Figure 22–1
.227 [R/S]	*a=?*	Intercept of first straight line in Figure 22–1
.557 [R/S]	*PWF=?*	
[←]	*1,821.0000*	This is the last PWF for the first straight line in Figure 22–1
[R/S]	*BO=?*	
1.27 [R/S]	*UO=?*	
.6 [R/S]	*CT=?*	
2 [EEX] [CHS] 5[R/S]	*DP/Q Σ KH*	KH, K, SKIN, DPSKIN, and %EFF printed

DRAWDOWN

```
OIL: YES
PI=2,906.0000 PSI

PWF=2,023.0000 PSI
QO=1,580.0000 BBL/DAY
DP/QO=0.5589

PWF=1,968.0000 PSI
DP/QO=0.5937

PWF=1,941.0000 PSI
DP/QO=0.6108

PWF=1,892.0000 PSI
QO=1,490.0000 BBL/DAY
DP/QO=0.6805

PWF=1,882.0000 PSI
DP/QO=0.6872

PWF=1,873.0000 PSI
DP/QO=0.6933

PWF=1,867.0000 PSI
DP/QO=0.6973

PWF=1,853.0000 PSI
QO=1,440.0000 BBL/DAY
DP/QO=0.7313

PWF=1,843.0000 PSI
DP/QO=0.7382

PWF=1,834.0000 PSI
DP/QO=0.7444

PWF=1,830.0000 PSI
DP/QO=0.7472

PWF=1,827.0000 PSI
QO=1,370.0000 BBL/DAY
DP/QO=0.7876

PWF=1,821.0000 PSI
DP/QO=0.7920


TIME1=1.0000 HR
QO1=1,580.0000 BBL/DAY
Σ1=0.0000

TIME2=1.5000 HR
Σ2=0.1761

TIME3=1.8900 HR
Σ3=0.2765

TIME4=2.4000 HR
Σ4=0.3802

TIME5=3.0000 HR
QO5=1,490.0000 BBL/DAY
Σ5=0.5193

TIME6=3.4500 HR
Σ6=0.5690

TIME7=3.9800 HR
Σ7=0.6241

TIME8=4.5000 HR
Σ8=0.6732

TIME9=4.8000 HR
Σ9=0.6994

TIME10=5.5000 HR
QO10=1,440.0000 BBL/DAY
Σ10=0.7870

TIME11=6.0500 HR
Σ11=0.8193

TIME12=6.5500 HR
Σ12=0.8485

TIME13=7.0000 HR
Σ13=0.8739

TIME14=7.2000 HR
Σ14=0.8849

TIME15=7.5000 HR
QO15=1,370.0000 BBL/DAY
Σ15=0.9737

TIME16=8.9500 HR
Σ16=1.0091


CORR: NO
PBP=2,100.0000 PSI
%POR=12.0000
H=40.0000 FT
RW=0.3000 FT

PI=2,906.0000 PSI
b=0.2270
a=0.5570
BO=1.2700
UO=0.6000 CP
CT=2.0000E-5 1/PSI

KH=545.8203 MD*FT
K=13.6455 MD
SKIN=-2.6954
DPSKIN=-729.2587 PSI
%EFF=167.2128
```

User Instructions

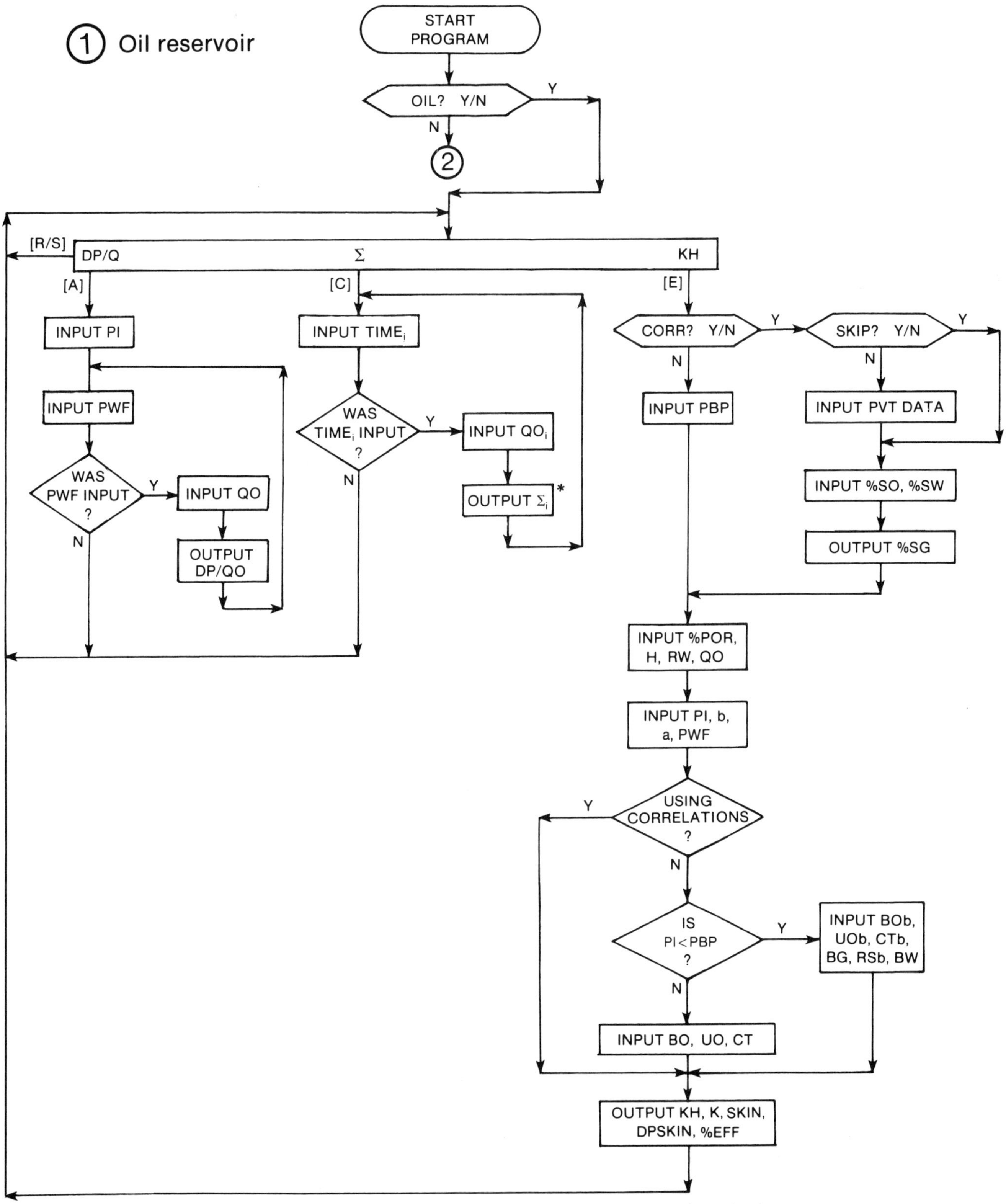

Note: for TIME$_i$, QO$_i$, and Σ_i, i = 1, 2, 3,..., n, where n is the number of TIME values input by the user.

* Tones will sound while Σ_i is calculated.

User Instructions

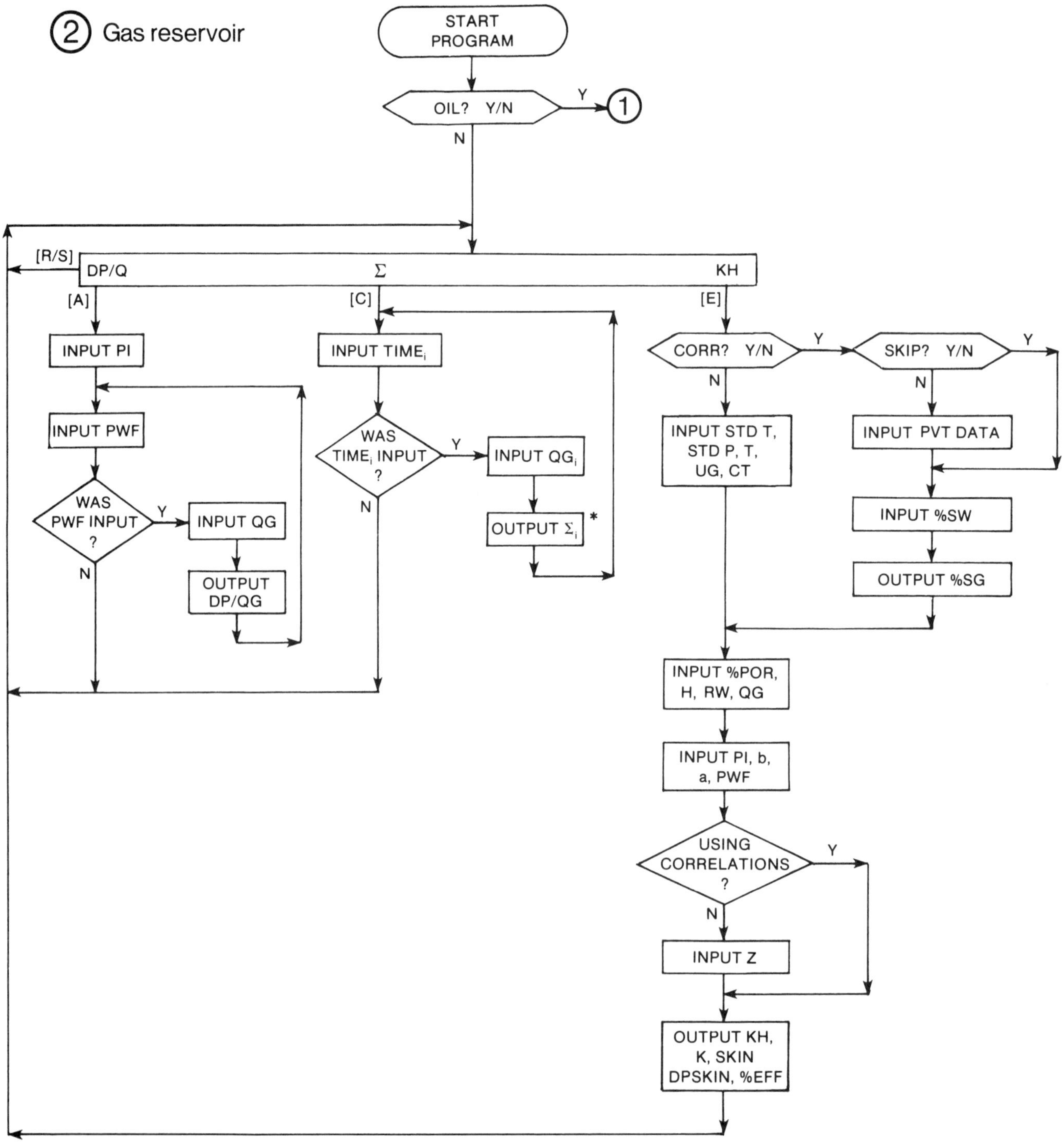

Note: for $TIME_i$, QG_i, and Σ_i, i = 1, 2, 3,..., n, where n is the number of TIME values input by the user.

* Tones will sound while Σ_i is calculated.

General Information

Memory Requirements

Program length: 1317 bytes (6 cards)
Minimum size: 089*
Minimum hardware: 41C + quad memory module or 41CV

*This size will allow 20 TIME values and 4 different QO or QG values. To accommodate v TIME values and w different QO or QG values, use size 61 + v + 2w.

Hidden Options

None

Pac Subroutines Called

TITLE, Y/N?, OUT, ITcPc, STDTP, W8, GASG, T, %NACL, IN, INU, %POR, CUOd, CBOb, CRSb, CUOb, CBO, CUO, CON, CUG, CCT, CCTb, OUTU, INK, OUTK

Registers

03	Oil or gas flow rate units
04	Oil or gas flow rate units
06	Scratch
07	Scratch
08	Pressure units
09	Pressure units
17	PI (PSI)
26	Z
27	TIME (HR)
28	H (FT)
29	K (MD)
31	RW (FT)
32	BO, BOb
33	UG, UOd, UO, UOb (CP)
34	CT, CTb (1/PSI)
35	PI (PSI)
42	QG (MCF/DAY), QO (BBL/DAY)
43	Time units
44	Time units
46	b*
47	a*
48	PWF (PSI)
50	Pointer
51	Pointer
52	Number of QO or QG values
53	Counter
54	Pointer
55	Pointer
56	Scratch
57	Counter
58	QO_0 (BBL/DAY) (=0)
59	$TIME_0$ (HR) (=0)
60	QO1 (BBL/DAY) (=1,580)
61	N1† (=4)
62	TIME1 (HR) (=1.00)
63	TIME2 (HR) (=1.50)
64	TIME3 (HR) (=1.89)
65	TIME4 (HR) (=2.40)
66	QO5 (BBL/DAY) (=1,490)
67	N5 (=5)
68	TIME5 (HR) (=3.00)
69	TIME6 (HR) (=3.45)
70	TIME7 (HR) (=3.98)
71	TIME8 (HR) (=4.50)
72	TIME9 (HR) (=4.80)
73	QO10 (BBL/DAY) (=1,440)
74	N10 (=5)
75	TIME10 (HR) (=5.50)
76	TIME11 (HR) (=6.05)
77	TIME12 (HR) (=6.55)
78	TIME13 (HR) (=7.00)
79	TIME14 (HR) (=7.20)
80	QO15 (BBL/DAY) (=1,370)
81	N15 (=2)
82	TIME15 (HR) (=7.50)
83	TIME16 (HR) (=9.60)

Registers 30, 36–41, 45, and 49 unused

Registers 60–83 illustrate the data storage for the example problem. The data structure is as follows:

$Rate_i$
Number of TIME values at $rate_i$
TIME values
$Rate_{i+1}$
Number of TIME values at $rate_{i+1}$
TIME values
etc.

Flags

00	Set:	Oil or gas flow rate units not yet input.
	Clear:	Oil or gas flow rate units have been input.
01	Set:	Use Pac correlations to estimate PVT properties.
	Clear:	Input PVT properties.
02	Set:	Skip input of PVT data.
	Clear:	Allow input of PVT data.
03	Set:	PI ≥ PBP.
	Clear:	PI < PBP.

*Physically, the intercept, a, has units of PSI*DAY/MCF or KPA*DAY/SCM for gas reservoirs, and PSI*DAY/BBL or KPA*DAY/M3 for oil reservoirs. The units of the slope, b, are the same units per cycle.

†N_i is the number of TIME values at QO_i or QG_i.

04	Set:	Time units not yet input.
	Clear:	Time units have been input.
05	Set:	Pressure units not yet input.
	Clear:	Pressure units have been input.
07	Set:	Oil reservoir.
	Clear:	Gas reservoir.

Program Listing

```
 01♦LBL "DRAW"
"DRAWDOWN" 89
XROM "TITLE" FC?C 25
PROMPT SF 00 SF 04
SF 05 SF 06 SF 27
"OIL" 7 XROM "Y/N?"
"KPA" ASTO 08 "M3"
FC? 07 "SCM" "⊦/DAY"
ASTO 03 ASHF ASTO 04
"HR" ASTO 43 CLA
ASTO 09 ASTO 44

 29♦LBL 15
"DP/Q Σ KH" PROMPT
GTO 15

 33♦LBL A
 34♦LBL B
XEQ 27

 36♦LBL 14
ADV XEQ 26 FS? 22
GTO 00 FC? 23 GTO 15

 43♦LBL 00
XEQ 20 RCL 35 RCL 48
- X<>Y / "DP/QO"
FC? 07 "DP/QG"
XROM "OUT" GTO 14

 55♦LBL C
GTO 17

 57♦LBL E
ADV "CORR" 1
XROM "Y/N?" FC? 01
GTO 01 "SKIP" 2
XROM "Y/N?" FS? 02
GTO 00 XROM "ITcPc"
XROM "STDTP" FS? 07
XROM "W8" FC? 07
XROM "GASG" FC? 07
XROM "T" XROM "%NACL"
```

```
 78♦LBL 00
19 FC? 07 20 STO 00
"%SO" FS? 07 XROM "IN"
CLX FC? 07 STO 20
"%SW" XROM "IN" RCL 20
+ 100 X<>Y - "%SG"
XROM "OUT" GTO 03

 99♦LBL 01
FS? 07 GTO 02
XROM "STDTP" XROM "T"
32 STO 00 "CP"
ASTO 01 CLA ASTO 02
ASTO Z "PA*S" ASTO Y
"UG" XROM "INU"
"1/PSI" ASTO 01 CLA
ASTO 02 ASTO Z "1/KPA"
ASTO Y "CT" XROM "INU"
GTO 03

125♦LBL 02
13 STO 00 "PSI"
ASTO 01 CLA ASTO 02
ASTO Z "KPA" ASTO Y
"PBP" XROM "INU"

137♦LBL 03
XROM "%POR" 27 STO 00
"FT" ASTO 01 CLA
ASTO 02 ASTO Z "M"
ASTO Y "H" XROM "INU"
30 STO 00 "M" ASTO Y
CLA ASTO Z "RW"
XROM "INU" XEQ 20 ADV
XEQ 27 STO 17 XEQ 25
FC? 07 GTO 16 RCL 17
RND RCL 14 RND CF 03
X<=Y? SF 03 FS? 01
GTO 04 31 STO 00 "BO"
FC? 03 "⊦b" XROM "IN"
"CP" ASTO 01 CLA
ASTO 02 ASTO Z "PA*S"
ASTO Y "UO" FC? 03
```

Program Listing (cont.)

```
"⊦b" XROM "INU"
"1/PSI" ASTO 01 CLA
ASTO 02 ASTO Z "1/KPA"
ASTO Y "CT" FC? 03
"⊦b" XROM "INU" 162.6
RCL 46 / STO 45
GTO 06

207♦LBL 04
XROM "CUOd" STO 33
162.6 XEQ 09 RCL 05
ST/ 45 FS? 03 GTO 05
RCL 17 XROM "CBOb"
STO 32 RCL 17
XROM "CRSb" RCL 33
XROM "CUOb" STO 33
GTO 06

225♦LBL 05
RCL 14 XROM "CBOb"
XROM "CBO" STO 32
RCL 13 RCL 33
XROM "CUOb" XROM "CUO"
STO 33

235♦LBL 06
RCL 32 GTO 10

238♦LBL 16
25 STO 00 "Z" FC? 01
XROM "IN" RCL 17 1/X

246♦LBL 07
XEQ 08 RCL 06 RCL 07
FS? 01 CZ FC? 01
RCL 26 RCL 23 *
RCL 22 "F-R" CON /
GTO 10

261♦LBL 08
162600 "FT3-BBL" CON
*

266♦LBL 09
RCL 46 / RCL 16 "F-R"
CON STO 05 * STO 45
FC? 01 RTN RCL 05
RCL 10 / STO 06
RCL 17 RCL 11 /
STO 07 FS? 07 RTN
XROM "CUG" STO 33 RTN
```

```
290♦LBL 10
RCL 33 * ST* 45
RCL 34 FC? 01 GTO 00
RCL 06 RCL 07 RCL 17
FC? 07 XROM "CCT"
FC? 07 GTO 00 FS? 03
XROM "CCT" FC? 03
XROM "CCTb"

308♦LBL 00
RCL 18 100 / *
RCL 31 X↑2 * RCL 45
RCL 28 / STO 29
RCL 33 / X<>Y / LOG
RCL 47 RCL 46 / X<>Y
- 3.2275 + 1.1513 *
X<> 45 ADV "MD*FT"
ASTO 01 CLA ASTO 02
ASTO Z "MD*M" ASTO Y
"KH" XROM "OUTU"
RCL 29 "MD" ASTO 01
ASTO Y CLA ASTO 02
ASTO Z "K" XROM "OUTU"
RCL 45 "SKIN"
XROM "OUT" 87 %
RCL 46 * RCL 42 *
"DPSKIN" XEQ 29 RCL 48
RCL 35 - / 1 + 100
* "%EFF" XROM "OUT"
ADV GTO 15

377♦LBL 17
CLX STO 58 STO 59 1
STO 52 STO 61 1.1
STO 53 60 STO 50 62
STO 51 RCL 62 STO 27
FS?C 04 SF 08 ADV
XEQ 23 CF 08 FS? 22
GTO 01 FC? 23 GTO 15

401♦LBL 01
STO 62 RCL 60 STO 42
FS?C 00 SF 08 XEQ 21
CF 08 STO 60 RCL 62
LOG * STO 56 GTO 19

415♦LBL 18
ADV XEQ 23 FS? 22
GTO 02 FC? 23 GTO 15
```

Program Listing (cont.)

```
422♦LBL 02
XEQ 21  RND  RCL IND 50
RND  X=Y?  GTO 03
RCL 42  STO IND 51
RCL 51  STO 50  1
ST+ 52  +  0  STO IND Y
2  ST+ 51

440♦LBL 03
RCL 27  STO IND 51
RCL 50  1  +  ISG IND X
CLD  60  STO 54  58
STO 55  CLX  STO 56
RCL 52  1 E3  +  LASTX
/  STO 57

460♦LBL 13
TONE 5  RCL IND 54
RCL IND 55  -  RCL 54  1
-  RCL IND X  RCL IND 51
X<>Y  -  LOG  R↑  *
ST+ 56  RCL 54  STO 55
1  +  RCL IND X  2  +
ST+ 54  ISG 57  GTO 13

486♦LBL 19
ISG 51  CLD  RCL IND 51
STO 27  RCL 56
RCL IND 50  /  "Σ"
XEQ 06  TONE 9
XROM "OUT"  ISG 53
GTO 18

500♦LBL 20
41  STO 00  FS?C 00
SF 08  "QO"  FC? 07
"QG"  XEQ 22  CF 08  RTN

511♦LBL 21
41  STO 00  "QO"  FC? 07
"QG"  XEQ 06

518♦LBL 22
ASTO T  "BBL"  FC? 07
"MCF"  "⊦/DAY"  ASTO 01
ASHF  ASTO 02  CLA
ARCL T  RCL 04  RCL 03
RCL Z  XROM "INK"  RDN
STO 03  X<>Y  STO 04  R↑
RTN

539♦LBL 23
26  STO 00  "TIME"
XEQ 06  XEQ 05
XROM "INK"  RDN  STO 43
X<>Y  STO 44  R↑  RTN

552♦LBL 24
XEQ 05  XROM "OUTK"  RDN
STO 43  X<>Y  STO 44  R↑
RTN

561♦LBL 05
ASTO T  "HR"  ASTO 01
CLA  ASTO 02  ARCL T
RCL 44  RCL 43  RCL Z
RTN

572♦LBL 06
STO 05  CLST  FS? 41  1
+  FS? 40  2  +  5  /
FS? 39  1  +  FS? 38  2
+  FS? 37  4  +  FS? 36
8  +  FS? 29  CHS
RCL 53  FIX 0  CF 29
ARCL X  X<>Y  X<0?
SF 29  ENTER↑  FRC  5  *
FIX IND Y  X=0?
SCI IND Y  1  X=Y?
ENG IND Z  RCL 05  RTN

616♦LBL 25
45  STO 00  "b"
XROM "IN"  "a"
XROM "IN"

623♦LBL 26
47  STO 00  "PWF"
XEQ 28  RTN

629♦LBL 27
FS?C 05  SF 08  34
STO 00  "PI"

635♦LBL 28
XEQ 07  XROM "INK"  RDN
STO 08  X<>Y  STO 09  R↑
CF 08  RTN

645♦LBL 29
XEQ 07  XROM "OUTK"  RDN
STO 08  X<>Y  STO 09  R↑
RTN

654♦LBL 07
ASTO T  "PSI"  ASTO 01
CLA  ASTO 02  ARCL T
RCL 09  RCL 08  RCL Z
END
```

Section 7
Waterflooding

23. FWVSW — Fractional Flow Calculations

Calculates dFW/dSW, %SWSZ, and %SWBT from the KO/KW correlation, predicts frontal advance, and provides an estimate of overall reservoir recovery.

24. 5SPOT — Five-Spot Waterflood Performance

Predicts reservoir rate-time performance for a five-spot pattern during all four stages of a waterflood.

25. SUMWF — Combined Performance for Stratified Waterfloods

Predicts performance of different layers of a stratified waterflood, given the performance of the base layer.

26. INJ — Water Injectivity for Regular Patterns with Unit Mobility Ratio

Estimates individual well injectivity or pressure drop for various patterns.

27. CUTCUM — Forecasting Mature Waterfloods

Predicts future performance of fully developed waterfloods and calculates the water-oil relative permeability ratio from production data.

23. FWVSW — Fractional Flow Calculations

FWVSW is a useful utility given the availability of KO/KW data, preferably from field performance. The program CUTCUM provides one method of performing the calculations of KO/KW for mature waterfloods. The KO/KW curve can often be represented in the form:

$$KO/KW = ae^{b\frac{\%SW}{100}}$$

The values of *a* and *b* are calculated easily given two representative points on the KO/KW versus water saturation curve (see Figure 23–1). FWVSW calculates FW as a function of %SW and the remaining calculations of Section E.1 of Craig. Since the programs FWVSW, 5SPOT, and SUMWF all perform the calculations of Appendix E of Craig, refer to these examples. The values of %SWSZ (the water saturation at the upstream end of the stabilized zone) and %SWBT (the average water saturation behind the flood front at and prior to breakthrough) are calculated iteratively and preserved by the program for use in 5SPOT (see Figure 23–2). The graphical method to determine these values is preferred when a portion of the KO/KW curve cannot be represented adequately by the exponential fit of *a* and *b*.

The program also provides a method of estimating ultimate waterflood recovery (Sections E.2 and E.3). This method is not recommended except as a quick estimate when only limited core data are available.

Equations

KO/KW equations:

$$KO/KW = ae^{b\frac{\%SW}{100}}$$

a and *b* are calculated from two %SW, KO/KW points as follows:

$$a = \frac{KO/KW}{e^{b\frac{2ND\%SW}{100}}}$$

$$b = 100\,\frac{\ln\frac{KO/KW2}{KO/KW1}}{2ND\%SW - 1ST\%SW}$$

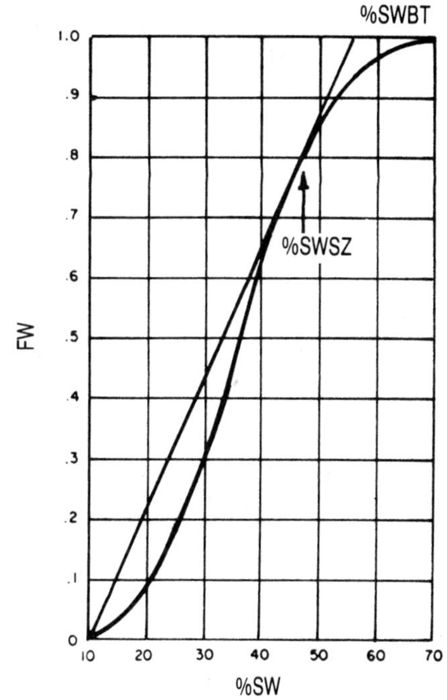

Figure 23–1 *(after Craig)*

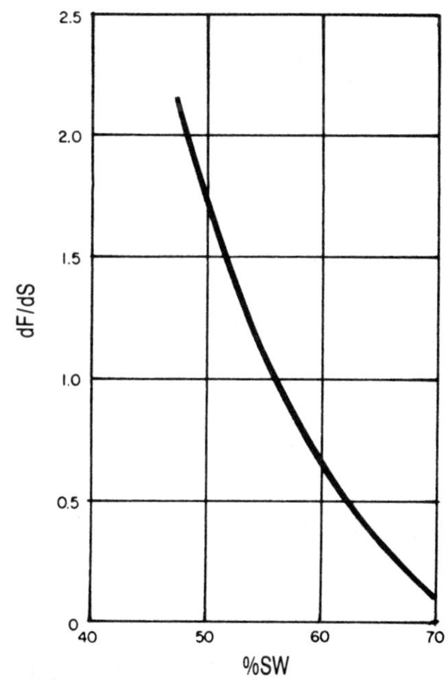

Figure 23–2 *(after Craig)*

Fractional flow equations:

$$FW = \frac{1}{1 + \frac{UW}{UO}\frac{KO}{KW}}$$

$$\frac{dFW}{dSW} \text{ (slope of fractional flow curve) } = b(FW^2 - FW)$$

To solve for %SWSZ:

$$f(\%SWSZ) = \frac{b(FW^2 - FW)}{(\%SWSZ - \%SWC)} - \frac{100\ FW}{(\%SWSZ - \%SWC)^2}$$

FWVSW attempts to minimize f(%SWSZ) using the regula falsi method. The new %SWSZ is predicted as:

$$\%SWSZ_{new} = \frac{\%SWSZ_{left}\ f(\%SWSZ_{right}) - \%SWSZ_{right}\ f(\%SWSZ_{left})}{f(\%SWSZ_{right}) - f(\%SWSZ_{left})}$$

$$\%SWBT = \frac{\%SWSZ - \%SWC}{FW} + \%SWC$$

$$FW2 = \frac{1}{1 + \frac{UW}{UO} ae^{b\frac{\%SW2}{100}}}$$

$$QI = \frac{1}{dFW/dSW}$$

$$\%SWAVG = \%SW2 + 100\ QI(1 - FW2)$$

Waterflood recovery equations:

$$M = \frac{UO}{UW}\ \frac{1}{KO/KW}$$

$$\%EV = 100\frac{1 - V^2}{M}$$

$$\%RF = 100 - \frac{\left[\frac{\%EV\ \%SOAVG}{BOWF} + \frac{(100 - \%EV)\ \%SOI}{BOWF}\right]}{\frac{\%SOI}{BOI}}$$

$$\%SOI = 100 - \%SWC$$

$$\%SOAVG = 100 - \%SWAVG$$

$$\%RFWF = \%RF - \%OILRF$$

Frontal advance equations:

$$X = \frac{IW'\ T}{POR\ H\ WIDTH}\ \frac{dFW}{dSW}$$

$$\frac{dFW}{dSW} \text{ evaluated at } \%SWSZ$$

$$IW' = IW \text{ in FT3/DAY}$$

$$POR = \frac{\%POR}{100}$$

Nomenclature

Symbol	*Variable Name*	*Input or Output*	*English Units*	*SI Units*
a	Intercept of straight-line portion of KO/KW curve	I,O	–	–
BOI	Initial oil formation volume factor	I	–	–
BOWF	Oil formation volume factor at the beginning of waterflooding	I	–	–
b	Slope of straight-line portion of KO/KW curve	I,O	–	–
FW	Fraction of total flowing stream composed of water	O	–	–
FW MAX	Maximum fraction of total flowing stream composed of water	I	–	–
FW2	Exit-end fraction of total flowing stream composed of water	O	–	–
H	Formation thickness*	I	FT	M
IW	Water injection rate	I	BBL/ DAY	M3/ DAY
KO/KW	Water-oil relative permeability ratio	I	–	–
KO/KW1	Water-oil relative permeability ratio at 1ST%SW	I	–	–

Symbol	Variable Name	Input or Output	English Units	SI Units
KO/KW2	Water-oil relative permeability ratio at 2ND%SW	I	–	–
M	Water-oil mobility ratio	O	–	–
QI	Pore volume of cumulative injected water	O	–	–
T†	Time	I	DAY	DAY
V	Permeability variation	I	–	–
WIDTH	Front width	I	FT	M
X	Frontal advance	O	FT	M
%EV	Percent volumetric sweep efficiency	O	–	–
%OILRF	Volume percent primary oil recovery	I	–	–
%RF	Volume percent ultimate oil recovery	O	–	–
%RFWF	Volume percent ultimate oil recovery after waterflooding	O	–	–
%SOc	Volume percent critical oil saturation	I	–	–
%SOI	Volume percent initial oil saturation	I	–	–
%SWAVG	Volume percent average water saturation	O	–	–
%SWBT	Volume percent average water saturation behind the flood front at and prior to breakthrough	O	–	–
%SWC	Volume percent connate water saturation	I	–	–
%SW-MAX	Upper limit of saturation range in which %SWSZ will be calculated	I	–	–
%SWMIN	Lower limit of saturation range in which %SWSZ will be calculated	I	–	–
%SWSZ	Volume percent water saturation at the upstream end of the stabilized zone	O	–	–
%SW2	Volume percent exit-end water saturation	O	–	–

Symbol	Variable Name	Input or Output	English Units	SI Units
1ST-%SW	First volume percent water saturation value used for calculating *a* and *b*	I	–	–
2ND-%SW	Second volume percent water saturation value used for calculating *a* and *b*	I	–	–

*In the case of deviated wells or slanted beds, use the true vertical thickness instead of the measured thickness of the formation.

†The units for these variables are saved by the program.

Yes/No Questions

FIT? Yes: Calculate *a* and *b* from two %SW KO/KW points.
No: Input a and b.

Example 1

This data is from Craig's example in Appendix E. Table 23-1 is KO/KW and %SW data for this example. Fit a KO/KW curve through the values at %SW = 40 and %SW = 65 to estimate *a* and *b*. Calculate %SWSZ and %SWBT and generate a table equivalent to Craig's Table E.3. Also, estimate ultimate waterflood recovery based on a permeability variation (V) of 0.5 and economic limiting value of FW MAX = 0.98. The initial oil saturation was 90%, the critical oil saturation is 25%, BOI was 1.29, the current value of BO is 1.2, and the primary recovery was 10.4%. Note that the calculated table differs from Craig's table due to the use of an exponential equation for KO/KW.

Table 23–1

%SW	KRO	KRW	KO/KW (= KRO/KRW)
10	1.000	0.000	∞
30	0.373	0.070	5.3286
40	0.210	0.169	1.2426
45	0.148	0.226	0.6549
50	0.100	0.300	0.3333
55	0.061	0.376	0.1622
60	0.033	0.476	0.0693
65	0.012	0.600	0.0200
70	0.000	0.740	0.0000

Keystrokes (SIZE>=051)	Display	Comments
[XEQ] [ALPHA] FWVSW [ALPHA]	*FIT? Y/N:*	Last character is Y or N
Y [R/S]	*1ST%SW=?*	
40 [R/S]	*KO/KW1=?*	
1.2426 [R/S]	*2ND%SW=?*	
65 [R/S]	*KO/KW2=?*	
.02 [R/S]	*%SOc=?*	a and b printed
25 [R/S]	*%SWC=?*	%SWC = 100 − %SOI
10 [R/S]	*UO=?*	
1 [R/S]	*UW=?*	
.5 [R/S]	*%SWMIN=?*	
35 [R/S]	*%SWMAX=?*	
70 [R/S]	*RCVRY ADV*	%SWSZ, FW, and %SWBT printed, followed by table of %SW2, FW2, QI, and %SWAVG
[A]	*FW MAX=?*	Waterflood recovery option
.98 [R/S]	*KO/KW=?*	%SW2 and %SWAVG printed
2.5 [R/S]	*V=?*	
.5 [R/S]	*BOI=?*	
1.29 [R/S]	*BOWF=?*	
1.2 [R/S]	*%OILRF=?*	
10.4 [R/S]	*RCVRY ADV*	M, %EV, %RF, and %RFWF printed

```
FW VS %SW

FIT: YES
1ST%SW=40.0000
KO/KW1=1.2426
2ND%SW=65.0000
KO/KW2=0.0200
a=919.6308
b=-16.5169

%SOc=25.0000
%SWC=10.0000
UO=1.0000 CP
UW=0.5000 CP
%SWMIN=35.0000
%SWMAX=70.0000

%SWSZ=46.9981
FW=0.8364
%SWBT=54.2340

%SW2=49.4981
FW2=0.8854
QI=0.5968
%SWAVG=56.3360

%SW2=51.9981
FW2=0.9211
QI=0.8333
%SWAVG=58.5709

%SW2=54.4981
FW2=0.9464
QI=1.1930
%SWAVG=60.8956

%SW2=56.9981
FW2=0.9639
QI=1.7380
%SWAVG=63.2795

%SW2=59.4981
FW2=0.9758
QI=2.5627
%SWAVG=65.7027

%SW2=61.9981
FW2=0.9838
QI=3.8096
%SWAVG=68.1519

%SW2=64.4981
FW2=0.9893
QI=5.6945
%SWAVG=70.6183

%SW2=66.9981
FW2=0.9929
QI=8.5432
%SWAVG=73.0960

%SW2=69.4981
FW2=0.9953
QI=12.8484
%SWAVG=75.5813

%SW2=71.9981
FW2=0.9969
QI=19.3547
%SWAVG=78.0715

%SW2=74.4981
FW2=0.9979
QI=29.1874
%SWAVG=80.5651

%SW2=75.0000
FW2=0.9981
QI=31.6997
%SWAVG=81.0660

FW MAX=0.9800
%SW2=60.6811
%SWAVG=66.8591
KO/KW=2.5000
V=0.5000
BOI=1.2900
BOWF=1.2000
%OILRF=10.4000

M=0.8000
%EV=93.7500
%RF=56.1703
%RFWF=45.7703
```

Example 2

Repeat Example 1 but use a = 460 and b = −21.5 to determine the %SW — FW table. Notice the effect that the KO/KW coefficients have on the values in the table.

Keystrokes	Display	Comments
[XEQ] [ALPHA] FWVSW [ALPHA]	*FIT? Y/N:Y*	
N [R/S]	*a=?*	
460 [R/S]	*b=?*	
21.5 [CHS] [R/S]	*%SOc=?*	
[R/S]	*%SWC=?*	
[R/S]	*UO=?*	
[R/S]	*UW=?*	
[R/S]	*%SWMIN=?*	
40 [R/S]	*%SWMAX=?*	
70 [R/S]	*NO ROOT*	Root not in specified range
[R/S]	*%SWMIN=?*	
30 [R/S]	*%SWMAX=?*	
90 [R/S]	*RCVRY ADV*	%SWSZ, FW, and %SWBT printed, followed by table of %SW2, QI, and %SWAVG
[A]	*FW MAX=?*	%SW2 and %SWAVG printed
.98 [R/S]	*KO/KW=?*	
[R/S]	*V=?*	
[R/S]	*BOI=?*	
[R/S]	*BOWF=?*	
[R/S]	*%OILRF=?*	
[R/S]	*RCVRY ADV*	M, %EV, %RF, and %RFWF printed

```
FW VS %SW

FIT: NO
a=460.0000
b=-21.5000

%SWMIN=40.0000
%SWMAX=70.0000
%SWMIN=30.0000
%SWMAX=90.0000

%SWSZ=31.1973
FW=0.7806
%SWBT=37.1541

%SW2=33.6973
FW2=0.8590
QI=0.3840
%SWAVG=39.1121

%SW2=36.1973
FW2=0.9125
QI=0.5824
%SWAVG=41.2946

%SW2=38.6973
FW2=0.9469
QI=0.9257
%SWAVG=43.6091

%SW2=41.1973
FW2=0.9683
QI=1.5155
%SWAVG=46.0008

%SW2=43.6973
FW2=0.9812
QI=2.5261
%SWAVG=48.4375

%SW2=46.1973
FW2=0.9890
QI=4.2568
%SWAVG=50.9005

%SW2=48.6973
FW2=0.9935
QI=7.2197
%SWAVG=53.3789

%SW2=51.1973
FW2=0.9962
QI=12.2915
%SWAVG=55.8662

%SW2=53.6973
FW2=0.9978
QI=20.9732
%SWAVG=58.3589

%SW2=56.1973
FW2=0.9987
QI=35.8340
%SWAVG=60.8546

%SW2=58.6973
FW2=0.9992
QI=61.2716
%SWAVG=63.3520

%SW2=61.1973
FW2=0.9996
QI=104.8138
%SWAVG=65.8506

%SW2=63.6973
FW2=0.9997
QI=179.3459
%SWAVG=68.3497

%SW2=66.1973
FW2=0.9998
QI=306.9244
%SWAVG=70.8492

%SW2=68.6973
FW2=0.9999
QI=525.3034
%SWAVG=73.3489

%SW2=71.1973
FW2=0.9999
QI=899.1125
%SWAVG=75.8488

%SW2=73.6973
FW2=1.0000
QI=1,538.9379
%SWAVG=78.3486

%SW2=75.0000
FW2=1.0000
QI=2,036.3219
%SWAVG=79.6513

FW MAX=0.9800
%SW2=43.3949
%SWAVG=48.1410

M=0.8000
%EV=93.7500
%RF=35.2099
%RFWF=24.8099
```

Example 3

From the end of Example 2, calculate displacement of the flood front at T = 1, 10, 20, 30, 60, and 180 days. The other data required are:

%POR = 20
H = 5 FT
IW = 100 BBL/DAY
WIDTH = 333 FT

Keystrokes	Display	Comments
[E]	*%POR=?*	Frontal advance option
20 [R/S]	*H=?*	
5 [R/S]	*WIDTH=?*	
333 [R/S]	*IW=?*	
100 [R/S]	*T=?*	
1 [R/S]	*T=?*	X printed
10 [R/S]	*T=?*	X printed
20 [R/S]	*T=?*	X printed
30 [R/S]	*T=?*	X printed
60 [R/S]	*T=?*	X printed
180 [R/S]	*T=?*	X printed

```
%POR=20.0000
H=5.0000 FT
WIDTH=333.0000 FT
IW=100.0000 BBL/DAY

T=1.0000 DAY
X=6.2077 FT

T=10.0000 DAY
X=62.0772 FT

T=20.0000 DAY
X=124.1544 FT

T=30.0000 DAY
X=186.2316 FT

T=60.0000 DAY
X=372.4632 FT

T=180.0000 DAY
X=1,117.3896 FT
```

General Information

Memory Requirements
Program length: 892 bytes (4 cards)
Minimum size: 051
Minimum hardware: 41C + 2 memory modules

Hidden Options
None

Pac Subroutines Called
TITLE, Y/N?, IN, OUT, INU, %POR, CON, INK, OUTU

Registers

03	Time units
04	Time units
06	UO/UW
07	WIDTH (FT)
08	IW (BBL/DAY)
09	T (DAY)
20	%SOc
21	%SWC
26	BOWF
27	BOI
28	H (FT)
29	%OILRF
30	%SW2
32	UO (CP)
33	UW (CP)
35	KO/KW
36	%SWMIN
37	%SWMAX
38	%SWSZ
39	%SWBT
40	%SWAVG
41	1ST%SW
42	KO/KW1
43	2ND%SW
44	KO/KW2
45	FW2, FW MAX
46	FW at %SWSZ
47	a
48	b
49	Scratch
50	V

Registers 10–17, 19, 22–25, 31, and 34 unused

Flags

04 Set: Calculate *a* and *b* from two %SW, KO/KW points.
Clear: Input *a* and *b*.

User Instructions

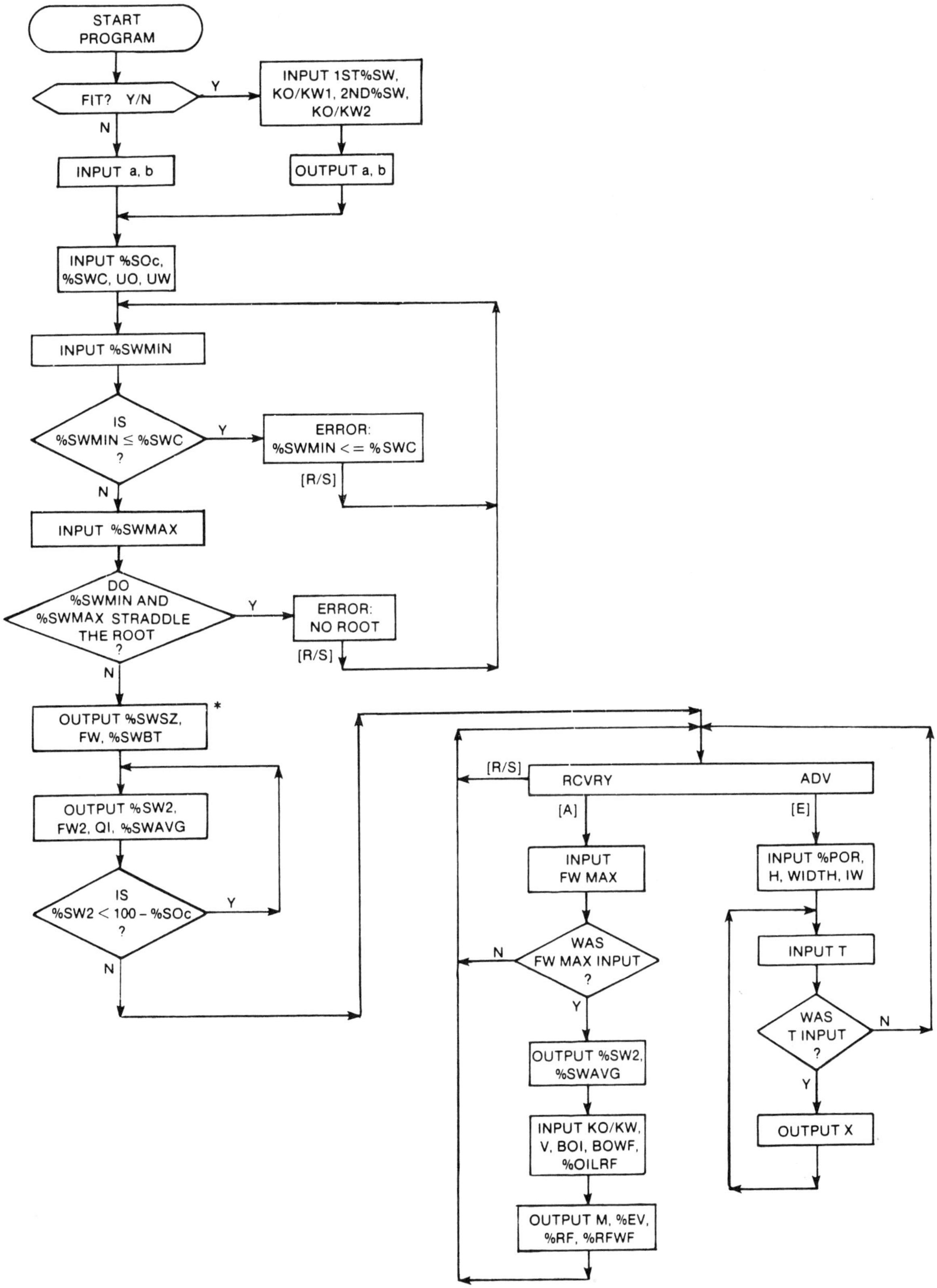

* Tones will sound while %SWSZ is calculated iteratively.

Program Listing

```
 01♦LBL "FWVSW"
"FW VS %SW" 51
XROM "TITLE" FC?C 25
PROMPT SF 27 "DAY"
ASTO 03 CLA ASTO 04
"FIT" 4 XROM "Y/N?"
FS? 04 GTO 00 46
STO 00 "a" XROM "IN"
"b" XROM "IN" GTO 01

 24♦LBL 00
40 STO 00 "1ST%SW"
XROM "IN" "KO/KW1"
XROM "IN" "2ND%SW"
XROM "IN" "KO/KW2"
XROM "IN" RCL 42 / LN
RCL 43 RCL 41 - /
100 X<>Y * STO 48
LASTX RCL 43 * E↑X
RCL 44 X<>Y / STO 47
"a" XROM "OUT" RCL 48
"b" XROM "OUT"

 59♦LBL 01
ADV 19 STO 00 "%SOc"
XROM "IN" "%SWC"
XROM "IN" 31 STO 00
"CP" ASTO 01 CLA
ASTO 02 ASTO Z "PA*S"
ASTO Y "UO" XROM "INU"
STO 06 "PA*S" ASTO Y
CLA ASTO Z "UW"
XROM "INU" ST/ 06
XEQ 12 ADV RCL 38
STO 30 "%SWSZ"
XROM "OUT" RCL 46 "FW"
XROM "OUT" RCL 39
"%SWBT" XROM "OUT" 100
RCL 20 - STO 49
GTO 02

103♦LBL 14
RCL 30 "%SW2"
XROM "OUT" XEQ 08
STO 45 "FW2"
XROM "OUT" X↑2 LASTX
- RCL 48 * 1/X "QI"
XROM "OUT" RCL 30
XEQ 09

121♦LBL 02
ADV RCL 49 RND RCL 30
2.5 + RND X<Y?
GTO 03 X<>Y RCL 30
RND X=Y? GTO 15
RCL 49 RND

138♦LBL 03
LASTX STO 30 GTO 14

142♦LBL 15
"RCVRY   ADV" PROMPT
GTO 15

146♦LBL A
147♦LBL B
44 STO 00 "FW MAX"
XROM "IN" FC? 22
GTO 15 1/X 1 -
RCL 06 * RCL 47 / LN
RCL 48 / 100 *
"%SW2" XROM "OUT"
XEQ 09 34 STO 00
"KO/KW" XROM "IN" 49
STO 00 "V" XROM "IN"
26 STO 00 "BOI"
XROM "IN" 25 STO 00
"BOWF" XROM "IN" 28
STO 00 "%OILRF"
XROM "IN" ADV RCL 06
RCL 35 / "M"
XROM "OUT" 1 RCL 50
X↑2 - X<>Y / 100 *
"%EV" XROM "OUT"
STO 49 100 STO 00 -
RCL 00 RCL 40 -
RCL 00 RCL 21 - /
RCL 49 * - RCL 27 *
RCL 26 / RCL 00 +
"%RF" XROM "OUT"
RCL 29 - "%RFWF"
XROM "OUT" ADV GTO 15

233♦LBL 09
1 RCL 45 - LASTX X↑2
LASTX - RCL 48 * /
100 * + STO 40
"%SWAVG" XROM "OUT"
RTN

251♦LBL D
252♦LBL E
XROM "%POR" 27 STO 00
"FT" ASTO 01 CLA
ASTO 02 ASTO Z "M"
ASTO Y "H" XROM "INU"
6 STO 00 "M" ASTO Y
CLA ASTO Z "WIDTH"
XROM "INU" "BBL/DAY"
ASTO 01 ASHF ASTO 02
"M3/DAY" ASTO Y CLA
ASTO Z "IW" XROM "INU"
RCL 46 X↑2 LASTX -
RCL 48 * * RCL 18
100 / / RCL 28 /
RCL 07 / "BBL-FT3"
CON STO 49

301♦LBL 13
ADV "DAY" ASTO 01 CLA
ASTO 02 RCL 04 RCL 03
8 STO 00 "T"
XROM "INK" RDN STO 03
X<>Y STO 04 R↑ CF 08
FS? 22 GTO 04 FC? 23
GTO 15

323♦LBL 04
RCL 49 * "FT" ASTO 01
CLA ASTO 02 ASTO Z
"M" ASTO Y "X"
XROM "OUTU" GTO 13

336♦LBL 12
35 STO 00 "%SWMIN"
XROM "IN" CLA ARCL 05
"⊢<=%SWC" RCL 21 X<>Y
X<=Y? GTO 05 "%SWMAX"
XROM "IN" STO 02
XEQ 07 STO 05 RCL 36
STO 30 XEQ 07 STO 01
RCL 05 * X<=0? GTO 11
"NO ROOT"

362♦LBL 05
TONE 3 PROMPT GTO 12

366♦LBL 11
TONE 5 RCL 30 RCL 05
* RCL 02 RCL 01 * -
RCL 05 RCL 01 - /
STO 38 XEQ 07 STO 39
ABS 1 E-4 X<=Y?
GTO 05 RCL 38 RCL 21
- RCL 46 / RCL 21 +
STO 39 TONE 9 RTN

396♦LBL 05
RCL 05 RCL 39 * X>0?
GTO 06 RCL 39 STO 01
RCL 38 STO 30 GTO 11

407♦LBL 06
RCL 39 STO 05 RCL 38
STO 02 GTO 11

413♦LBL 07
STO 38 XEQ 08 STO 46
X↑2 LASTX - RCL 48 *
RCL 46 RCL 38 RCL 21
- ST/ Z X↑2 / 100 *
- RTN

433♦LBL 08
1 % RCL 48 * E↑X
RCL 47 * RCL 06 / 1
+ 1/X END
```

24. 5SPOT — Five-Spot Waterflood Performance

5SPOT performs most of the calculations required for Section E.5 of Craig. Basically, the Craig, Geffen, and Morse method is used to relate oil recovery and WOR to cumulative injected water, and the correlation of Caudle and Witte is used for calculating five-spot water injection rates. Five-spot flood performances approximate many other patterns.

Reservoir heterogeneities should be reflected by using stratified reservoir predictions. Craig's Table 7.1 indicates the minimum number of layers required. 5SPOT performs the calculations for one layer, and SUMWF adjusts the other performances appropriately.

These calculations are divided into the four stages Craig mentions for the waterflood:

Stage	Description
1	Prior to interference
2	From interference to fillup
3	From fillup to breakthrough
4	After water breakthrough

Typically, stage four is most of the time of a flood; however, the timing of stages 1–3 have a clear and certain effect on the economics of a flood. This technique usually is used to predict the performance of proposed floods. The program CUTCUM can be used for waterfloods later in stage four.

5SPOT uses correlations to replace Craig's Figure E.5 (GAMMA vs M and %EA), E.6 (%EABT vs M), and E.7 (C vs M) and Slider's correlation for KRW vs %SW (typical values for the coefficients *c* and *d* are given in Table 24–1). The use of these correlations greatly simplifies calculations, since the only value the user must look up at each point in stage four is QI/QIB (Craig's Table E.9, reproduced here in Table

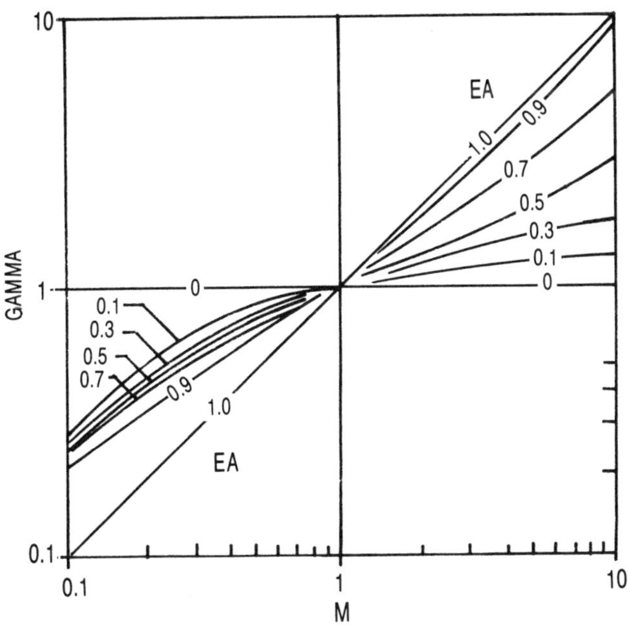

Figure 24–1 *(after Craig)*

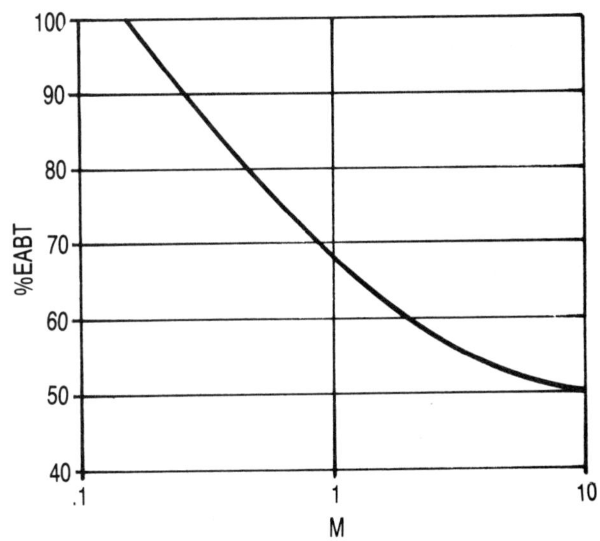

Figure 24–2 *(after Craig)*

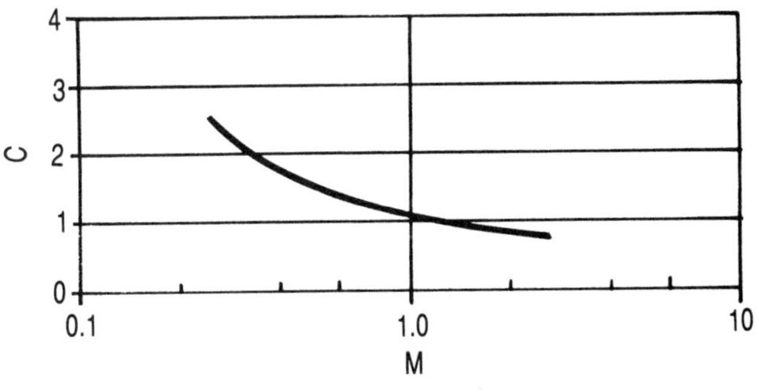

Figure 24–3 *(after Craig)*

24-2). If desired, the user can optionally select to input the values that the correlations would calculate. This will duplicate Craig's calculations more closely.

Another correlation is used as well: Ershaghi and Omoregie's exponential KO/KW correlation (see FWVSW and CUTCUM). This is used to calculate %SW2 from FW2, as in FWVSW. If KO/KW data over the range %SWBT to $\%SW_{el}$ (economic limit) is well approximated by this KO/KW correlation, use it. FWVSW can be used to determine the *a* and *b* coefficients of the correlation and how well it models the data. If data is not well represented by this exponential, select the option to input SW2 and FW2. (You will have to plot graphs equivalent to Figure 23-1, FW vs %SW, and Figure 23-2, dFW/dSW vs %SW).

Table 24-1 (after Slider)

Formation	*c*	*d*
Unconsolidated sandstones (well sorted)	3	0
Unconsolidated sandstones (poorly sorted)	2	1.5
Cemented sandstones, oolitic limestones, and vugular limestones	2	2

Table 24-2
QI/QIB For Various Values of %EABT

WI/WIB	*%EABT* 50.	51.	52.	53.	54.	55.	56.	57.	58.	59.
1.0	1.000	1.000	1.000	1.000	1.000	1.000	1.000	1.000	1.000	1.000
1.2	1.190	1.191	1.191	1.191	1.191	1.191	1.191	1.191	1.192	1.192
1.4	1.365	1.366	1.366	1.367	1.368	1.368	1.369	1.369	1.370	1.370
1.6	1.529	1.530	1.531	1.532	1.533	1.535	1.536	1.536	1.537	1.538
1.8	1.684	1.686	1.688	1.689	1.691	1.693	1.694	1.696	1.697	1.699
2.0	1.832	1.834	1.837	1.839	1.842	1.844	1.846	1.849	1.851	1.853
2.2	1.974	1.977	1.981	1.984	1.987	1.990	1.993	1.996	1.999	2.001
2.4	2.111	2.115	2.119	2.124	2.127	2.131	2.135	2.139	2.142	2.146
2.6	2.244	2.249	2.254	2.259	2.264	2.268	2.273	2.277	2.282	2.286
2.8	2.373	2.379	2.385	2.391	2.397	2.402	2.407	2.413	2.418	2.422
3.0	2.500	2.507	2.513	2.520	2.526	2.533	2.539	2.545	2.551	2.556
3.2	2.623	2.631	2.639	2.646	2.653	2.660	2.667	2.674	2.681	2.687
3.4	2.744	2.752	2.761	2.770	2.778	2.786	2.793	2.801	2.808	2.816
3.6	2.862	2.872	2.881	2.891	2.900	2.909	2.917	2.926	2.934	2.942
3.8	2.978	2.989	3.000	3.010	3.020	3.030	3.039	3.048	3.057	3.066
4.0	3.093	3.105	3.116	3.127	3.138	3.149	3.159	3.169	3.179	3.189
4.2	3.205	3.218	3.231	3.243	3.254	3.266	3.277	3.288	3.299	3.309
4.4	3.316	3.330	3.343	3.357	3.369	3.382	3.394	3.406	3.417	3.428
4.6	3.426	3.441	3.455	3.469	3.483	3.496	3.509	3.521	3.534	3.546
4.8	3.534	3.550	3.565	3.580	3.594	3.609	3.622	3.636	3.649	
5.0	3.641	3.657	3.674	3.689	3.705	3.720	3.735			
5.2	3.746	3.764	3.781	3.798	3.814	3.830				
5.4	3.851	3.869	3.887	3.905	3.922					
5.6	3.954	3.973	3.993	4.011						
5.8	4.056	4.077	4.097							
6.0	4.157	4.179								
6.2	4.257									
Values of WI/WIB at which %EA = 100	6.164	5.944	5.732	5.527	5.330	5.139	4.956	4.779	4.608	4.443

Table 24-2 (cont.)

WI/WIB	*%EABT* 60.	61.	62.	63.	64.	65.	66.	67.	68.	69.
1.0	1.000	1.000	1.000	1.000	1.000	1.000	1.000	1.000	1.000	1.000
1.2	1.192	1.192	1.192	1.192	1.192	1.192	1.193	1.193	1.193	1.193
1.4	1.371	1.371	1.371	1.372	1.372	1.373	1.373	1.373	1.374	1.374
1.6	1.539	1.540	1.541	1.542	1.543	1.543	1.544	1.545	1.546	1.546
1.8	1.700	1.702	1.703	1.704	1.706	1.707	1.708	1.709	1.710	1.711
2.0	1.855	1.857	1.859	1.861	1.862	1.864	1.866	1.868	1.869	1.871
2.2	2.004	2.007	2.009	2.012	2.014	2.016	2.019	2.021	2.023	2.025
2.4	2.149	2.152	2.155	2.158	2.161	2.164	2.167	2.170	2.173	2.175
2.6	2.290	2.294	2.298	2.301	2.305	2.308	2.312	2.315	2.319	2.322
2.8	2.427	2.432	2.436	2.441	2.445	2.449	2.453	2.457	2.461	2.465
3.0	2.562	2.567	2.572	2.577	2.582	2.587	2.592	2.597	2.601	2.606
3.2	2.693	2.700	2.705	2.711	2.717	2.723	2.728	2.733	2.738	2.744
3.4	2.823	2.830	2.836	2.843	2.849	2.855	2.862	2.867	2.873	
3.6	2.950	2.957	2.965	2.972	2.979	2.986	2.993			
3.8	3.075	3.083	3.091	3.099	3.107					
4.0	3.198	3.207	3.216	3.225						
4.2	3.319	3.329								
4.4	3.439									
Values of WI/WIB at which %EA = 100	4.285	4.132	3.984	3.842	3.704	3.572	3.444	3.321	3.203	3.088

WI/WIB	*%EABT* 70.	71.	72.	73.	74.	75.	76.	77.	78.	79.
1.0	1.000	1.000	1.000	1.000	1.000	1.000	1.000	1.000	1.000	1.000
1.2	1.193	1.193	1.193	1.193	1.193	1.193	1.193	1.194	1.194	1.194
1.4	1.374	1.375	1.375	1.375	1.376	1.376	1.376	1.377	1.377	1.377
1.6	1.547	1.548	1.548	1.549	1.550	1.550	1.551	1.551	1.552	1.552
1.8	1.713	1.714	1.715	1.716	1.717	1.718	1.719	1.720	1.720	1.721
2.0	1.872	1.874	1.875	1.877	1.878	1.880	1.881	1.882	1.884	1.885
2.2	2.027	2.029	2.031	2.033	2.035	2.037	2.039	2.040	2.042	2.044
2.4	2.178	2.180	2.183	2.185	2.188	2.190	2.192	2.195	2.197	
2.6	2.325	2.328	2.331	2.334	2.337	2.340				
2.8	2.469	2.473	2.476	2.480						
3.0	2.610	2.614								
Values of WI/WIB at which %EA = 100	2.978	2.872	2.769	2.670	2.575	2.483	2.394	2.309	2.226	2.147

WI/WIB	*%EABT* 80.	81.	82.	83.	84.	85.	86.	87.	88.	89.
1.0	1.000	1.000	1.000	1.000	1.000	1.000	1.000	1.000	1.000	1.000
1.2	1.194	1.194	1.194	1.194	1.194	1.194	1.194	1.194	1.194	1.194
1.4	1.377	1.378	1.378	1.378	1.378	1.379	1.379	1.379	1.379	1.379
1.6	1.553	1.553	1.554	1.555	1.555	1.555	1.556	1.556	1.557	1.557
1.8	1.722	1.723	1.724	1.725	1.725	1.726	1.727	1.728		
2.0	1.886	1.887	1.888	1.890						
2.2	2.045									
Values of WI/WIB at which %EA = 100	2.070	1.996	1.925	1.856	1.790	1.726	1.664	1.605	1.547	1.492

WI/WIB	*%EABT* 90.	91.	92.	93.	94.	95.	96.	97.	98.	99.
1.0	1.000	1.000	1.000	1.000	1.000	1.000	1.000	1.000	1.000	1.000
1.2	1.194	1.195	1.195	1.195	1.195	1.195	1.195	1.195	1.195	1.195
1.4	1.380	1.380	1.380	1.380	1.381					
1.6	1.558									
Values of WI/WIB at which %EA = 100	1.439	1.387	1.338	1.290	1.244	1.199	1.157	1.115	1.075	1.037

Equations

$$KRW = c \left(\frac{\%SWAVG - \%SWC}{100 - \%SWC} \right)^{d}$$

$$M = \frac{KRW}{UW} \frac{UO}{KRO}$$

$$C = 0.62 + \frac{0.46836}{M}^{*}$$

$$\%SGI* = C(\%SOI - \%SOBT)$$

$$\%SOI = 100 - \%SWC - \%SGI$$

$$\%SOBT = 100 - \%SWBT$$

$$VP \text{ (pore volume)} = 7758 \text{ POR AREA H}$$

$$OIPWF = \frac{VP\ \%SOI}{100\ BOI}$$

$$IBASE = \frac{0.001538\ H\ K\ KRO\ DP}{UO \left(\log \frac{D}{RW} - 0.2688 \right)}$$

(D is distance from injector to producer, $\sqrt{2}$ REI)

Prior to interference:

$$WII' = \pi \text{ H POR SGI REI}^2$$

$$R = \left(\frac{WI'}{\pi \text{ H POR (SWBT} - \text{SWC)}} \right)^{1/2}$$

$$IW = \frac{7.07\ (10^{-3})\ H\ K\ DP}{\frac{UW}{KRW} \ln \frac{R}{RW} + \frac{UO}{KRO} \ln \left(\frac{\%SWBT - \%SWC}{\%SGI} \right)^{1/2}}$$

T is calculated at the ith WI as follows:

$$T_i = T_{i-1} + \frac{WI_i - WI_{i-1}}{(IW_i + IW_{i-1})/2}, \quad T_0 = 0,$$

$$WI_0 = \text{BEG WI}, \quad IW_0 = IW_1$$

WII' = WII in FT3

WI' = WI in FT3

*Unpublished correlation, D.N. Meehan, Champlin Petroleum Company.

Interference to fillup:

$$WIF = VP\ SGI$$

$$\%EAF = \frac{100\ \%SGI}{\%SWBT - \%SWC}$$

If $M \geq 1$, $GAMMA = M^{\%EA/100^{*}}$

If $M < 1$, $GAMMA = D + EM + F \log M$

$$D = 1.412546 - 0.9881\ EA$$

$$E = -0.40586 + 0.985715\ EA$$

$$F = 1.094353 - 0.812527\ EA$$

If %EA = 0, GAMMA = 1

If %EA = 1, GAMMA = M

GAMMAF = GAMMA at %EA = %EAF

IWF = GAMMAF IBASE

Fillup to breakthrough:

$$\%EABT = 70.4\ M^{-0.1736^{*}}$$

$$WIBT = VP\ EABT\ (SWBT - SWC)$$

$$\%EA = \frac{100\ WI}{VP\ \%SWBT - \%SWC}$$

IW = GAMMA IBASE

QO = IW/BOWF

$$NP = \frac{WI - WIF}{BOWF}$$

$$\%OILRF = 100 \frac{NP}{OIPWF}$$

After water breakthrough:

WI = WI/WIB WIBT

$$\%EA = 27.49 \ln \frac{WI}{WIBT} + \%EABT$$

If $\%EA \geq 100$, $QI/QIB_i =$

$$QI/QIB_{i-1} + \frac{WI\ INC}{WIBT} \frac{\%EABT}{100}$$

If %EA < 100, QI/QIB is read or interpolated from Table 24-2 at WI/WIB and %EABT

QIBT = (%SWBT − %SWC)/100

QI = QI/QIB QIBT

$$dF/dS = 1/QI = b(FW2^2 - FW2)$$

$$FW2 = \frac{1 + \sqrt{1 + \frac{4}{b}\frac{dF}{dS}}}{2}$$

$$FW2 = \frac{1}{1 + \frac{UO}{UW}\frac{KO}{KW}}, \quad \frac{KO}{KW} = a\,e^{b\frac{\%SW2}{100}}$$

$$\%SW2 = \frac{100}{b}\ln\left[\left(\frac{1}{FW2} - 1\right)\frac{UW}{UO}\frac{1}{a}\right]$$

$$\%SWAVG = \%SW2 + 100\,QI\,(1 - FW2)$$

$$WOR = BOWF\left[\frac{1}{FO2 + (1 - FO2)A} - 1\right]$$

$$FO2 = 1 - FW2$$

If $\%EA < 100$, A =

$$27.49 \cdot \frac{1}{WI/WIB}\,\frac{\%SWSZ - \%SWC}{\%EABT(\%SWBT - \%SWC)}$$

If $\%EA \geq 100$, A = 0

$$NP = \left[\frac{(\%SWAVG - \%SWC)EA - \%SGI}{100 - \%SWC - \%SGI}\right] OIPWF$$

$$QO = \left[\frac{FO2 + (1 - FO2)A}{BOWF}\right] IW$$

$$POR = \frac{\%POR}{100}$$

$$SGI = \frac{\%SGI}{100}$$

$$SWC = \frac{\%SWC}{100}$$

$$EA = \frac{\%EA}{100}$$

$$SWBT = \frac{\%SWBT}{100}$$

Nomenclature

Symbol	*Variable Name*	*Input or Output*	*English Units*	*SI Units*
AREA	Five-spot pattern area	I	ACRE	M2
a	Intercept of straight-line portion of KO/KW curve	I	–	–
BEG WI*	Beginning cumulative injected water volume of table	I	BBL	M3
BEG-W/W	Beginning WI/WIB of table	I	–	–
BOWF	Oil formation volume factor at the beginning of waterflooding	I	–	–
b	Slope of straight-line portion of KO/KW curve	I	–	–
C	Coefficient used to determine whether prediction method is valid	I,O	–	–
c	First coefficient of KRW correlation	I	–	–
DP	Pressure difference between injection and producing well	I	PSI	KPA
d	Second coefficient of KRW correlation	I	–	–
dF/dS	Slope of fractional flow curve	O	–	–
END WI*	Ending cumulative injected water volume of table	O	BBL	M3
END-W/W	Ending WI/WIB of table	I	–	–
FW2	Exit-end fraction of total flowing stream composed of water	I	–	–
GAMMA	Conductance ratio	I,O	–	–
GAM-MAF	Conductance ratio at fillup	I,O	–	–
H	Thickness of a single layer of the formation†	I	FT	M
IBASE*	Base water injection rate	O	BBL/DAY	M3/DAY
IW*	Water injection rate	O	BBL/DAY	M3/DAY

Symbol	Variable Name	Input or Output	English Units	SI Units
IWF*	Water injection rate at fillup	O	BBL/DAY	M3/DAY
K	Permeability	I	MD	MD
KRO	Relative permeability to oil	I	–	–
KRW	Relative permeability to water	I,O	–	–
M	Water-oil mobility ratio	O	–	–
NP*	Cumulative oil production	O	BBL	M3
OIPWF*	Oil in place at the beginning of waterflooding	O	BBL	M3
QI/QIB	QI divided by QIBT; cumulative injected water volume divided by injected water volume at breakthrough (both volumes in water-contacted pore volumes)	I,O	–	–
QO*	Oil producing rate	O	BBL/DAY	M3/DAY
R	Outer radius of waterflood front prior to interference	O	FT	M
REI	Half the distance between adjacent injectors	I	FT	M
RW	Wellbore radius	I	FT	M
T*	Time	O	DAY	DAY
TF*	Time to fillup	O	DAY	DAY
WI*	Cumulative injected water volume	O	BBL	M3
WIBT*	Cumulative injected water volume at breakthrough	O	BBL	M3
WIF*	Cumulative injected water volume at fillup	O	BBL	M3
WII*	Cumulative injected water volume at interference	O	BBL	M3
WI/WIB	WI divided by WIBT	O	–	–
WI INC*	Cumulative injected water volume increment of table	I	BBL	M3
W/WINC	WI/WIB increment of table	I	–	–
WOR	Water-oil ratio	O	–	–
%EA	Percent areal sweep efficiency	I,O	–	–
%EABT	Percent areal sweep efficiency at breakthrough	I,O	–	–
%EAF	Percent areal sweep efficiency at fillup	O	–	–
%OILRF	Percent oil recovery factor	O	–	–
%SGI	Volume percent initial gas saturation	I	–	–
%SGI*	Volume percent maximum initial gas saturation at which prediction method is valid	I	–	–
%SW-AVG	Volume percent average water saturation	O	–	–
%SWBT	Volume percent average water saturation behind the flood front at and prior to breakthrough	I	–	–
%SWC	Volume percent connate water saturation	I	–	–
%SWSZ	Volume percent water saturation at the upstream end of the stabilized zone	I	–	–
%SW2	Volume percent exit-end water saturation	I	–	–

*The units for these variables are saved by the program.
†In the case of deviated wells or slanted beds, use the true vertical thickness instead of the measured thickness of the layer.

Yes/No Questions

None

Example 1

A certain oil reservoir's performance is being used as a model for a field being considered for waterflood.

The relevant data for the proposed flood is summarized here:

Well spacing	80 ACRE
%SWC	10%
Current gas saturation	9.5%
Oil viscosity	1.0 CP
Water viscosity	0.5 CP
Reservoir pressure	2,200 PSI
Permeability variation, V	0.756
Water-oil relative permeability ratio characteristics	a = 460.0 b = −21.5
Current oil recovery, %OIP	5.6%
Oil formation volume factor at discovery (= bubble point)	1.381
current	1.305
Pattern	20-acre five spot
Effective well bore radius	0.35 FT

Layer	Upper	Lower
Average porosity	14%	18%
Average connate water saturation	10%	10%
Net thickness	18 FT	27 FT
Absolute permeability	19 MD	22 MD
Distance between injectors	933.4 FT	
Pressure drop between producers and injectors	1,500 PSI	

The values of *a* and *b* were determined by analogy with another reservoir using the program CUTCUM. These values of *a* and *b* were used in Example 2 of FWVSW to calculate %SWSZ = 31.2 and %SWBT = 37.2. Not all values of *a* and *b* will result in realistic %SWSZ and %SWBT values. If the correlations are not representative of the data, the user must input KRO and KRW values.

To estimate KRW, use c = 5.5 and d = 2.93. Also, KRO = 0.67 ahead of the flood front, and correlations provided by the program should be used whenever possible for this example. Output just enough points to define the flood. Forecast to an ultimate ratio of injected water to water injected at breakthrough of 10, scaling the values appropriately. Make the prediction for the upper layer.

Keystrokes (SIZE> =065)	Display	Comments
[XEQ] [ALPHA] 5SPOT [ALPHA]	*%POR=?*	
14 [R/S]	*%SWC=?*	
10 [R/S]	*UO=?*	
1 [R/S]	*UW=?*	
.5 [R/S]	*BOWF=?*	
1.305 [R/S]	*AREA=?*	
20 [R/S]	*H=?*	
18 [R/S]	*REI=?*	
933.4 [ENTER↑] 2 [÷]	*466.7000*	REI is one-half the distance between injectors
[R/S]	*RW=?*	
.35 [R/S]	*DP=?*	
1500 [R/S]	*%SWSZ=?*	
31.2 [R/S]	*%SWBT=?*	
37.2 [R/S]	*c=?*	
5.5 [R/S]	*d=?*	
2.93 [R/S]	*KRO=?*	KRW printed
.67 [R/S]	*K=?*	
19 [R/S]	*C=?*	M and C printed
[R/S]	*%SGI=?*	
9.5 [R/S]	*BEG WI=?*	%SGI*, OIPWF, and IBASE printed. Prior to interference stage starts and WII printed
5000 [R/S]	*WI INC=?*	END WI printed
5000 [R/S]	*%EABT=?*	Table printed. Interference to fillup stage starts and WIF, %EAF, GAMMAF, and TF printed. Fillup to breakthrough stage starts and %EABT printed
[R/S]	*WI INC=?*	WIBT and END WI printed
10000 [R/S]	*ENDW/W=?*	Table printed. After water breakthrough stage starts and BEGW/W printed
4 [R/S]	*W/WINC=?*	
1 [R/S]	*a=?*	WI/WIB, WI, %EA, and QI/QIB printed. Coefficients for KO/KW correlation, used to calculate %SWAVG and WOR, must be input the first time the correlation is used
460 [R/S]	*b=?*	
21.5 [CHS] [R/S]	*QI/QIB=?*	%SWAVG, WOR, KRW, M, GAMMA, IW, T, QO, NP, %OILRF, WI/WIB, WI, and %EA printed. QI/QIB interpolated from Table 24–2 at WI/WIB=1 and %EABT=79.6
1.886 [R/S]	*ENDW/W=?*	Once %EA ≥100, QI/QIB is calculated and not input
10 [R/S]	*W/WINC=?*	
2 [R/S]	*ENDW/W=?*	Table printed

FIVE SPOT

```
%POR=14.0000
%SWC=10.0000
UO=1.0000 CP
UW=0.5000 CP
BOWF=1.3050
AREA=20.0000 ACRE
H=18.0000 FT
REI=466.7000 FT
RW=0.3500 FT
DP=1,500.0000 PSI
%SWSZ=31.2000
%SWBT=37.2000
c=5.5000
d=2.9300
KRW=0.1651
KRO=0.6700
K=19.0000 MD

M=0.4928
C=1.5704
%SGI=9.5000
%SGI*=27.7960
OIPWF=241,204.9656 BBL
IBASE=175.8169 BBL/DAY


PRIOR TO INTERFERENCE

WII=29,176.4258 BBL
BEG WI=5,000.0000 BBL
END WI=29,176.4258 BBL
WI INC=5,000.0000 BBL

WI=5,000.0000 BBL
R=114.1784 FT
IW=198.0439 BBL/DAY
T=25.2469 DAY

WI=10,000.0000 BBL
R=161.4727 FT
IW=187.3083 BBL/DAY
T=51.1972 DAY

WI=15,000.0000 BBL
R=197.7629 FT
IW=181.5513 BBL/DAY
T=78.3078 DAY

WI=20,000.0000 BBL
R=228.3569 FT
IW=177.6767 BBL/DAY
T=106.1453 DAY

WI=25,000.0000 BBL
R=255.3108 FT
IW=174.7834 BBL/DAY
T=134.5173 DAY

WI=29,176.4258 BBL
R=275.8133 FT
IW=172.8349 BBL/DAY
T=158.5461 DAY


INTERFERENCE TO FILLUP

WIF=37,147.0629 BBL
%EAF=34.9265
GAMMAF=0.7880
IWF=138.5403 BBL/DAY
TF=209.7424 DAY


FILLUP TO BREAKTHROUGH

%EABT=79.6021
WIBT=84,663.1055 BBL
END WI=84,663.1055 BBL
WI INC=10,000.0000 BBL

WI=47,147.0629 BBL
%EA=44.3287
GAMMA=0.7642
IW=134.3642 BBL/DAY
T=283.0281 DAY
QO=102.9611 BBL/DAY
NP=7,662.8352 BBL
%OILRF=3.1769

WI=57,147.0629 BBL
%EA=53.7309
GAMMA=0.7405
IW=130.1882 BBL/DAY
T=358.6275 DAY
QO=99.7610 BBL/DAY
NP=15,325.6705 BBL
%OILRF=6.3538

WI=67,147.0629 BBL
%EA=63.1331
GAMMA=0.7167
IW=126.0121 BBL/DAY
T=436.6914 DAY
QO=96.5610 BBL/DAY
NP=22,988.5058 BBL
%OILRF=9.5307

WI=77,147.0629 BBL
%EA=72.5353
GAMMA=0.6930
IW=121.8360 BBL/DAY
T=517.3860 DAY
QO=93.3609 BBL/DAY
NP=30,651.3410 BBL
%OILRF=12.7076

WI=84,663.1055 BBL
%EA=79.6021
GAMMA=0.6751
IW=118.6973 BBL/DAY
T=579.8808 DAY
QO=90.9558 BBL/DAY
NP=36,410.7606 BBL
%OILRF=15.0954


AFTER WATER BREAKTHROUGH

BEGW/W=1.0000
ENDW/W=4.0000
W/WINC=1.0000

WI/WIB=1.0000
WI=84,663.1055 BBL
%EA=79.6021
QI/QIB=1.0000
a=460.0000
b=-21.5000
%SWAVG=43.6120
WOR=1.7358
KRW=0.3069
M=0.9163
GAMMA=0.9561
IW=168.0927 BBL/DAY
T=579.8808 DAY
QO=55.2793 BBL/DAY
NP=51,704.4104 BBL
%OILRF=21.4359

WI/WIB=2.0000
WI=169,326.2110 BBL
%EA=98.6567
QI/QIB=1.8860
```

```
%SWAVG=47.0907
WOR=4.5774
KRW=0.4096
M=1.2227
GAMMA=1.2194
IW=214.3980 BBL/DAY
T=1,022.5745 DAY
QO=36.4471 BBL/DAY
NP=81,178.0346 BBL
%OILRF=33.6552

WI/WIB=3.0000
WI=253,989.3164 BBL
QI/QIB=2.6820
%SWAVG=48.8778
WOR=17.7623
KRW=0.4702
M=1.4035
GAMMA=1.4035
IW=246.7590 BBL/DAY
T=1,389.7514 DAY
QO=12.9414 BBL/DAY
NP=88,025.6475 BBL
%OILRF=36.4941

WI/WIB=4.0000
WI=338,652.4219 BBL
QI/QIB=3.4780
%SWAVG=50.1634
WOR=23.8618
KRW=0.5172
M=1.5439
GAMMA=1.5439
IW=271.4391 BBL/DAY
T=1,716.5110 DAY
QO=10.7856 BBL/DAY
NP=91,877.9301 BBL
%OILRF=38.0912

ENDW/W=10.0000
W/WINC=2.0000

WI/WIB=6.0000
WI=507,978.6329 BBL
QI/QIB=5.0701
%SWAVG=51.9949
WOR=36.0358
KRW=0.5894
M=1.7594
GAMMA=1.7594
IW=309.3235 BBL/DAY
T=2,299.6278 DAY
QO=8.2838 BBL/DAY
NP=97,365.5183 BBL
%OILRF=40.3663

WI/WIB=8.0000
WI=677,304.8438 BBL
QI/QIB=6.6621
%SWAVG=53.3049
WOR=48.1976
KRW=0.6449
M=1.9250
GAMMA=1.9250
IW=338.4557 BBL/DAY
T=2,822.4177 DAY
QO=6.8371 BBL/DAY
NP=101,290.8387 BBL
%OILRF=41.9937

WI/WIB=10.0000
WI=846,631.0548 BBL
QI/QIB=8.2542
%SWAVG=54.3257
WOR=60.3546
KRW=0.6905
M=2.0611
GAMMA=2.0611
IW=362.3677 BBL/DAY
T=3,305.6385 DAY
QO=5.8769 BBL/DAY
NP=104,349.5279 BBL
%OILRF=43.2618
```

Example 2

In this example, use the input data from Craig's example. Set flags 00, 01, and 02 early in the program to override the use of the correlations.

Keystrokes	Display	Comments
[XEQ] [ALPHA] 5SPOT [ALPHA]	*%POR=?*	
20 [R/S]	*%SWC=?*	
10 [R/S]	*UO=?*	
1 [R/S]	*UW=?*	
.5 [R/S]	*BOWF=?*	
1.2 [R/S]	*AREA=?*	
40 [R/S]	*H=?*	
[//] [SF] 00 [//] [SF] 01 [//] [SF] 02		No correlations will be used
	REI=?	
5 [R/S]	*RW=?*	
1 [R/S]	*DP=?*	
3000 [R/S]	*%SWSZ=?*	
46.9 [R/S]	*%SWBT=?*	

Keystrokes	Display	Comments
56.3 [R/S]	*KRW=?*	
.4 [R/S]	*KRO=?*	
1 [R/S]	*K=?*	
31.5 [R/S]	*C=?*	M and C printed
1.18 [R/S]	*%SGI=?*	
15 [R/S]	*BEG WI=?*	%SGI*, OIPWF, and IBASE printed. Prior to interference stage starts and WII printed
5000 [R/S]	*WI INC=?*	END WI printed
5000 [R/S]	*GAMMAF=?*	Table printed. Interference to fillup stage starts and WIF and %EAF printed
.96 [R/S]	*%EABT=?*	IWF and TF printed. Fillup to breakthrough stage starts and %EABT printed
71.7 [R/S]	*WI INC=?*	
20000 [R/S]	*GAMMA=?*	A larger increment is used here than in Example 1. GAMMA values are not sufficiently significant to justify more accurate interpolations
.94 [R/S]	*GAMMA=?*	IW, T, QO, NP, %OILRF, WI,and %EA printed
.92 [R/S]	*GAMMA=?*	
.91 [R/S]	*ENDW/W=?*	After water break-through stage starts and BEGW/W printed
2 [R/S]	*W/WINC=?*	
.4 [R/S]	*%SW2=?*	WI/WIB, WI%EA, QI/QIB, and dF/dS printed
47 [R/S]	*FW2=?*	
.8 [R/S]	*KRW=?*	%SWAVG and WOR printed
.4 [R/S]	*GAMMA=?*	M printed
.91 [R/S]	*QI/QIB=?*	IW, T, QO, NP, % OILRF, WI/WIB, WI and %EA printed
1.375 [R/S]	*%SW2=?*	dF/dS printed
50.7 [R/S]	*FW2=?*	
.87 [R/S]	*KRW=?*	%SWAVG and WOR printed

Keystrokes	Display	Comments
.45 [R/S]	*GAMMA=?*	M printed
.96 [R/S]	*QI/QIB=?*	IW, T, QO, NP, %OILRF, WI/WIB, WI, and %EA printed
1.715 [R/S]	*%SW2=?*	dF/dS printed
53.4 [R/S]	*FW2=?*	
.905 [R/S]	*KRW=?*	%SWAVG and WOR printed
.5 [R/S]	*GAMMA=?*	M printed
1 [R/S]	*QI/QIB=?*	IW, T, QO, NP, % OILRF, WI/WIB, WI and %EA printed
1.875 [R/S]	*%SW2=?*	dF/dS printed
54.3 [R/S]	*FW2=?*	
.92 [R/S]	*KRW=?*	%SWAVG and WOR printed
.51 [R/S]	*GAMMA=?*	M printed
1.02 [R/S]	*ENDW/W=?*	IW, T, QO, NP, and %OILRF printed
10 [R/S]	*W/WINC=?*	
2 [R/S]	*%SW2=?*	Once %EA $\geq$ 100, QI/QIB is calculated and not input
59.7 [R/S]	*FW2=?*	
.963 [R/S]	*KRW=?*	
.6 [R/S]	*GAMMA=?*	
1.2 [R/S]	*%SW2=?*	
62.2 [R/S]	*FW2=?*	
.98 [R/S]	*KRW=?*	
.635 [R/S]	*GAMMA=?*	
1.27 [R/S]	*%SW2=?*	
63.7 [R/S]	*FW2=?*	
.985 [R/S]	*KRW=?*	
.67 [R/S]	*GAMMA=?*	
1.34 [R/S]	*%SW2=?*	
65 [R/S]	*FW2=?*	
.99 [R/S]	*KRW=?*	
.69 [R/S]	*GAMMA=?*	
1.38 [R/S]	*ENDW/W=?*	

The values for QI/QIB begin to diverge from the values in Craig. When the value for areal sweep efficiency exceeds 100%, the increase in QI/QIBT is proportional to the increase in WI/WIBT. In this example the jump from WI/WIBT = 2 to WI/WIBT = 4 was sufficiently large to induce some error, which propagates through the remainder of the example. A smaller value of W/WINC as %EA approaches 100% would be appropriate if it were desired to match Craig's example more accurately.

```
FIVE SPOT

%POR=20.0000
%SWC=10.0000
UO=1.0000 CP
UW=0.5000 CP
BOWF=1.2000
AREA=40.0000 ACRE
H=5.0000 FT
REI=660.0000 FT
RW=1.0000 FT
DP=3,000.0000 PSI
%SWSZ=46.9000
%SWBT=56.3000
KRW=0.4000
KRO=1.0000
K=31.5000 MD

M=0.8000
C=1.2055
C=1.1800
%SGI=15.0000
%SGI*=36.9340
OIPWF=193,959.1837 BBL
IBASE=269.0246 BBL/DAY


PRIOR TO INTERFERENCE

WII=36,560.4448 BBL
BEG WI=5,000.0000 BBL
END WI=36,560.4448 BBL
WI INC=5,000.0000 BBL

WI=5,000.0000 BBL
R=138.9243 FT
IW=496.3001 BBL/DAY
T=10.0745 DAY

WI=10,000.0000 BBL
R=196.4687 FT
IW=466.2889 BBL/DAY
T=20.4632 DAY

WI=15,000.0000 BBL
R=240.6240 FT
IW=450.3585 BBL/DAY
T=31.3725 DAY

WI=20,000.0000 BBL
R=277.8487 FT
IW=439.7002 BBL/DAY
T=42.6077 DAY

WI=25,000.0000 BBL
R=310.6443 FT
IW=431.7742 BBL/DAY
T=54.0825 DAY

WI=30,000.0000 BBL
R=340.2937 FT
IW=425.5072 BBL/DAY
T=65.7473 DAY

WI=35,000.0000 BBL
R=367.5592 FT
IW=420.3487 BBL/DAY
T=77.5697 DAY

WI=36,560.4448 BBL
R=375.6636 FT
IW=418.9117 BBL/DAY
T=81.2883 DAY


INTERFERENCE TO FILLUP

WIF=46,550.2041 BBL
%EAF=32.3974
GAMMAF=0.9600
IWF=258.2636 BBL/DAY
TF=110.7925 DAY


FILLUP TO BREAKTHROUGH

%EABT=73.1806
%EABT=71.7000
WIBT=103,022.1187 BBL
END WI=103,022.1187 BBL
WI INC=20,000.0000 BBL

WI=66,550.2041 BBL
%EA=46.3167
GAMMA=0.9400
IW=252.8831 BBL/DAY
T=189.0479 DAY
QO=210.7359 BBL/DAY
NP=16,666.6667 BBL
%OILRF=8.5929

WI=86,550.2041 BBL
%EA=60.2361
GAMMA=0.9200
IW=247.5026 BBL/DAY
T=268.9863 DAY
QO=206.2522 BBL/DAY
NP=33,333.3333 BBL
%OILRF=17.1857

WI=103,022.1187 BBL
%EA=71.7000
GAMMA=0.9100
IW=244.8124 BBL/DAY
T=335.9024 DAY
QO=204.0103 BBL/DAY
NP=47,059.9288 BBL
%OILRF=24.2628


AFTER WATER BREAKTHROUGH

BEGW/W=1.0000
ENDW/W=2.0000
W/WINC=0.4000

WI/WIB=1.0000
WI=103,022.1187 BBL
%EA=71.7000
QI/QIB=1.0000
dF/dS=2.1598
%SW2=47.0000
FW2=0.8000
%SWAVG=56.2600
WOR=1.5000
KRW=0.4000
M=0.8000
```

```
GAMMA=0.9100
IW=244.8124 BBL/DAY
T=335.9024 DAY
QO=90.6725 BBL/DAY
NP=46,985.7588 BBL
%OILRF=24.2246

WI/WIB=1.4000
WI=144,230.9662 BBL
%EA=80.9496
QI/QIB=1.3750
dF/dS=1.5708
%SW2=50.7000
FW2=0.8700
%SWAVG=58.9761
WOR=2.5513
KRW=0.4500
M=0.9000
GAMMA=0.9600
IW=258.2636 BBL/DAY
T=499.7300 DAY
QO=68.8457 BBL/DAY
NP=63,737.5427 BBL
%OILRF=32.8613

WI/WIB=1.8000
WI=185,439.8138 BBL
%EA=87.8583
QI/QIB=1.7150
dF/dS=1.2594
%SW2=53.4000
FW2=0.9050
%SWAVG=60.9434
WOR=3.6264
KRW=0.5000
M=1.0000
GAMMA=1.0000
IW=269.0246 BBL/DAY
T=656.0348 DAY
QO=55.7397 BBL/DAY
NP=76,957.8490 BBL
%OILRF=39.6773

WI/WIB=2.0000
WI=206,044.2374 BBL
%EA=90.7546
QI/QIB=1.8750
dF/dS=1.1519
%SW2=54.3000
FW2=0.9200
%SWAVG=61.2450
WOR=4.2407
KRW=0.5100
M=1.0200
GAMMA=1.0200
IW=274.4051 BBL/DAY
T=731.8659 DAY
QO=50.4354 BBL/DAY
NP=81,481.4849 BBL
%OILRF=42.0096

ENDW/W=10.0000
W/WINC=2.0000

WI/WIB=4.0000
WI=412,088.4749 BBL
QI/QIB=3.3090
dF/dS=0.6527
%SW2=59.7000
FW2=0.9630
%SWAVG=65.3686
WOR=31.2324
KRW=0.6000
M=1.2000
GAMMA=1.2000
IW=322.8295 BBL/DAY
T=1,421.8603 DAY
QO=9.9539 BBL/DAY
NP=104,398.2666 BBL
%OILRF=53.8249

WI/WIB=6.0000
WI=618,132.7121 BBL
QI/QIB=4.7430
dF/dS=0.4554
%SW2=62.2000
FW2=0.9800
%SWAVG=66.5920
WOR=58.8000
KRW=0.6350
M=1.2700
GAMMA=1.2700
IW=341.6612 BBL/DAY
T=2,042.0172 DAY
QO=5.6944 BBL/DAY
NP=107,562.0515 BBL
%OILRF=55.4560

WI/WIB=8.0000
WI=824,176.9496 BBL
QI/QIB=6.1770
dF/dS=0.3497
%SW2=63.7000
FW2=0.9850
%SWAVG=67.9899
WOR=78.8000
KRW=0.6700
M=1.3400
GAMMA=1.3400
IW=360.4929 BBL/DAY
T=2,628.9090 DAY
QO=4.5062 BBL/DAY
NP=111,177.2140 BBL
%OILRF=57.3199

WI/WIB=10.0000
WI=1,030,221.187 BBL
QI/QIB=7.6110
dF/dS=0.2838
%SW2=65.0000
FW2=0.9900
%SWAVG=68.5239
WOR=118.8000
KRW=0.6900
M=1.3800
GAMMA=1.3800
IW=371.2539 BBL/DAY
T=3,192.0662 DAY
QO=3.0938 BBL/DAY
NP=112,558.1168 BBL
%OILRF=58.0319
```

User Instructions

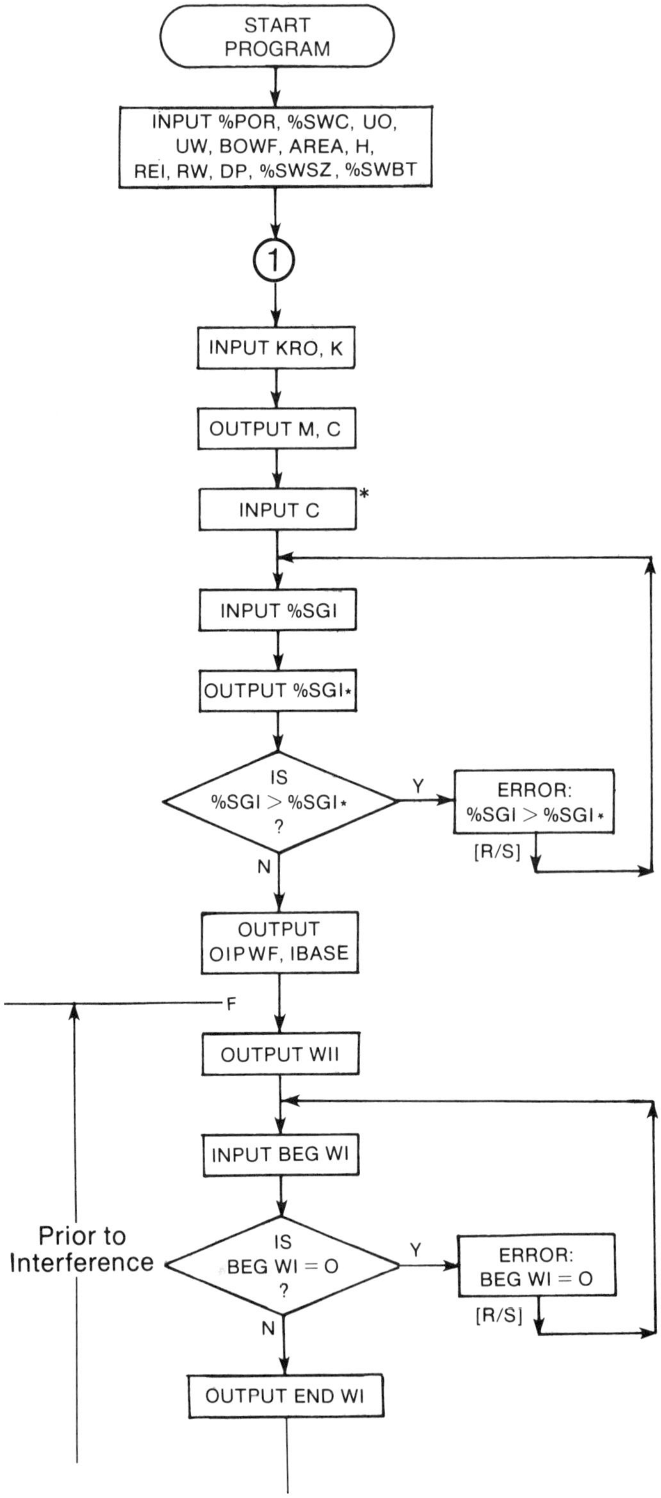

* This option is provided in case you know the value for C.

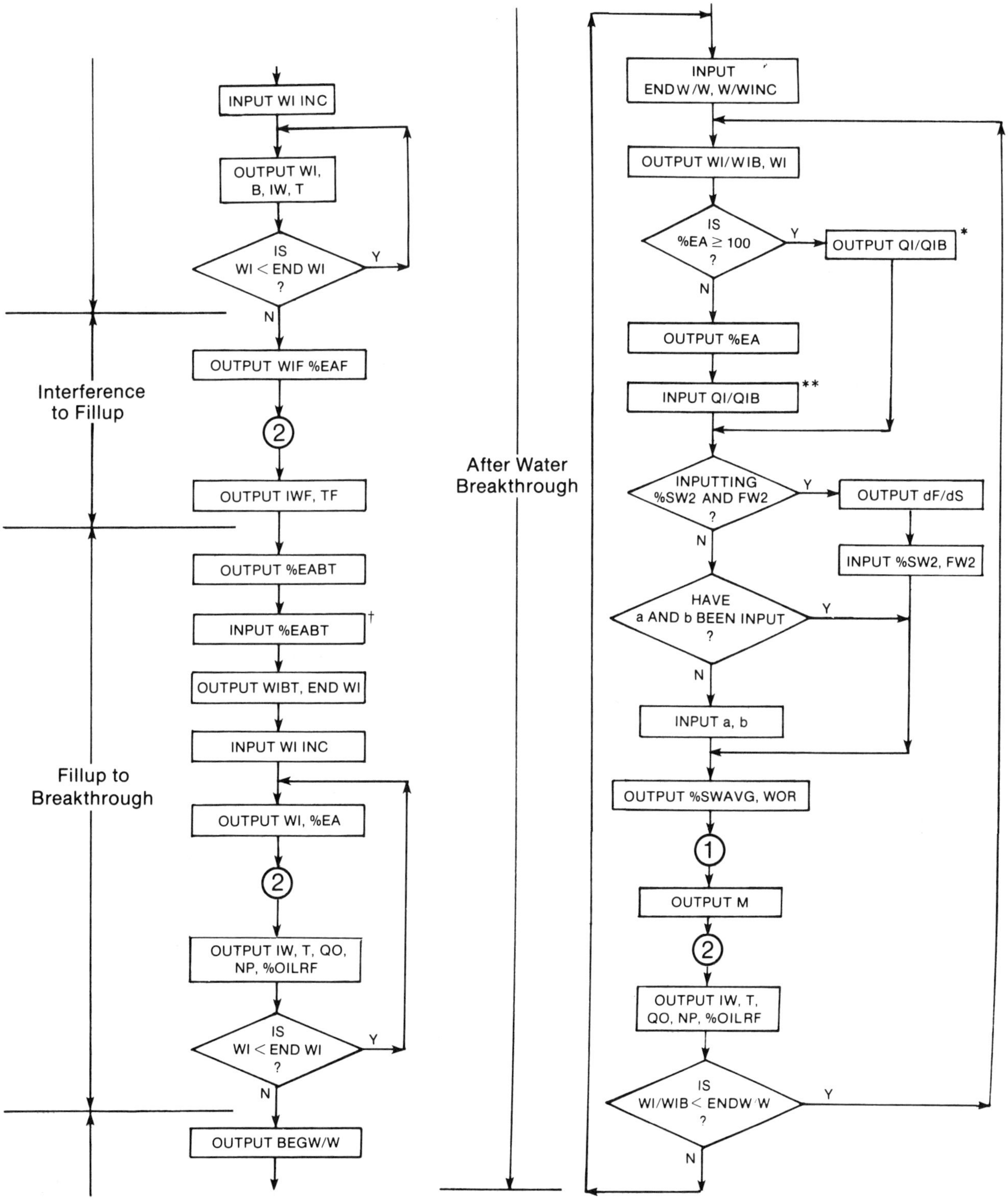

† This option is provided in case you know the value for %EABT.

**When WI/WIB = 1, QI/QIB = 1 and is output regardless of the value of %EA.

User Instructions

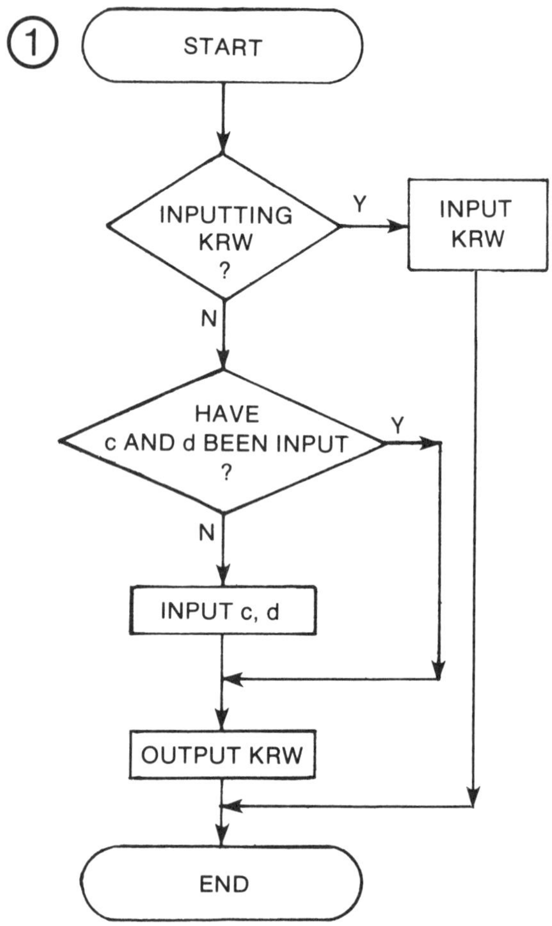

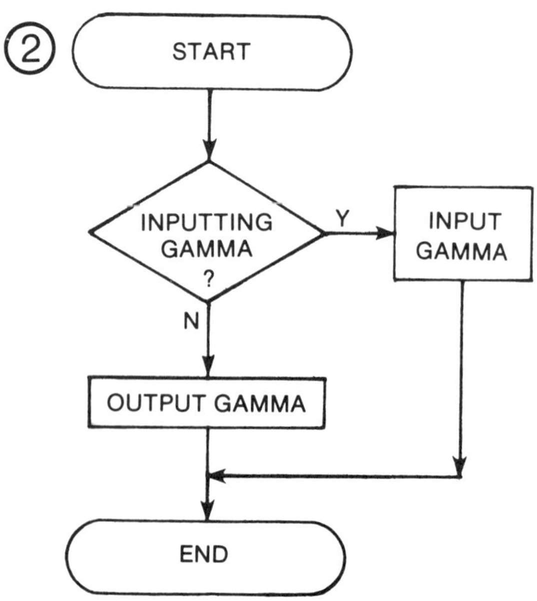

General Information

Memory Requirements

Program length: 1766 bytes* (8 cards)
Minimum size: 065*
Minimum hardware: 41C + quad memory module or 41CV

*5SPOT requires a total of 2,221 bytes of the 2,237 available, leaving no more than 4 key assignments other than the 5SPOT assignment.

Hidden Options

When 5SPOT is run, correlations will be used for GAMMA, KRW, %SW2, and FW2. To allow inputing GAMMA, set flag 00 ([//] [SF]00) any time after the program is run. To change back to using the correlation, clear flag 00 ([//] [CF]00) or run the program again. The flag 00 annunciator in the display will be off if the correlation is being used and on if GAMMA is to be input. Flag 01 is used in the same way to select calculating or inputting %SW2 and FW2, as is flag 02 for KRW.

Pac Subroutines Called

TITLE, %POR, IN, INU, OUT, CON, OUTU, INK, OUTK

Registers

03	Injected water or oil production units
04	Injected water or oil production units
06	Water injection rate or oil producing rate units
07	Water injection rate or oil producing rate units
08	Time units
09	Time units
17	DP (PSI)
18	%POR
20	%SGI
21	%SWC
26	BOWF
27	AREA (ACRE)
28	H (FT)
29	K (MD)
30	%SW2
31	RW (FT)
32	UO (CP)
33	UW (CP)
34	KRO
35	KRW
36	c
37	d
38	%SWSZ
39	%SWBT
40	%SWAVG
41	%SWBT — %SWC
42	C
43	VP, WI/WIB

44 %EABT
45 FW2
46 OIPWF (BBL)
47 a
48 b
49 WIBT (BBL), scratch
50 WII (BBL)
51 WIF (BBL)
52 BEG WI (BBL), BEGW/W
53 END WI (BBL), ENDW/W
54 WI INC (BBL), W/WINC
55 FO2, QI, scratch
56 T_{i-1} (DAY)
57 WI_{i-1} (BBL)
58 IW_{i-1} (BBL/DAY)
59 IBASE (BBL/DAY)
60 QI/QIB
61 M
62 GAMMA
63 REI (FT)
64 %EAF, %EA, scratch

Registers 10 – 16, 19, and 22 – 25 unused

Flags

00 Set: Input GAMMA.
Clear: Use correlation for GAMMA.

01 Set: Input %SW2 and FW2.
Clear: Use correlation for %SW2 and FW2.

02 Set: Input KRW.
Clear: Use correlation for KRW.

03 Set: a and b not yet input.
Clear: a and b have been input.

04 Set: c and d not yet input.
Clear: c and d have been input.

05 Set: Time units not yet input.
Clear: Time units have been input.

Flag 05 is also used after water breakthrough to control whether the first output of the forecast is the last WI/WIB or the last WI/WIB + W/WINC and to set QI/QIB to 1 if WI/WIB is 1.

Program Listing

```
 01♦LBL "5SPOT"
"FIVE SPOT"  65
XROM "TITLE"  FC?C 2
PROMPT  CF 00  CF 01
CF 02  SF 03  SF 04
SF 05  "M3"  ASTO 03
"⊢/DAY"  ASTO 06  "DAY"
ASTO 08  CLA  ASTO 04
ASTO 07  ASTO 09
XROM "%POR"  20  STO 00
"%SWC"  XROM "IN"  31
"CP"  ASTO 01  "PA*S"
XEQ 05  "UO"  XROM "INU"
"PA*S"  ASTO Y  CLA
ASTO Z  "UW"  XROM "INU"
25  STO 00  "BOWF"
XROM "IN"  26  "ACRE"
ASTO 01  "M2"  XEQ 05
"AREA"  XROM "INU"  27
XEQ 04  "H"  XROM "INU"
62  XEQ 04  "REI"
XROM "INU"  30  XEQ 04
```

Program Listing (cont.)

```
"RW"  XROM "INU"  16
"PSI"  ASTO 01  "KPA"
XEQ 05  "DP"  XROM "INU"
37  STO 00  "%SWSZ"
XROM "IN"  "%SWBT"
XROM "IN"  STO 40
RCL 21  -  STO 41
XEQ 31  33  STO 00
"KRO"  XROM "IN"  28
STO 00  "MD"  ASTO 01
ASTO Y  CLA  ASTO 02
ASTO Z  "K"  XROM "INU"
RCL 35  RCL 33  /
RCL 32  *  RCL 34  /
STO 61  ADV  "M"
XROM "OUT"  .46836  X<>Y
/  .62  +  STO 42  "C"
XROM "OUT"  41  STO 00
"C"  XROM "IN"

119♦LBL 14
19  STO 00  "%SGI"
XROM "IN"  RCL 41  X<>Y
-  RCL 42  *  "%SGI*"
XROM "OUT"  RCL 20
X<=Y?  GTO 00
"%SGI > %SGI*"  TONE 3
PROMPT  GTO 14

138♦LBL 00
RCL 27  RCL 28  *
RCL 18  %  "ACRE*FT-BBL"
CON  STO 43  100  RCL 21
-  RCL 20  -  %  RCL 26
/  STO 46  "OIPWF"
XEQ 24  .001538  RCL 28
*  RCL 29  *  RCL 34  *
RCL 17  *  2  SQRT
RCL 63  *  RCL 31  /
LOG  .2688  -  RCL 32  *
/  STO 59  "IBASE"
XEQ 28  CF 08  ADV  ADV
"PRIOR TO INTERF"
"⊢ERENCE"  FS? 55  PRA
ADV  PI  RCL 28  *
RCL 18  %  "FT3-BBL"
CON  STO 64  RCL 20  %
RCL 63  X↑2  *  STO 50
STO 53  "WII"  XEQ 24

207♦LBL 13
51  "BEG WI"  XEQ 22
X≠0?  GTO 01
"BEG WI = 0"  TONE 3
PROMPT  GTO 13

217♦LBL 01
XEQ 20  ADV  RCL 41
RCL 20  /  LN  RCL 32  *
RCL 34  /  2  /  STO 49
RCL 28  RCL 29  *
RCL 17  *  7.07 E-3  *
STO 55  CLX  STO 56
STO 57  RCL 52

243♦LBL 15
XEQ 23  RCL 64  /
RCL 41  /  SQRT  10  *
XEQ 04  "R"  XROM "OUTU"
RCL 31  /  LN  RCL 33  *
RCL 35  /  RCL 49  +
RCL 55  X<>Y  /  FS? 05
STO 58  XEQ 25  CF 08
XEQ 12  X<Y?  GTO 02
X<>Y  RCL 52  RND  X=Y?
GTO 06  RCL 53  RND

281♦LBL 02
LASTX  GTO 15

284♦LBL 04
"FT"  ASTO 01  "M"

288♦LBL 05
STO 00  ASTO Y  CLA
ASTO 02  ASTO Z  RTN

295♦LBL 06
ADV  "INTERFERENCE TO"
"⊢ FILLUP"  FS? 55  PRA
ADV  RCL 43  RCL 20  %
STO 51  STO 52  "WIF"
XEQ 24  RCL 20  RCL 41
/  100  *  STO 64
"%EAF"  XROM "OUT"
"GAMMAF"  XEQ 30  RCL 59
*  "IWF"  XEQ 28  "TF"
XEQ 26  ADV  ADV
"FILLUP TO BREAK"
"⊢THROUGH"  FS? 55  PRA
ADV  RCL 61  -.1736  Y↑X
70.4  *  STO 44  "%EABT"
XROM "OUT"  43  STO 00
"%EABT"  XROM "IN"
```

Program Listing (cont.)

```
RCL 43  X<>Y  %  RCL 41
%  STO 49  STO 53
"WIBT"  XEQ 24  RCL 51
STO 52  XEQ 20  XEQ 12
GTO 07

358♦LBL 16
XEQ 23  1 E4  *  RCL 43
/  RCL 41  /  STO 64
"%EA"  XROM "OUT"
XEQ 29  RCL 59  *
XEQ 25  RCL 58  RCL 26
/  "QO"  XEQ 28  RCL 52
RCL 51  -  RCL 26  /
XEQ 11

384♦LBL 07
X<Y?  GTO 08  X<>Y
RCL 52  RND  X=Y?
GTO 09  RCL 53  RND

394♦LBL 08
LASTX  GTO 16

397♦LBL 09
ADV  "AFTER WATER BRE"
"⊢AKTHROUGH"  FS? 55
PRA  ADV  1  STO 43
STO 60  "BEGW/W"
XROM "OUT"  SF 05

410♦LBL 17
52  STO 00  "ENDW/W"
XROM "IN"  "W/WINC"
XROM "IN"  RCL 43
FS? 05  ADV  FS? 05
GTO 18  XEQ 12  GTO 32

424♦LBL 18
STO 43  "WI/WIB"
XROM "OUT"  RCL 49  *
XEQ 23  RCL 43  LN
27.49  *  RCL 44  +  100
X<=Y?  GTO 10  X<>Y
STO 64  "%EA"
XROM "OUT"  FS? 05
GTO 00  59  STO 00
"QI/QIB"  XROM "IN"
GTO 01

451♦LBL 10
STO 64  RCL 52  RCL 57
-  RCL 49  /  RCL 44  %
ST+ 60

461♦LBL 00
RCL 60  "QI/QIB"
XROM "OUT"

465♦LBL 01
RCL 41  %  STO 55
FS? 01  GTO 02  FC? 03
GTO 01  46  STO 00  "a"
XROM "IN"  "b"
XROM "IN"  CF 03

480♦LBL 01
RCL 55  ENTER↑  1/X  4
*  RCL 48  /  1  +  SQRT
1  X<>Y  -  2  /  STO 55
*  LASTX  1  LASTX  -  /
RCL 33  *  RCL 32  /
RCL 47  /  LN  RCL 48  /
100  *  GTO 03

515♦LBL 02
1/X  "dF/dS"  XROM "OUT"
29  STO 00  "%SW2"
XROM "IN"  44  STO 00
"FW2"  XROM "IN"  1
X<>Y  -  ENTER↑  X<> 55
*  RCL 30

534♦LBL 03
X<>Y  100  *  +  STO 40
"%SWAVG"  XROM "OUT"
RCL 38  RCL 21  -
RCL 44  /  RCL 41  /
RCL 43  /  27.49  *  1
RCL 55  -  *  RCL 55  +
LASTX  RCL 64  100  X=Y?
R↑  R↑  STO 55  1/X  1
-  RCL 26  *  "WOR"
XROM "OUT"  XEQ 31
RCL 33  /  RCL 32  *
RCL 34  /  STO 61  "M"
XROM "OUT"  XEQ 29
RCL 59  *  XEQ 25
RCL 58  RCL 26  /
RCL 55  *  "QO"  XEQ 28
RCL 43  STO 52  RCL 40
RCL 21  -  RCL 64  %
RCL 20  -  100  LASTX  -
RCL 21  -  /  RCL 46  *
XEQ 11

612♦LBL 32
X<Y?  GTO 04  X<>Y
RCL 52  X=Y?  GTO 17
RCL 53  RND

621♦LBL 04
LASTX  GTO 18

624♦LBL 11
"NP"  XEQ 24  RCL 46  /
100  *  "%OILRF"
XROM "OUT"

633♦LBL 12
ADV  RCL 53  RND  RCL 52
RCL 54  +  RND  RTN

642♦LBL 20
RCL 53  "END WI"  XEQ 24

646♦LBL 21
53  "WI INC"

649♦LBL 22
STO 00  XEQ 05
XROM "INK"  RDN  STO 06
X<>Y  STO 07  R↑  RTN

659♦LBL 23
STO 52  "WI"

662♦LBL 24
XEQ 05  XROM "OUTK"  RDN
STO 06  X<>Y  STO 07  R↑
RTN

671♦LBL 05
ASTO T  "BBL"  ASTO 01
CLA  ASTO 02  ARCL T
RCL 07  RCL 06  RCL Z
RTN

682♦LBL 25
"IN"  XEQ 28  "T"

686♦LBL 26
ENTER↑  X<> 58  +  2  /
RCL 52  ENTER↑  X<> 57
-  X<>Y  /  ST+ 56
RCL 56  FS?C 05  SF 08

702♦LBL 27
ASTO T  "DAY"  ASTO 01
CLA  ASTO 02  ARCL T
RCL 09  RCL 08  RCL Z
XROM "OUTK"  RDN  STO 08
X<>Y  STO 09  R↑  RTN

719♦LBL 28
ASTO T  "BBL/DAY"
ASTO 01  ASHF  ASTO 02
CLA  ARCL T  RCL 04
RCL 03  RCL Z
XROM "OUTK"  RDN  STO 03
X<>Y  STO 04  R↑  RTN

737♦LBL 29
"GAMMA"

739♦LBL 30
61  STO 00  FS? 00
XROM "IN"  FS? 00  RTN
RCL 64  1  %  STO 00  1
RCL 61  X<Y?  GTO 06
RCL 00  Y↑X  XROM "OUT"
RTN

758♦LBL 06
1.094353  RCL 00
.812527  *  -  RCL 61
LOG  *  RCL 00  .985715
*  .40586  -  RCL 61  *
+  1.412546  RCL 00
.9881  *  -  +  1
RCL 00  X≠0?  RDN  X<>Y
RCL 61  RCL 00  1  X=Y?
ENTER↑  R↑  XROM "OUT"
RTN

794♦LBL 31
FC? 02  FC? 04  GTO 07
35  STO 00  "c"
XROM "IN"  "d"
XROM "IN"  CF 04

805♦LBL 07
34  STO 00  "KRW"
FS? 02  XROM "IN"
FS? 02  RTN  RCL 40
RCL 21  -  100  LASTX  -
/  RCL 37  Y↑X  RCL 36
*  STO 35  XROM "OUT"
END
```

25. SUMWF — Combined Performance for Stratified Waterfloods

SUMWF calculates the performance of other layers of a stratified waterflood that may differ in thickness, porosity, or permeability from the base layer. The calculated value for the total performance is merely the sum of the individual layer performances. Other assumptions include that each layer have identical initial saturation and water-oil relative permeability characteristics and minimal producing rates prior to fillup. If these assumptions are not valid, the performance must be calculated for each layer individually using 5SPOT. The composite performance will still be the sum of the individual layer performances. However, since the calculated oil producing rates for the layers are different than those for the base layer, the summation is performed most easily graphically, as shown in Figure 25–1.

Equations

$$T_i = T1 \frac{K1}{K_i} \frac{\%POR_i}{\%POR1}$$

$$QO_i = QO1 \frac{K_i}{K1} \frac{H_i}{H1}$$

$$IW_i = \frac{K_i}{K1} \frac{H_i}{H1}$$

Note: For K_i, H_i, $\%POR_i$, T_i, QO_i, and IW_i, i = 1, 2, 3, . . ., n where n is the number of K values input by the user.

Nomenclature

Symbol	*Variable Name*	*Input or Output*	*English Units*	*SI Units*
H_i	Thickness of layer i*	I	FT	M
IW_i†	Water injection rate of layer i	I,O**	BBL/DAY	M3/DAY
K_i	Permeability layer i	I	MD	MD
QO_i†	Oil producing rate of layer i	I,O**	BBL/DAY	M3/DAY
T_i†	Time to inject into layer i	I,O**	DAY	DAY
$\%POR_i$	Percent porosity of layer i	I	-	-

*In the case of deviated wells or slanted beds, use the true vertical thickness instead of the measured thickness.
†The units for these variables are saved by the program.
**T1, QO1, and IW1 are input; the remaining T, QO, IW values are output.

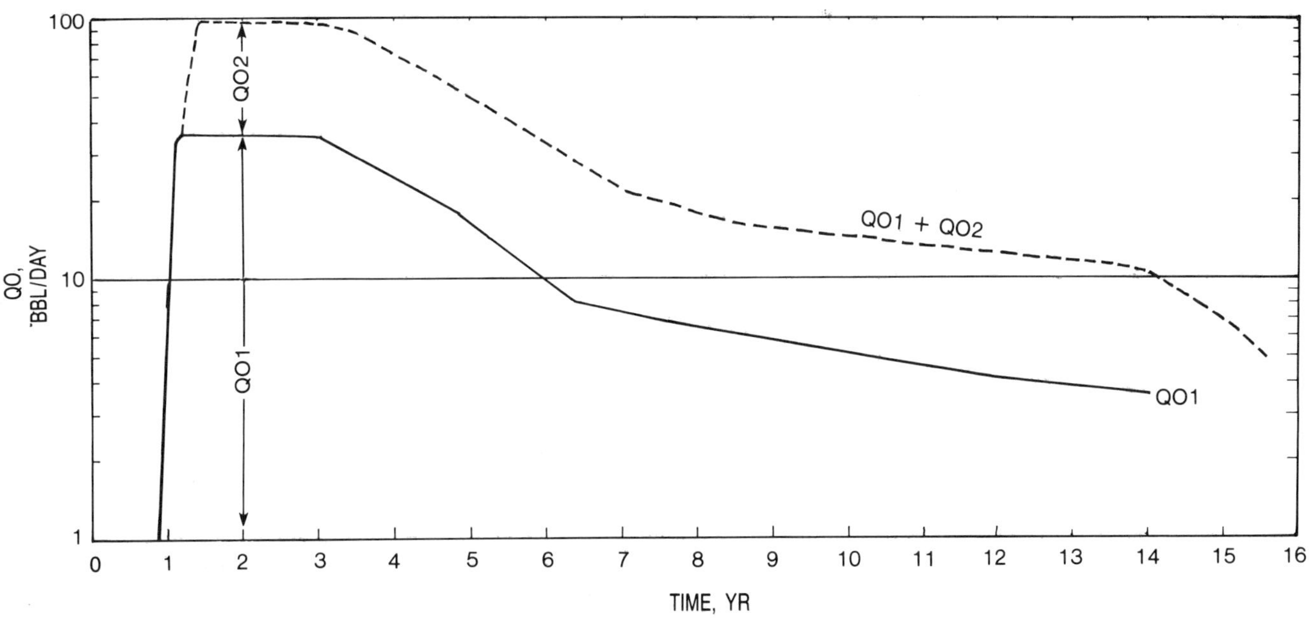

Figure 25–1 Composite waterflood performance

Yes/No Questions

EDIT? Yes: Allow editing of K, H, and %POR values.
No: No editing necessary.

Example

An oil reservoir similar to Example 1 of 5SPOT has two layers with the following properties:

Layer	*Upper*	*Lower*
Porosity	14%	18%
%SWI	36%	36%
H	18 FT	27 FT
K	19 MD	22 MD
%SGI at start of waterflood	4.5%	4.5%

Based on the following summary of the calculated performance, calculate the performance of the second layer and combine the two layers' performances. This is illustrated in Figure 25-1.

TIME (DAY)	*QO (BBL/DAY)*	*QW (BBL/DAY)*
106	-	90.8
148	-	88.6
203	-	49.6
405	35.7	49.2
814	35.2	48.6
1,132	34.8	48.1
1,791	17.6	109.8
2,314	8.2	131.0
2,777	6.8	141.0
3,624	5.1	157.0
4,397	4.2	170.0
5,113	3.6	181.0

Keystrokes (SIZE>=061)	Display	Comments
[XEQ] [ALPHA] SUMWF [ALPHA]	*K1=?*	
19 [R/S]	*H1=?*	
18 [R/S]	*%POR1=?*	
14 [R/S]	*K2=?*	
22 [R/S]	*H2=?*	
27 [R/S]	*%POR2=?*	
18 [R/S]	*K3=?*	
[R/S]	*EDIT? Y/N:N*	
[R/S]	*T1=?*	
106 [R/S]	*QO1=?*	
0 [R/S]	*IW1=?*	
90.8 [R/S]	*T1=?*	T2, QO2, and IW2 printed
[/] [FIX] 1	*106.0*	Set display format to match accuracy expected
148 [R/S]	*QO1=?*	
[R/S]	*IW1=?*	
88.6 [R/S]	*T1=?*	T2, QO2, and IW2 printed
203 [R/S]	*QO1=?*	
[R/S]	*IW1=?*	
49.6 [R/S]	*T1=?*	T2, QO2, and IW2 printed
405 [R/S]	*QO1=?*	
35.7 [R/S]	*IW1=?*	
49.2 [R/S]	*T1=?*	
814 [R/S]	*QO1=?*	
35.2 [R/S]	*IW1=?*	
48.6 [R/S]	*T1=?*	
1132 [R/S]	*QO1=?*	
34.8 [R/S]	*IW1=?*	
48.1 [R/S]	*T1=?*	
1791 [R/S]	*QO1=?*	
17.6 [R/S]	*IW1=?*	
109.8 [R/S]	*T1=?*	
2314 [R/S]	*QO1=?*	
8.2 [R/S]	*IW1=?*	
131 [R/S]	*T1=?*	
2777 [R/S]	*QO1=?*	
6.75 [R/S]	*IW1=?*	
141 [R/S]	*T1=?*	
3624 [R/S]	*QO1=?*	
5.13 [R/S]	*IW1=?*	
157 [R/S]	*T1=?*	
4397 [R/S]	*QO1=?*	
4.22 [R/S]	*IW1=?*	
170 [R/S]	*T1=?*	
5113 [R/S]	*QO1=?*	
3.6 [R/S]	*IW1=?*	
181 [R/S]	*T1=?*	
[/] [FIX] 4	*5113.0000*	Back to FIX 4

```
WF LAYERS

K1=19.0000 MD
H1=18.0000 FT
%POR1=14.0000

K2=22.0000 MD
H2=27.0000 FT
%POR2=18.0000

T1=106.0000 DAY
QO1=0.0000 BBL/DAY
IW1=90.8000 BBL/DAY

T2=117.7013 DAY
QO2=0.0000 BBL/DAY
IW2=157.7053 BBL/DAY

T1=148.0 DAY
IW1=88.6 BBL/DAY

T2=164.3 DAY
QO2=0.0 BBL/DAY
IW2=153.9 BBL/DAY

T1=203.0 DAY
IW1=49.6 BBL/DAY

T2=225.4 DAY
QO2=0.0 BBL/DAY
IW2=86.1 BBL/DAY
```

```
T1=405.0 DAY              T2=2,569.4 DAY
QO1=35.7 BBL/DAY          QO2=14.2 BBL/DAY
IW1=49.2 BBL/DAY          IW2=227.5 BBL/DAY

T2=449.7 DAY              T1=2,777.0 DAY
QO2=62.0 BBL/DAY          QO1=6.8 BBL/DAY
IW2=85.5 BBL/DAY          IW1=141.0 BBL/DAY

T1=814.0 DAY              T2=3,083.6 DAY
QO1=35.2 BBL/DAY          QO2=11.7 BBL/DAY
IW1=48.6 BBL/DAY          IW2=244.9 BBL/DAY

T2=903.9 DAY              T1=3,624.0 DAY
QO2=61.1 BBL/DAY          QO1=5.1 BBL/DAY
IW2=84.4 BBL/DAY          IW1=157.0 BBL/DAY

T1=1,132.0 DAY            T2=4,024.1 DAY
QO1=34.8 BBL/DAY          QO2=8.9 BBL/DAY
IW1=48.1 BBL/DAY          IW2=272.7 BBL/DAY

T2=1,257.0 DAY            T1=4,397.0 DAY
QO2=60.4 BBL/DAY          QO1=4.2 BBL/DAY
IW2=83.5 BBL/DAY          IW1=170.0 BBL/DAY

T1=1,791.0 DAY            T2=4,882.4 DAY
QO1=17.6 BBL/DAY          QO2=7.3 BBL/DAY
IW1=109.8 BBL/DAY         IW2=295.3 BBL/DAY

T2=1,988.7 DAY            T1=5,113.0 DAY
QO2=30.6 BBL/DAY          QO1=3.6 BBL/DAY
IW2=190.7 BBL/DAY         IW1=181.0 BBL/DAY

T1=2,314.0 DAY            T2=5,677.4 DAY
QO1=8.2 BBL/DAY           QO2=6.3 BBL/DAY
IW1=131.0 BBL/DAY         IW2=314.4 BBL/DAY
```

User Instructions

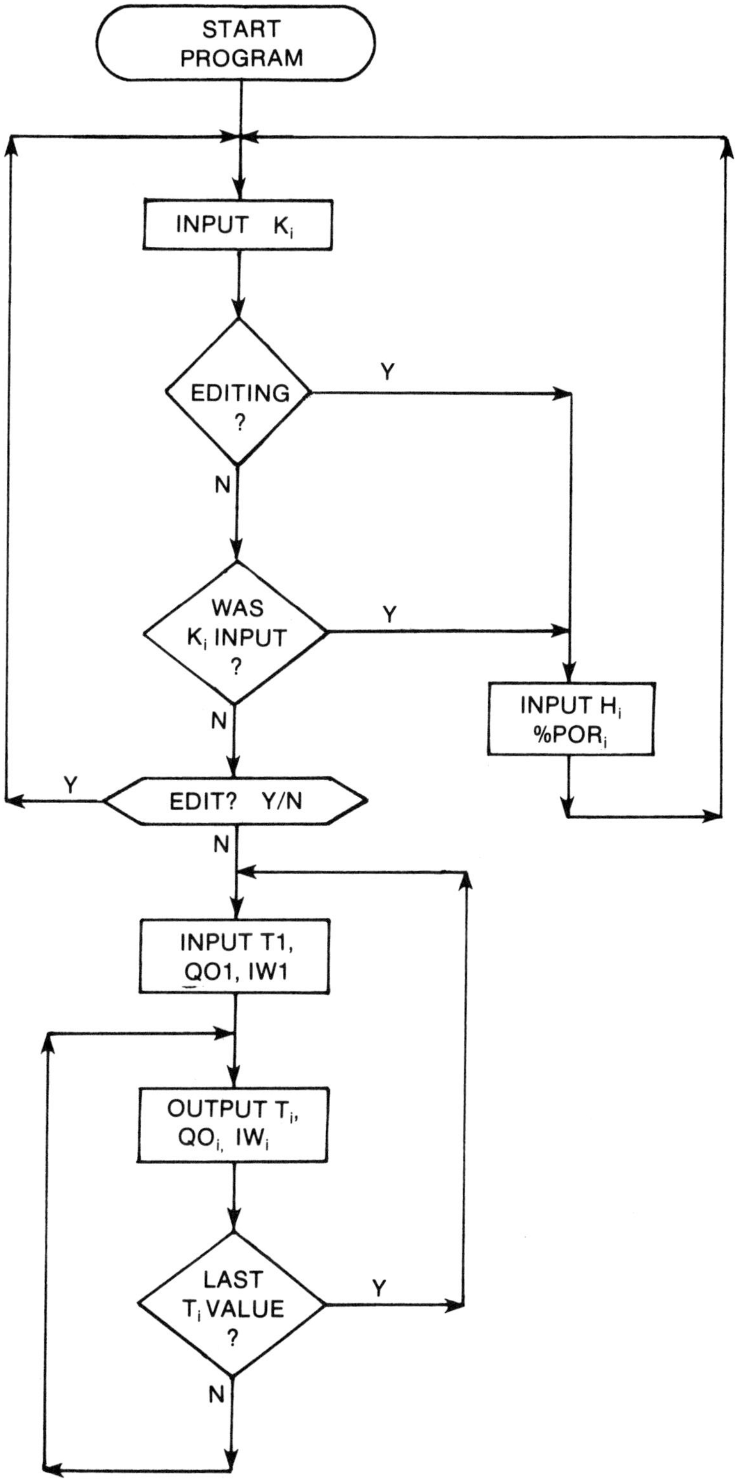

Note: for K_i, H_i, %POR_i, T_i, QO_i, and IW_i, i = 1, 2, 3,..., n, where n is the number of K_i values input by the user.

General Information

Memory Requirements

Program length: 439 bytes (2 cards)
Minimum size: 061*
Minimum hardware: 41C + 1 memory module

*This size will allow up to 10 K, H, %POR values. To accommodate v K, H, and %POR values, use size 31 + 3v.

Hidden Options

None

Pac Subroutines Called

TITLE, IN, Y/N?, INU, INK, OUTK

Registers

03	Time units
04	Time units
06	Oil producing rate or water injection rate units
07	Oil producing rate or water injection rate units
08	Pointer
09	Pointer
26	Scratch
27	T1 (DAY)
28	QO1 (BBL/DAY)
29	IW1 (BBL/DAY)
30	K1 (MD)
31	H1 (FT)
32	%POR1
33	K2 (MD)
34	H2 (FT)
35	%POR2
36	K3 (MD)
37	H3 (FT)
38	%POR3
	etc.

Flags

03 Set: Allow editing of K, H, and %POR values.
Clear: No editing necessary.

Program Listing

```
 01♦LBL "SUMWF"
"WF LAYERS" 61
XROM "TITLE" FC?C 25
PROMPT "DAY" ASTO 03
"M3/DAY" ASTO 06 CLA
ASTO 04 ASTO 07 CF 03
1.1 STO 08

 17♦LBL 14
29 STO 00

 20♦LBL 13
FC? 03 ADV XEQ 02
FC? 03 FS? 22 GTO 00
FC? 23 GTO 01

 29♦LBL 00
XEQ 03 "%POR" XEQ 18
XROM "IN" ISG 08
GTO 13

 36♦LBL 01
RCL 08 INT 1 - 1 E3
+ LASTX / STO 08
STO 09 "EDIT" 3
XROM "Y/N?" FS? 03
GTO 14

 52♦LBL 12
RCL 09 STO 08 26
STO 00 ADV XEQ 05
XEQ 08 CF 08 XEQ 15
ISG 08

 63♦LBL 11
ADV RCL 08 INT 3 *
27 + RCL 27 RCL IND Y
RCL 30 / STO 26 ISG Z
CLD RCL IND Z *
RCL 31 / ISG Z CLD
X<> 26 RCL IND Z /
RCL 32 * / XEQ 06
RCL 28 RCL 26 *
XEQ 09 RCL 29 RCL 26
* XEQ 16 ISG 08
GTO 11 GTO 12

102♦LBL 02
"K" XEQ 18 ASTO T
"MD" ASTO 01 GTO 04

109♦LBL 03
```

Program Listing (cont.)

```
"H" XEQ 18 ASTO T
"FT" ASTO 01 "M"

116♦LBL 04
ASTO Y CLA ASTO 02
ASTO Z ARCL T
XROM "INU" RTN

124♦LBL 05
"T1" XEQ 07 XROM "INK"
RDN STO 03 X<>Y
STO 04 R↑ RTN

134♦LBL 06
"T" XEQ 18 XEQ 07
XROM "OUTK" RDN STO 03
X<>Y STO 04 R↑ RTN

145♦LBL 07
ASTO T "DAY" ASTO 01
CLA ASTO 02 ARCL T
RCL 04 RCL 03 RCL Z
RTN

156♦LBL 08
"Q01" XEQ 17
XROM "INK" GTO 10

161♦LBL 09
"Q0" XEQ 18 XEQ 17
XROM "OUTK" GTO 10

167♦LBL 15
"IW1" XEQ 17
XROM "INK" GTO 10

172♦LBL 16
"IW" XEQ 18 XEQ 17
XROM "OUTK"

177♦LBL 10
RDN STO 06 X<>Y
STO 07 R↑ RTN

184♦LBL 17
ASTO T "BBL/DAY"
ASTO 01 ASHF ASTO 02
CLA ARCL T RCL 07
RCL 06 RCL Z RTN

196♦LBL 18
STO 05 CLST FS? 41 1
+ FS? 40 2 + 5 /
FS? 39 1 + FS? 38 2
+ FS? 37 4 + FS? 36
8 + FS? 29 CHS
RCL 08 FIX 0 CF 29
ARCL X X<>Y X<0?
SF 29 ENTER↑ FRC 5 *
FIX IND Y X=0?
SCI IND Y 1 X=Y?
ENG IND Z RCL 05 END
```

26. INJ — Water Injectivity for Regular Patterns with Unit Mobility Ratio

When designing a waterflood, it is extremely important to be able to estimate the injectivity of water into the reservoir. This is best determined by actual tests; however, an estimate of injectivity frequently is required before a pilot can be approved. Water injectivity varies as a function of the pattern selected for the flood. The location of existing wells is a major consideration in selecting waterflood patterns. Other factors affecting the selection of waterflood patterns include the well spacing, response time, estimated flood life, water availability, and the productivity and injectivity of the reservoir. The life of the flood will be affected strongly by the injectivity into the reservoir. Muskat and Deppe developed empirical methods for estimating injectivity. Because of the heterogeneity in most reservoirs which are candidates for waterfloods, this method should be considered only to provide a reasonable first estimate of injectivity to design pilot tests or filtering prospects.

INJ lets the user calculate the differential pressure for a given injection rate or the injection rate resulting from an input differential pressure. A common assumption in waterflood design and frequently a good operating procedure is to pump the producing wells down, i.e., maintain a low pressure at the producing wells. It is often desirable to maintain injection pressures below the fracture pressure of the formation. With the fracture pressure estimated, INJ will provide a reasonable estimate of injectivity that can be used as a preliminary estimate for forecasting the flood life. The five different patterns whose injectivity can be calculated are shown in Figure 26–1.

Equations

$$IW = \frac{A\ K\ KRO\ H\ DP}{UO\left[\left(\log\frac{d}{RW} - B + C\frac{a}{d}\right)\cdot\left(\frac{E + RATIO}{2 + RATIO}\right) - \frac{D}{2 + RATIO}\right]}$$

Pattern	*A*	*B*	*C*	*D*	*E*
Line drive	0.001538	0.9020	0.682	0	0
Five spot	0.001538	0.2688	0	0	0
Seven spot	0.002051	0.2472	0	0	0
Inverted nine spot (corner well)	0.001538	0.1183	0	0	1
Inverted nine spot (side well)	0.003076	0.1183	0	0.301	3

Nomenclature

Symbol	*Variable Name*	*Input or Output*	*English Units*	*SI Units*
a	Pattern dimension (default = 2 FT)	I	FT	M
DP*	Pressure difference between injection and producing well	I,O	PSI	KPA
d	Pattern dimension	I	FT	M
H	Formation thickness†	I	FT	M
IW*	Water injection rate	I,O	BBL/DAY	M3/DAY
K	Permeability	I	MD	MD
KRO	Relative permeability to oil	I	–	–
RATIO	Ratio of corner to side well producing rate (default = 2)	I	–	–
RW	Effective wellbore radius	I	FT	M

*The units for these variables are saved by the program.
†In the case of deviated wells or slanted beds, use the true vertical thickness instead of the measured thickness of the formation.

Yes/No Questions

None

Example

Estimate the injection rate of the various patterns if the well spacing is 10 acres. Producing well pressures are estimated to be 300 PSI. The maximum desired bottom hole injection pressure is 1,600 PSI, so the maximum differential pressure between injection and producing wells is 1,300 PSI. Use the pattern dimensions shown in Table 26–1 for the 10-acre spacing. Other data are as follows:

UO = 3.8 CP
K = 92 MD
KRO = 0.62
H = 38 FT
RW = 4 IN

A. LINE DRIVE $\left(\frac{a}{d} \geq 1\right)$

Direct

$$IW = \frac{0.001538 \text{ K KRO H DP}}{UO\left(\log \frac{d}{RW} + 0.682\ \frac{a}{d} - 0.902\right)}$$

Staggered

$$IW = \frac{0.001538 \text{ K KRO H DP}}{UO\left(\log \frac{d}{RW} + 0.682\ \frac{a}{d} - 0.902\right)}$$

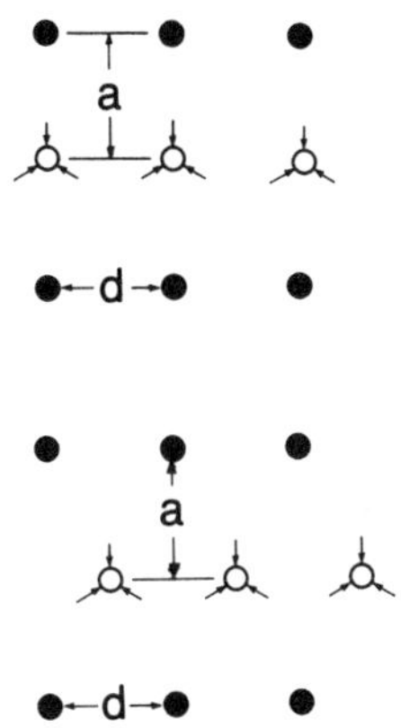

B. FIVE SPOT PATTERN

$$IW = \frac{0.001538 \text{ K KRO H DP}}{UO\left(\log \frac{d}{RW} - 0.2688\right)}$$

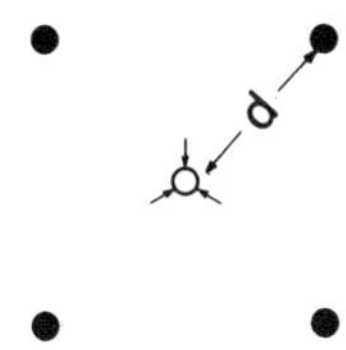

C. SEVEN SPOT PATTERN

$$IW = \frac{0.002051 \text{ K KRO H DP}}{UO\left(\log \frac{d}{RW} - 0.2472\right)}$$

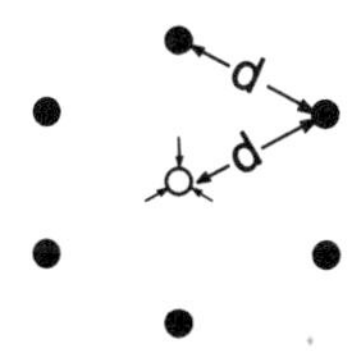

D. INVERTED NINE SPOT PATTERN

Corner Well

$$IW = \frac{0.001538 \text{ K KRO H DP}}{UO\left(\frac{1 + RATIO}{2 + RATIO}\right)\left(\log \frac{d}{RW} - 0.1183\right)}$$

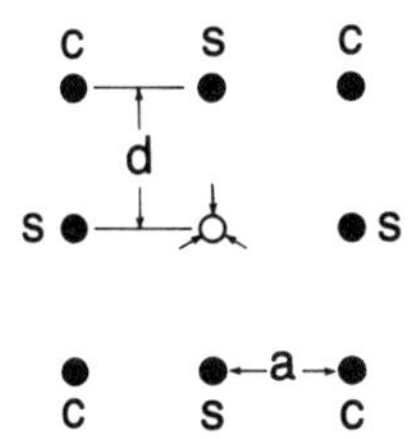

E. INVERTED NINE SPOT PATTERN

Side Well

$$IW = \frac{0.003076 \text{ K KRO H DP}}{UO\left[\left(\frac{3 + RATIO}{2 + RATIO}\right)\left(\log \frac{d}{RW} - 0.1183\right) - \frac{0.301}{2 + RATIO}\right]}$$

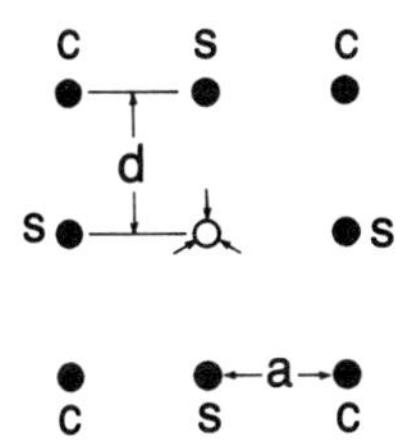

(injection well symbol) = Injection Well ● = Producing Well c = Corner Well s = Side Well

Figure 26–1 *(after Muskat)*

Table 26-1

Pattern	*a (FT)*	*d (FT)*	*Ratio*
Line drive	660	660	–
Five spot	–	933.4	–
Seven spot	–	933.4	–
Inverted nine spot (corner well)	–	660	1.5
Inverted nine spot (side well)	–	660	1.5

Keystrokes (SIZE> =040)	**Display**	**Comments**
[XEQ] [ALPHA] INJ [ALPHA]	*UO=?*	
3.8 [R/S]	*K=?*	
92 [R/S]	*KRO=?*	
.62 [R/S]	*H=?*	
38 [R/S]	*RW=?*	
4 [ALPHA] IN [R/S]	*LD 5 7 9C 9S*	
[A]	*d=?*	Line drive pattern
660 [R/S]	*a=?*	
660 [R/S]	*IW DP*	
[E]	*DP=?*	DP known
1300 [R/S]	*IW DP*	IW printed
[R/S]	*LD 5 7 9C 9S*	
[B]	*d=?*	Five-spot pattern
933.4 [R/S]	*IW DP*	
[E]	*DP=?*	
[R/S]	*IW DP*	IW printed
[R/S]	*LD 5 7 9C 9S*	
[C]	*d=?*	Seven-spot pattern
[R/S]	*IW DP*	
[E]	*DP=?*	
[R/S]	*IW DP*	IW printed
[R/S]	*LD 5 7 9C 9S*	
[D]	*d=?*	Inverted nine-spot pattern for corner well
660 [R/S]	*RATIO=?*	
1.5 [R/S]	*IW DP*	
[E]	*DP=?*	
[R/S]	*IW DP*	IW printed
[R/S]	*LD 5 7 9C 9S*	
[E]	*d=?*	Inverted nine-spot pattern for side well
[R/S]	*RATIO=?*	
1.5 [R/S]	*IW DP*	
[E]	*DP=?*	
[R/S]	*IW DP*	IW printed

```
WATER INJ

UO=3.8000 CP
K=92.0000 MD
KRO=0.6200
H=38.0000 FT
RW=4.0000 IN

LINE DRIVE

d=660.0000 FT
a=660.0000 FT
DP=1,300.0000 PSI
IW=370.6798 BBL/DAY


FIVE SPOT

d=933.4000 FT
IW=358.8163 BBL/DAY


SEVEN SPOT

IW=475.2696 BBL/DAY


INVERTED 9 SPOT, CORNER

d=660.0000 FT
RATIO=1.5000
IW=502.3466 BBL/DAY


INVERTED 9 SPOT, SIDE

RATIO=1.5000
IW=570.1620 BBL/DAY
```

User Instructions

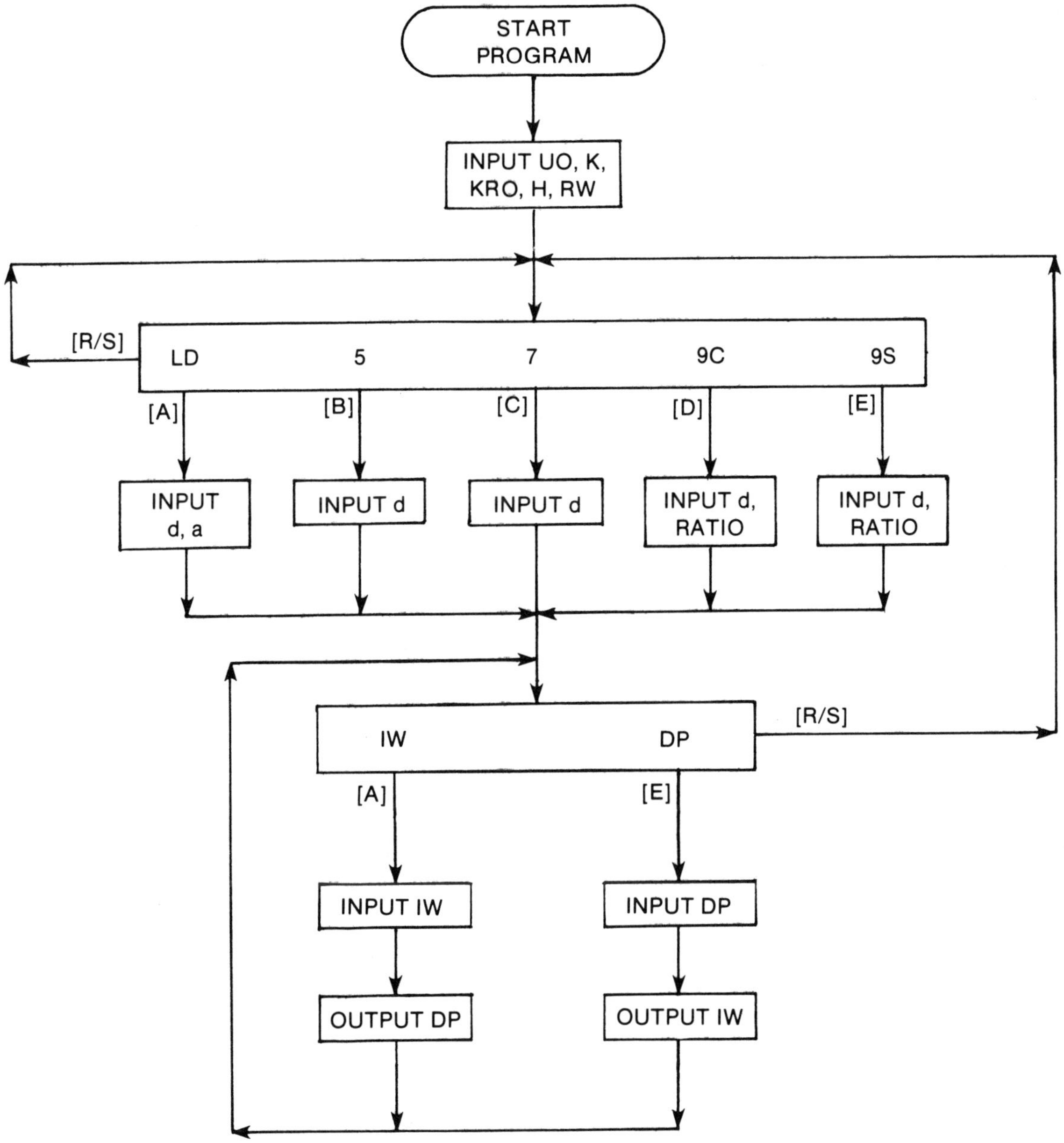

General Information

Memory Requirements

Program length: 572 bytes (3 cards)
Minimum size: 040
Minimum hardware: 41C + 1 memory module

Hidden Options

None

Pac Subroutines Called

TITLE, INU, IN, INK, OUTK

Registers

03	Pressure units
04	Pressure units
06	Water injection rate units
07	Water injection rate units
08	A
09	B
17	DP (PSI)
26	RATIO
27	IW (BBL/DAY)
28	H (FT)
29	K (MD)
30	d (FT)
31	RW (FT)
32	UO (CP)
34	KRO
35	d (FT)
36	a (FT)
37	C
38	D
39	E

Registers 10–16, 18–25, and 33 unused

Flags

None

Program Listing

```
 01♦LBL "INJ"
"WATER INJ"  40
XROM "TITLE"  FC?C 25
PROMPT  SF 27  "KPA"
ASTO 03  "M3/DAY"
ASTO 06  CLA  ASTO 04
ASTO 07  31  STO 00
"CP"  ASTO 01  CLA
ASTO 02  ASTO Z  "PA*S"
ASTO Y  "UO"  XROM "INU"
28  STO 00  "MD"
ASTO 01  ASTO Y  CLA
ASTO 02  ASTO Z  "K"
XROM "INU"  33  STO 00
"KRO"  XROM "IN"  27
"H"  XEQ 07  30  "RW"
XEQ 07

 46♦LBL 15
.001538  STO 08  CL
STO 37  STO 39  2
STO 38  STO 26  STO 36

 56♦LBL 00
"LD 5 7 9C 9S"  PROMPT
GTO 00

 60♦LBL A
"LINE DRIVE"  XEQ 06  35
"a"  XEQ 07  .682
STO 37  .902  GTO 02

 70♦LBL B
"FIVE SPOT"  XEQ 06
.2688  GTO 02

 75♦LBL C
"SEVEN SPOT"  XEQ 06
.002051  STO 08  .2472
GTO 02

 82♦LBL D
XEQ 04  "⊦, CORNER"
XEQ 06  1  STO 38
GTO 01

 89♦LBL E
XEQ 04  "⊦, SIDE"
XEQ 06  .003076  STO 08
3  STO 38  .301  STO 39

 99♦LBL 01
25  STO 00  "RATIO"
XROM "IN"  .1183

105♦LBL 02
STO 09

107♦LBL 03
"IW        DP"  PROMPT
GTO 15

111♦LBL 04
"INVERTED 9 SPOT"  RTN

114♦LBL a
XEQ 08  XEQ 05  RCL 27
X<>Y  /  STO 17  XEQ 12
CF 08  ADV  GTO 03

125♦LBL e
XEQ 11  XEQ 05  RCL 17
*  STO 27  XEQ 09  CF 08
ADV  GTO 03

135♦LBL 05
RCL 26  2  +  RCL 08  *
RCL 29  *  RCL 34  *
RCL 28  *  RCL 35
RCL 31  /  LOG  RCL 09
-  RCL 36  RCL 35  /
RCL 37  *  +  RCL 38
RCL 26  +  *  RCL 39  -
RCL 32  *  /  RTN

169♦LBL 06
ADV  FS? 55  PRA  ADV
34  "d"

176♦LBL 07
STO 00  ASTO T  "FT"
ASTO 01  "M"  ASTO Y
CLA  ASTO 02  ASTO Z
ARCL T  XROM "INU"  RTN

189♦LBL 08
26  STO 00  XEQ 10
XROM "INK"  RDN  STO 06
X<>Y  STO 07  R↑  RTN

200♦LBL 09
XEQ 10  XROM "OUTK"  RDN
STO 06  X<>Y  STO 07  R↑
RTN

209♦LBL 10
"BBL/DAY"  ASTO 01  ASHF
ASTO 02  "IW"  RCL 07
RCL 06  RCL Z  RTN

219♦LBL 11
16  STO 00  XEQ 13
XROM "INK"  RDN  STO 03
X<>Y  STO 04  R↑  RTN

230♦LBL 12
XEQ 13  XROM "OUTK"  RDN
STO 03  X<>Y  STO 04  R↑
RTN

239♦LBL 13
"PSI"  ASTO 01  CLA
ASTO 02  "DP"  RCL 04
RCL 03  RCL Z  END
```

27. CUTCUM — Forecasting Mature Waterfloods

The techniques previously presented in this section and most techniques related to forecasting waterfloods deal with predicting the behavior early in the life of a waterflood. CUTCUM simplifies the forecast for developed waterfloods using an equation that represents the waterflood process in a fully developed waterflood with stabilized operations. The program also permits calculation of water-oil relative permeability ratio KO/KW as a function of water saturation from production data.

CUTCUM allows calculation of a correlation variable, performs a linear regression of the cumulative recovery fraction versus this variable, extrapolates to any future value of water cut or cumulative production, and generates a simple rate-time forecast. It is based on a technique presented by Ershaghi and Omoregie.

Equations

Linear regression:

$$\%OILRF = 100\,(AX + B)$$

$$\%OILRF = 100\,\frac{NP}{N}$$

$$X = \frac{1}{FW} - \ln\left(\frac{1}{FW} - 1\right)$$

$$WOR = \frac{FW}{1 - FW}$$

$$B = \frac{\Sigma X^2\,\Sigma\,OILRF - \Sigma X\,\Sigma X\,OILRF}{n\Sigma X^2 - (\Sigma X)^2}$$

$$A = \frac{\Sigma\,OILRF - nB}{\Sigma X}$$

$$R\uparrow 2 = \frac{B\,\Sigma\,OILRF + A\,\Sigma X\,OILRF - (\Sigma\,OILRF)^2/n}{\Sigma\,OILRF^2 - (\Sigma\,OILRF)^2/n}$$

where n is the number of FW, %OILRF points input by the user.

KO/KW:

$$b = \frac{1}{A(SWI - 1)}$$

$$a = \frac{UO}{UW}\,e^{-b[B(1 - SWI) + SWI]}$$

$$KO/KW = a\,e^{b\frac{\%SW}{100}}$$

FW from NP:

FW is calculated iteratively using Newton's method as follows:

$$FW_{i+1} = FW_i + \frac{(X_i - X)FW_i^2\,(1 - FW_i)}{1 - 2FW_i}$$

X evaluated at the input NP is used as the initial guess.

Forecast:

$$QOAVG_i = \frac{QO_{i-1} + QO_i}{2}$$

$$T_i = T_{i-1} + DT_i$$

$$DT_i = \frac{DNP_i}{365\,QOAVG_i}$$

$$QO_i = (1 - FW_i)\,QO + QW$$

Nomenclature

Symbol	*Variable Name*	*Input or Output*	*English Units*	*SI Units*
A	Slope of %OILRF vs X correlation	I,O	-	-
a	Intercept of straight line portion of KO/KW curve	O	-	-
B	Intercept of %OILRF vs X correlation	I,O	-	-
BEG FW	Beginning FW of forecast	I,O	-	-
b	Slope of straight-line portion of KO/KW curve	O	-	-
DT*	Delta T - change in time	O	YR	YR

Symbol	Variable Name	Input or Output	English Units	SI Units
END FW	Ending FW of forecast	I,O	-	-
FW	Fraction of total flowing stream composed of water	I,O	-	-
FW INC	FW increment of forecast	I,O	-	-
KO/KW	Water-oil relative permeability ratio	O	-	-
N*	Initial oil in place	I	BBL	M3
NP*	Cumulative oil production	I,O	BBL	M3
QO*	Oil producing rate	O	BBL/DAY	M3/DAY
QOAVG*	Average oil producing rate	O	BBL/DAY	M3/DAY
QO+QW*	Total fluid producing rate	I	BBL/DAY	M3/DAY
R↑2	Coefficient of determination	O	-	-
T*	Time	O	YR	YR
WOR	Water-oil ratio	I,O	-	-
X	Correlation variable (function of FW)	I,O	-	-
%OILRF	Volume percent oil recovery factor	I,O	-	-
%SW	Volume percent water saturation	I	-	-
%SWI	Volume percent initial water saturation	I	-	-

*The units for these variables are saved by the program.

Yes/No Questions

None

Example 1

Table 27-1 and Figure 27-1 present the cumulative production for the Plum Nearly (3,800 FT) Field. Initial oil in place was 10.5 MM barrels. In this example, forecast the production to an economic limit water-oil ratio of 50:1. What will be the water cut when 1 MM barrels of waterflood oil have been produced? Generate a rate-time forecast for the production assuming a constant total fluid producing rate of 5,000 BBL/DAY. The graphs of forecast production are shown in Figures 27-2 and 27-3.

Table 27-1

FW	NP (BBL)
0.500	189,230
0.550	211,100
0.571	223,709
0.600	263,548
0.623	275,019
0.650	311,150
0.679	363,263

(Note that the A and B coefficients of this correlation can be input instead of calculated by the program.)

Keystrokes (SIZE> =055)	Display	Comments
[XEQ] [ALPHA] CUTCUM [ALPHA]	*N=?*	
10.5 [EEX] 6 [R/S]	*FIT A,B*	
[A]	*FW=?*	Regression option
.5 [R/S]	*NP=?*	X printed
189230 [R/S]	*FW=?*	
.55 [R/S]	*NP=?*	X printed
211100 [R/S]	*FW=?*	
.571 [R/S]	*NP=?*	X printed
223709 [R/S]	*FW=?*	
.6 [R/S]	*NP=?*	X printed
263548 [R/S]	*FW=?*	
.623 [R/S]	*NP=?*	X printed
275019 [R/S]	*FW=?*	
.65 [R/S]	*NP=?*	X printed
311150 [R/S]	*FW=?*	
.679 [R/S]	*NP=?*	X printed
363263 [R/S]	*FW=?*	
[R/S]	*WOR=?*	
[R/S]	*SW FW NP FOR*	A, B, and R↑2 printed
[B]	*FW=?*	Forecast to future WOR
[R/S]	*WOR=?*	
50 [R/S]	*SW FW NP FOR*	FW, %OILRF, and NP printed
[C]	*NP=?*	Forecast to future NP
[EEX] 6 [RS]	*SW FW NP FOR*	FW and X printed
[E]	*BEG FW=?*	Forecast option
.679 [R/S]	*END FW=?*	
50 [ENTER↑] 51 [÷]	*0.9804*	A WOR of 50 is 50/51 water or FW = 0.9804
[R/S]	*FW INC=?*	
.03 [R/S]	*QO+QW=?*	
5000 [R/S]	*SW FW NP FOR*	Forecast printed

```
FW VS NP

N=10,500,000.00 BBL

FW=0.5000
X=2.0000
NP=189,230.0000 BBL

FW=0.5500
X=2.0189
NP=211,100.0000 BBL

FW=0.5710
X=2.0372
NP=223,709.0000 BBL

FW=0.6000
X=2.0721
NP=263,548.0000 BBL

FW=0.6230
X=2.1074
NP=275,019.0000 BBL

FW=0.6500
X=2.1575
NP=311,150.0000 BBL

FW=0.6790
X=2.2219
NP=363,263.0000 BBL

A=0.0719
B=-0.1252
R↑2=0.9893

WOR=50.0000
FW=0.9804
%OILRF=22.9551
NP=2,410,287.134 BBL

NP=1,000,000.000 BBL
FW=0.8719
X=3.0645

BEG FW=0.6790
END FW=0.9804
FW INC=0.0300
QO+QW=5,000.0000 BBL/DAY
%OILRF=3.4636
NP=363,672.8740 BBL

FW=0.7090
%OILRF=4.0320
NP=423,359.1286 BBL
DNP=59,686.2546 BBL
QOAVG=1,530.0000 BBL/DAY
T=0.1069 YR
DT=0.1069 YR
QO=1,455.0000 BBL/DAY

FW=0.7390
%OILRF=4.7008
NP=493,582.1184 BBL
DNP=70,222.9899 BBL
QOAVG=1,380.0000 BBL/DAY
T=0.2463 YR
DT=0.1394 YR
QO=1,305.0000 BBL/DAY

FW=0.7690
%OILRF=5.4855
NP=575,977.0779 BBL
DNP=82,394.9595 BBL
QOAVG=1,230.0000 BBL/DAY
T=0.4298 YR
DT=0.1835 YR
QO=1,155.0000 BBL/DAY

FW=0.7990
%OILRF=6.4101
NP=673,061.2822 BBL
DNP=97,084.2044 BBL
QOAVG=1,080.0000 BBL/DAY
T=0.6761 YR
DT=0.2463 YR
QO=1,005.0000 BBL/DAY

FW=0.8290
%OILRF=7.5120
NP=788,761.9709 BBL
DNP=115,700.6886 BBL
QOAVG=930.0000 BBL/DAY
T=1.0169 YR
DT=0.3408 YR
QO=855.0000 BBL/DAY

FW=0.8590
%OILRF=8.8521
NP=929,470.8948 BBL
DNP=140,708.9239 BBL
QOAVG=780.0000 BBL/DAY
T=1.5112 YR
DT=0.4942 YR
QO=705.0000 BBL/DAY

FW=0.8890
%OILRF=10.5370
NP=1,106,390.013 BBL
DNP=176,919.1181 BBL
QOAVG=630.0000 BBL/DAY
T=2.2806 YR
DT=0.7694 YR
QO=555.0000 BBL/DAY

FW=0.9190
%OILRF=12.7778
NP=1,341,667.275 BBL
DNP=235,277.2620 BBL
QOAVG=480.0000 BBL/DAY
T=3.6235 YR
DT=1.3429 YR
QO=405.0000 BBL/DAY

FW=0.9490
%OILRF=16.0887
NP=1,689,314.227 BBL
DNP=347,646.9510 BBL
QOAVG=330.0000 BBL/DAY
T=6.5097 YR
DT=2.8862 YR
QO=255.0000 BBL/DAY

FW=0.9790
%OILRF=22.4620
NP=2,358,509.235 BBL
DNP=669,195.0090 BBL
QOAVG=180.0000 BBL/DAY
T=16.6953 YR
DT=10.1856 YR
QO=105.0000 BBL/DAY

FW=0.9804
%OILRF=22.9551
NP=2,410,287.134 BBL
DNP=51,777.8990 BBL
QOAVG=101.5196 BBL/DAY
T=18.0927 YR
DT=1.3973 YR
QO=98.0392 BBL/DAY
```

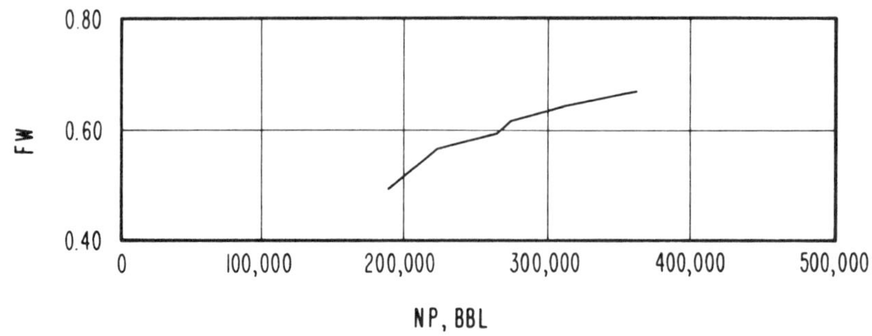

Figure 27–1

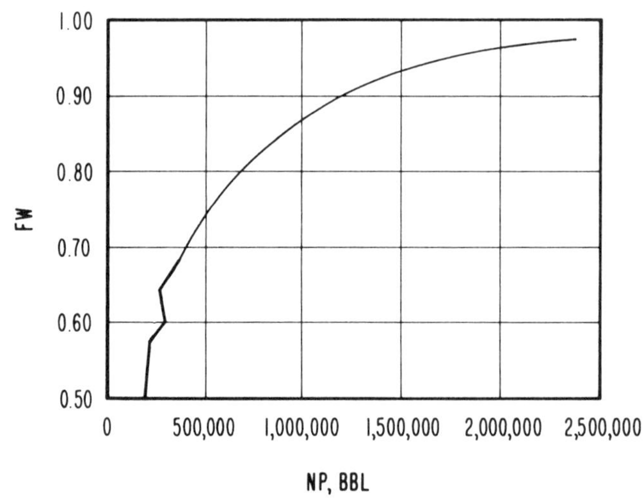

Figure 27–2

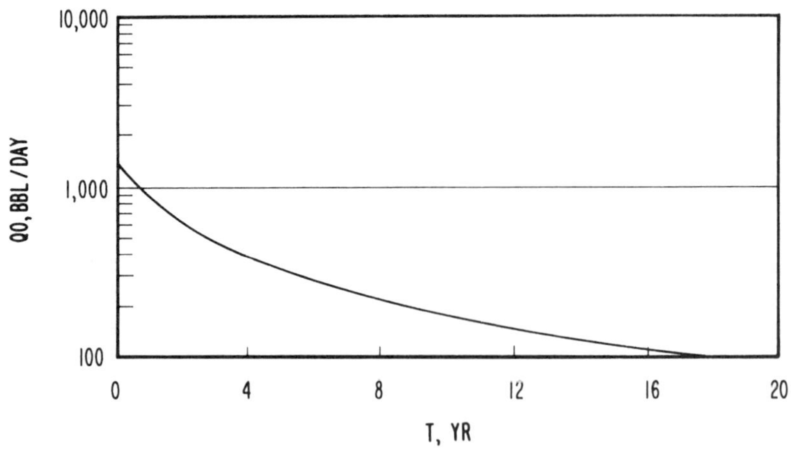

Figure 27–3

Example 2

Based on the following properties, generate KO/KW data from this performance data. A graph of calculated KO/KW data is given in Figure 27–4.

Oil viscosity = 3.5 CP
Water viscosity = 0.70 CP
Initial water saturation = 36%

Keystrokes	Display	Comments
[A]	*%SWI=?*	Water saturation option
36 [R/S]	*UO=?*	
3.5 [R/S]	*UW=?*	
.7 [R/S]	*%SW=?*	a and b printed
40 [R/S]	*%SW=?*	KO/KW printed
45 [R/S]	*%SW=?*	KO/KW printed
50 [R/S]	*%SW=?*	KO/KW printed
55 [R/S]	*%SW=?*	KO/KW printed
60 [R/S]	*%SW=?*	KO/KW printed
70 [R/S]	*%SW=?*	KO/KW printed

```
%SWI=36.0000
UO=3.5000 CP
UW=0.7000 CP
a=2,186.4191
b=-21.7249

%SW=40.0000
KO/KW=0.3679

%SW=45.0000
KO/KW=0.1242

%SW=50.0000
KO/KW=0.0419

%SW=55.0000
KO/KW=0.0141

%SW=60.0000
KO/KW=0.0048

%SW=70.0000
KO/KW=0.0005
```

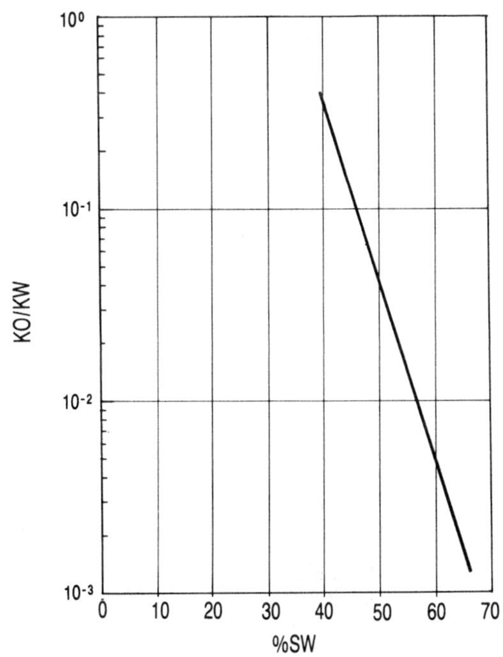

Figure 27–4

User Instructions

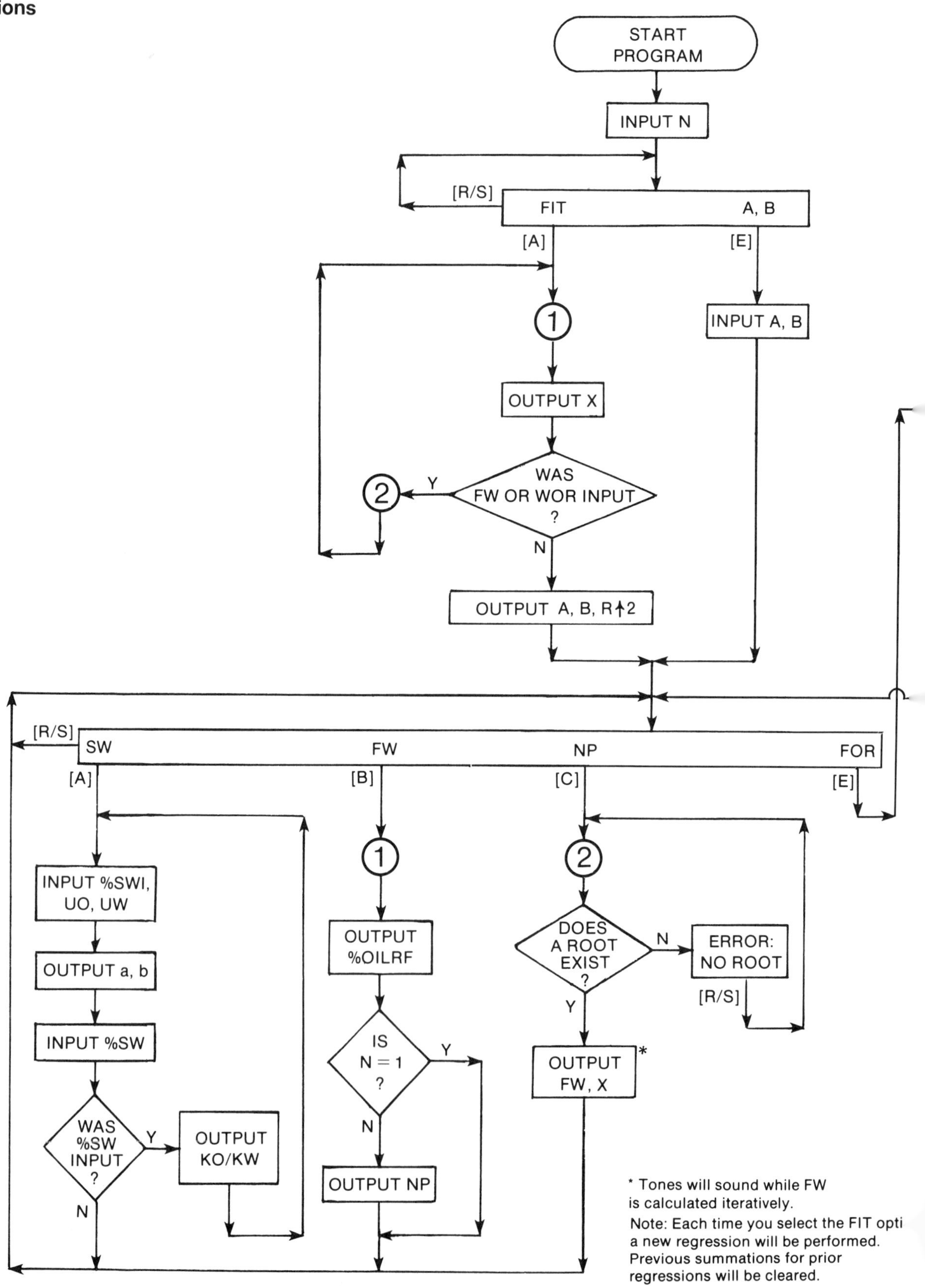

* Tones will sound while FW is calculated iteratively.

Note: Each time you select the FIT opti a new regression will be performed. Previous summations for prior regressions will be cleared.

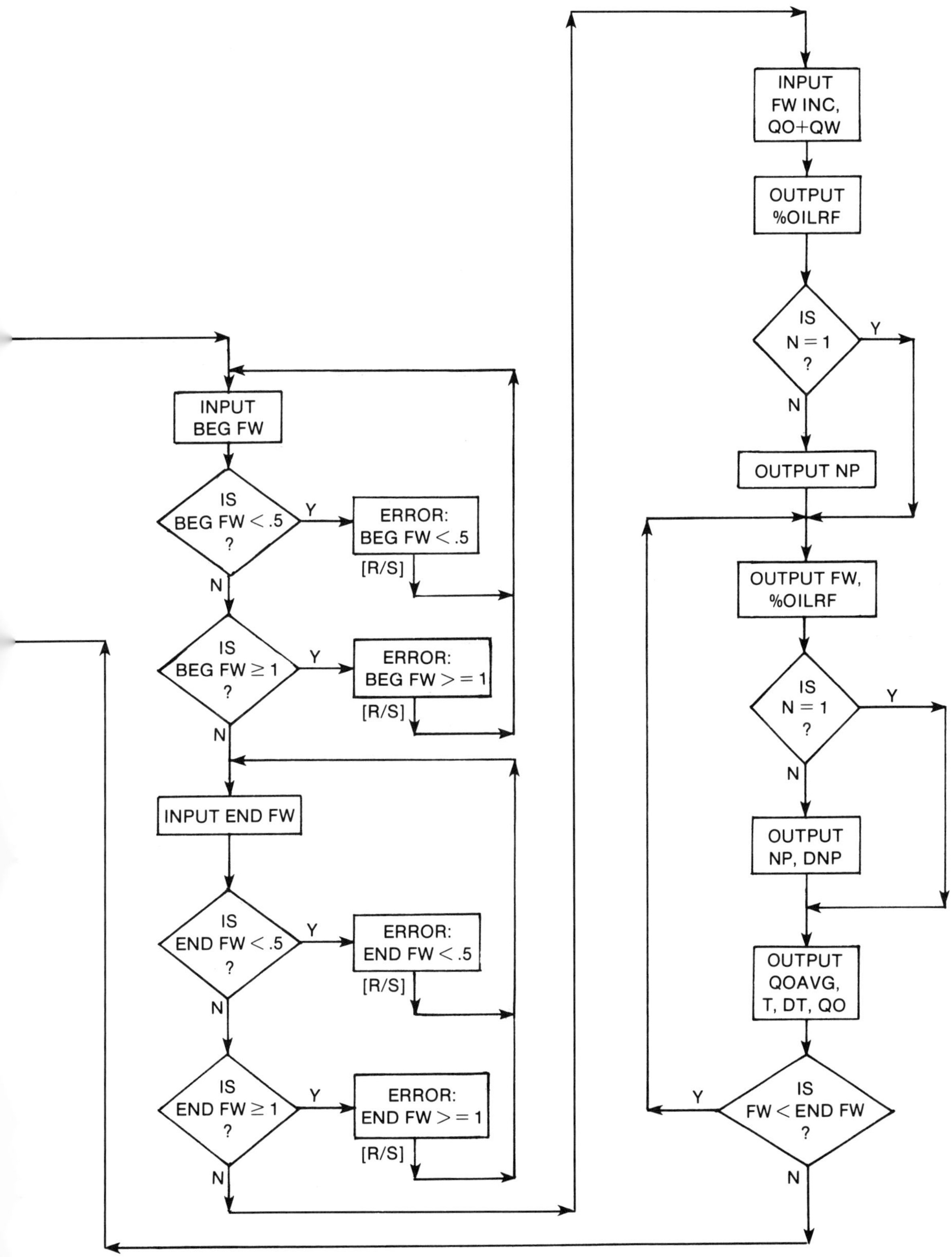
INPUT BEG FW
IS BEG FW < .5 ?
ERROR: BEG FW < .5
[R/S]
IS BEG FW ≥ 1 ?
ERROR: BEG FW >= 1
[R/S]
INPUT END FW
IS END FW < .5 ?
ERROR: END FW < .5
[R/S]
IS END FW ≥ 1 ?
ERROR: END FW >= 1
[R/S]
INPUT FW INC, QO+QW
OUTPUT %OILRF
IS N = 1 ?
OUTPUT NP
OUTPUT FW, %OILRF
IS N = 1 ?
OUTPUT NP, DNP
OUTPUT QOAVG, T, DT, QO
IS FW < END FW ?
Y
N

User Instructions

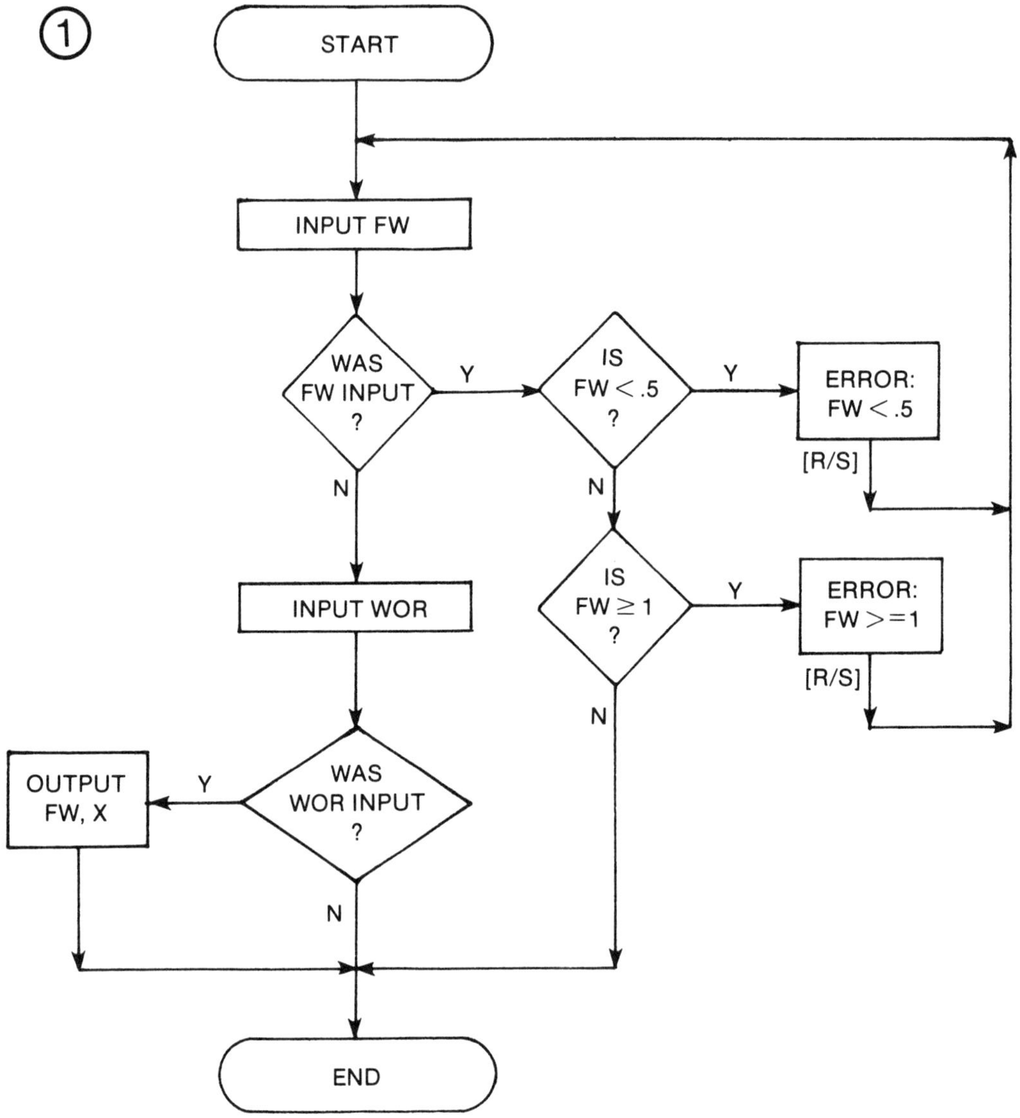

User Instructions

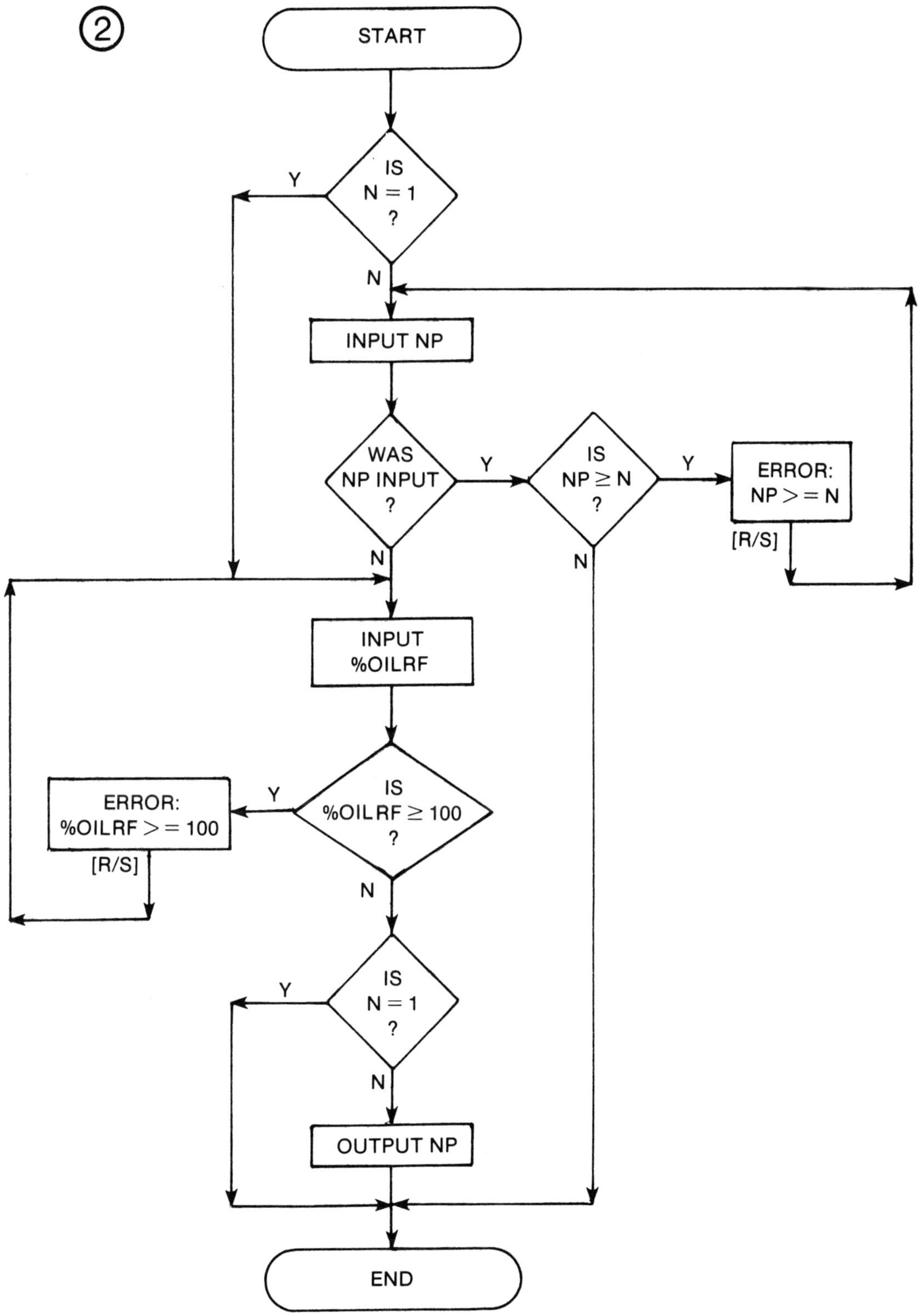

General Information

Memory Requirements

Program length:	1054 bytes (5 cards)
Minimum size:	055
Minimum hardware:	41C + quad memory module or 41CV*

*This program can be run in a 41C + 3 memory modules but will not leave a port available for a printer.

Hidden Options

None

Pac Subroutines Called

TITLE, IN, OUT, INU, OUTK, INK

Registers

03	Oil production units
04	Oil production units
06	Oil producing rate units
07	Oil producing rate units
08	Time units
09	Time units
21	%SWI
26	A
27	B
30	%SW
32	UO (CP)
33	UW (CP)
35	X
36	%OILRF
37	NP (BBL)
38	ΣX
39	ΣX^2
40	ΣOILRF
41	$\Sigma OILRF^2$
42	ΣX OILRF
43	n
44	WOR
45	FW, BEG FW
46	N (BBL)
47	a
48	b
49	END FW
50	FW INC
51	QO+QW (BBL/DAY)
52	NP_{i-1} (BBL)
53	QO_{i-1} (BBL/DAY)
54	T_{i-1} (YR)

Registers 10–20, 28, 29, 31, and 34 unused

Note: The summation registers have been moved to start at register 38.

Flags

00	Set:	Do not allow input of WOR.
	Clear:	Allow input of WOR.
01	Set:	Oil producing rate units not yet input.
	Clear:	Oil producing rate units have been input.
02	Set:	Time units not yet input.
	Clear:	Time units have been input.

Program Listing

```
 01♦LBL "CUTCUM"
"FW VS NP"  55
XROM "TITLE"  FC?C 25
PROMPT  CF 00  SF 01
SF 02  CF 03  SF 27
ΣREG 38  "M3"  ASTO 03
"⊦/DAY"  ASTO 06  "YR"
ASTO 08  CLA  ASTO 04
ASTO 07  ASTO 09  CLX
STO 54  45  "N"  XEQ 24
CF 08  RND  1  X=Y?
SF 03  ADV

 34♦LBL 14
"FIT        A,B"  PROMPT
GTO 14

 38♦LBL E
25  STO 00  "A"
XROM "IN"  "B"
XROM "IN"  ADV  GTO 15

 47♦LBL A
CLΣ

 49♦LBL 13
XEQ 17  FC? 22  GTO 00
"X"  XROM "OUT"  XEQ 22
RCL 36  1  %  RCL 35  Σ+
ADV  GTO 13

 63♦LBL 00
RCL 39  RCL 40  *
RCL 38  RCL 42  *  -
RCL 43  RCL 39  *
RCL 38  X↑2  -  /
STO 27  RCL 40  RCL 43
RCL Z  *  -  RCL 38  /
STO 26  "A"  XROM "OUT"
RCL 27  "B"  XROM "OUT"
RCL 27  RCL 40  *
RCL 26  RCL 42  *  +
RCL 40  X↑2  RCL 43  /
-  RCL 41  LASTX  -  /
"R↑2"  XROM "OUT"  ADV

111♦LBL 15
"SW FW NP FOR"  PROMPT
GTO 15

115♦LBL D
116♦LBL E
GTO 16

118♦LBL B
XEQ 17  XEQ 21  ADV
GTO 15

123♦LBL A
20  STO 00  "%SWI"
XROM "IN"  1  %  1  -
1/X  RCL 26  /  STO 48
31  STO 00  "CP"
ASTO 01  CLA  ASTO 02
ASTO Z  "PA*S"  ASTO Y
"UO"  XROM "INU"  "PA*S"
ASTO Y  CLA  ASTO Z
"UW"  XROM "INU"  100
RCL 21  -  RCL 27  *
RCL 21  +  RCL 48  %
CHS  E↑X  RCL 32  *
RCL 33  /  STO 47  "a"
XROM "OUT"  RCL 48  "b"
XROM "OUT"

174♦LBL 12
ADV  29  STO 00  "%SW"
XROM "IN"  FC? 22
GTO 15  1  %  RCL 48  *
E↑X  RCL 47  *  "KO/KW"
```

Program Listing (cont.)

```
XROM "OUT"  GTO 12

192♦LBL C
193♦LBL 11
XEQ 22  RCL 36  1  %
RCL 27  -  RCL 26  /
STO 00  .5  STO 45
XEQ 20  RCL 00  -  .995
STO 01  STO 45  X<>Y
XEQ 20  RCL 00  -  *
X<0?  GTO 10  "NO ROOT"
TONE 3  PROMPT  GTO 11

222♦LBL 10
TONE 5  RCL 01  1/X
RCL 01  1/X  1  -  LN  -
RCL 00  -  RCL 01  X↑2
*  1  RCL 01  -  *  1
RCL 01  2  *  -  /
ENTER↑  X<> 01  ST+ 01
/  ABS  1 E-4  X<=Y?
GTO 10  RCL 01  STO 45
TONE 9  XEQ 19  "X"
XROM "OUT"  ADV  GTO 15

263♦LBL 16
SF 00  44  "BEG FW"
XEQ 18  48  "END FW"
XEQ 18  CF 00  "FW INC"
XROM "IN"  FS?C 01
SF 08  "QO+QW"  XEQ 25
CF 08  1  RCL 45  -  *
STO 53  XEQ 20  XEQ 21
STO 52  GTO 01

288♦LBL 09
XEQ 19  XEQ 21  RCL 52
-  STO 52  "DNP"  FC? 03
XEQ 23  RCL 53  1
RCL 45  -  RCL 51  *
STO 53  +  2  /  "QOAVG"
XEQ 26  365  *  RCL 37
X<> 52  X<>Y  /  ENTER↑
X<> 54  +  FS?C 02
SF 08  "T"  XEQ 27
CF 08  X<> 54  "DT"
XEQ 27  RCL 53  "QO"
XEQ 26

329♦LBL 01
ADV  RCL 49  RCL 45
RCL 50  +  RND  X<Y?
GTO 02  X<>Y  RCL 45
X=Y?  GTO 03  RCL 49
RND

344♦LBL 02
LASTX  STO 45  GTO 09

348♦LBL 03
XEQ 20  GTO 15

351♦LBL 17
44  "FW"

354♦LBL 18
STO 00  XROM "IN"
FC? 22  GTO 05  CLA
ARCL 05  "⊢ < .5"  .5
X>Y?  GTO 04  CLA
ARCL 05  "⊢ >= 1"  X<>Y
1  X>Y?  GTO 20

372♦LBL 04
TONE 3  PROMPT  RCL 00
1  -  CLA  ARCL 05
GTO 18

381♦LBL 05
FS? 00  RTN  43  STO 00
"WOR"  XROM "IN"  1  -
1/X  1  X<>Y  -  STO 45
FC? 22  RTN

397♦LBL 19
RCL 45  "FW"  XROM "OUT"

401♦LBL 20
1  RCL 45  -  1/X  1  -
LN  RCL 45  1/X  +
STO 35  RTN

414♦LBL 21
RCL 26  *  RCL 27  +
100  *  STO 36  "%OILRF"
XROM "OUT"  GTO 00

425♦LBL 22
FS? 03  GTO 08  36  "NP"
XEQ 24  FC? 22  FS? 23
GTO 02

434♦LBL 08
35  STO 00  "%OILRF"
XROM "IN"  100  X>Y?
GTO 00  "%OILRF >=100"
TONE 3  PROMPT  GTO 08

446♦LBL 00
RCL 46  %  STO 37
FS? 03  RTN  "NP"

453♦LBL 23
XEQ 01  XROM "OUTK"  RDN
STO 03  X<>Y  STO 04  R↑
RTN

462♦LBL 24
STO 00  XEQ 01
XROM "INK"  RDN  STO 03
X<>Y  STO 04  R↑  RTN

472♦LBL 01
ASTO T  "BBL"  ASTO 01
CLA  ASTO 02  ARCL T
RCL 04  RCL 03  RCL Z
RTN

483♦LBL 02
RCL 46  /  100  *
STO 36  LASTX  X<=Y?
GTO 03  RCL 37  RTN

494♦LBL 03
"NP >= N"  TONE 3
PROMPT  GTO 22

499♦LBL 25
XEQ 04  XROM "INK"  RDN
STO 06  X<>Y  STO 07  R↑
RTN

508♦LBL 26
XEQ 04  XROM "OUTK"  RDN
STO 06  X<>Y  STO 07  R↑
RTN

517♦LBL 04
ASTO T  "BBL/DAY"
ASTO 01  ASHF  ASTO 02
CLA  ARCL T  RCL 07
RCL 06  RCL Z  RTN

529♦LBL 27
ASTO T  "YR"  ASTO 01
CLA  ASTO 02  ARCL T
RCL 09  RCL 08  RCL Z
XROM "OUTK"  RDN  STO 08
X<>Y  STO 09  R↑  END
```

Section 8
Well Log Analysis

28. RW — Calculating RW from SP

Calculates water resistivity from the SP log and PPM NaC1 equivalency.

29. XPLOT — Cross Plot Porosities

Calculates cross plot porosities for density-neutron or sonic-neutron logs with options to input densities and transit times or porosities. Also calculates RWa, M, and N for lithology identification.

30. H2OSAT — Water Saturation Calculations

Calculates water saturation for clean or shaly sands and carbonates. Shale content can be calculated by five options.

31. SLANT — True Stratigraphic and Vertical Thicknesses

Calculates true vertical thickness and true stratigraphic thickness for deviated wells and slanted beds.

28. RW — Calculating RW From SP

The SP log is a widely run tool that measures the spontaneous potential due to the combination of electrochemical and electrokinetic components. The electrokinetic component is generally quite small. The static SP (SSP) is proportional to the logarithm of the electrochemical activities of the formation water divided by the mud filtrate. The electrochemical activity of a solution is related to its salt content and resistivity. Therefore, a value of RW can be estimated from the SP log, and this process is simplified with the RW program.

However, the SP log has certain weaknesses in determining RW (Bateman and Konen). These are summarized as follows:

- the full SSP is not developed in thin beds
- shaly formations have a reduced SP deflection
- shallow invasion can cause distortions
- salinity changes can cause SP base-line shifts
- the electrokinetic component can be fairly large in depleted reservoirs or tight formations drilled with fresh mud
- resistivity of mud samples may not be representative
- the relationship between RMF and RMFE or RWE and RW may not be valid
- the SP curve is often noisy or affected by magnetism

The SP is more reliable in low to moderate resistivities rather than high resistivities. The SP is frequently the only source of RW for exploratory wells or wells that produce no water. Drillstem tests generally do not give valid RW values due to filtrate invasion. Water samples from wells that have produced large amounts of water are the best source of RW. Another excellent source of RW is the RWa calculation in a 100% water-saturated formation.

Equations

$$\text{RMF75 (RMF at 75 F)} = \text{RMF}\,\frac{\text{TMF} + 7}{82}$$

If RMF75 > 0.1,

$$\text{RMFE (equivalent RMF)} = 0.85\ \text{RMF75}$$

If RMF75 ≤ 0.1,

$$\text{RMFE} = \frac{146\ \text{RMF75} - 5}{337\ \text{RMF75} + 77}$$

$$\text{RWE (equivalent RW)} = \frac{\text{RMFE}}{10^{\text{SP}/(60 + \text{T}/7.5)}}$$

If RWE ≥ 0.12,

$$\text{RW75 (RW at 75 F)} = -0.58 + 10^{(0.69\ \text{RWE} - 0.24)}$$

If RWE < 0.12,

$$\text{RW75} = \frac{77\ \text{RWE} + 5}{146 - 337\ \text{RWE}}$$

$$\text{RW} = \text{RW75}\,\frac{82}{\text{T} + 7}$$

$$\text{PPM} = 10^{\text{A}}$$

$$\text{A} = \frac{3.562 - \log(\text{RW75} - 0.0123)}{0.955}$$

If T unknown:

$$\text{T} = \text{SURF T} + \text{T GRAD}\,\frac{\text{DEPTH}}{100}$$

Nomenclature

Symbol	*Variable Name*	*Input or Output*	*English Units*	*SI Units*
DEPTH	Depth	I	FT	M
RMF	Mud filtrate resistivity	I	OHM*M*	OHM*M*
RW	Water resistivity	O	OHM*M*	OHM*M*
SP	Spontaneous potential	I	MV*	MV*
SURF T	Surface temperature	I	F	C
TMF	Mud filtrate temperature	I	F	C
T GRAD	Temperature gradient	I	F/100 FT*	F/100 FT*

*These units are not allowed by the unit management system in the Pac. Special input and output subroutines were included in this program to provide these units for annotation purposes only.

Yes/No Questions

None

Example

RW requires the user first to input the resistivity and temperature of the mud filtrate. The formation temperature can be input as a temperature or as a surface temperature and temperature gradient in conjunction with depth. For this example, use RMF = 0.15 OHM*M at TMF = 110 F. Note that the SP sign convention is that normal SP deflection (positive) is to the left (RMF > RW).

First calculate RW with T = 150 F and SP = 40 MV. Then assume the surface temperature is 70 F and the temperature gradient is 1.5 F/100 FT and calculate RW at the following depths:

DEPTH (FT)	SP (MV)
3,950	18
4,725	36
4,728	39

Finally, calculate the PPM NaCl equivalency at the 4,728 FT depth.

Keystrokes (SIZE> =028)	Display	Comments
[XEQ] [ALPHA] RW [ALPHA]	*RMF=?*	
.15 [R/S]	*TMF=?*	
110 [R/S]	*T DEPTH PPM*	
[A]	*T=?*	T known
150 [R/S]	*SP=?*	
40 [R/S]	*T DEPTH PPM*	RW printed
[C]	*SURF T=?*	Depth known. SURF T and T GRAD must be input the first time the DEPTH option is used
70 [R/S]	*T GRAD=?*	
1.5 [R/S]	*DEPTH=?*	
3950 [R/S]	*SP=?*	T printed
18 [R/S]	*T DEPTH PPM*	RW printed
[C]	*DEPTH=?*	
4725 [R/S]	*SP=?*	T printed
36 [R/S]	*T DEPTH PPM*	RW printed
[C]	*DEPTH=?*	
4728 [R/S]	*SP=?*	
39 [R/S]	*T DEPTH PPM*	RW printed
[E]	*T DEPTH PPM*	PPM printed

```
RW FROM SP

RMF=0.1500 OHM*M
TMF=110.0000 F

T=150.0000 F
SP=40.0000 MV
RW=0.0389 OHM*M

SURF T=70.0000 F
T GRAD=1.5000 F/100 FT
DEPTH=3,950.0000 FT
T=129.2500 F
SP=18.0000 MV
RW=0.0721 OHM*M

DEPTH=4,725.0000 FT
T=140.8750 F
SP=36.0000 MV
RW=0.0440 OHM*M

DEPTH=4,728.0000 FT
T=140.9200 F
SP=39.0000 MV
RW=0.0416 OHM*M

PPM=97,529.3979
```

General Information

Memory Requirements

Program length: 462 bytes (3 cards)
Minimum size: 028
Minimum hardware: 41C + 1 memory module

Hidden Options

None

Pac Subroutines Called

TITLE, INU, T, OUT, OUTU

Registers

03 RMF75 (OHM*M)
04 SP (MV)
06 SURF T (F)
07 T GRAD (F/100 FT)
08 DEPTH (FT)
09 RW75 (OHM*M)
26 RMF (OHM*M)
27 TMF (F)
Registers 10–15 and 17–25 unused

Flags

None

User Instructions

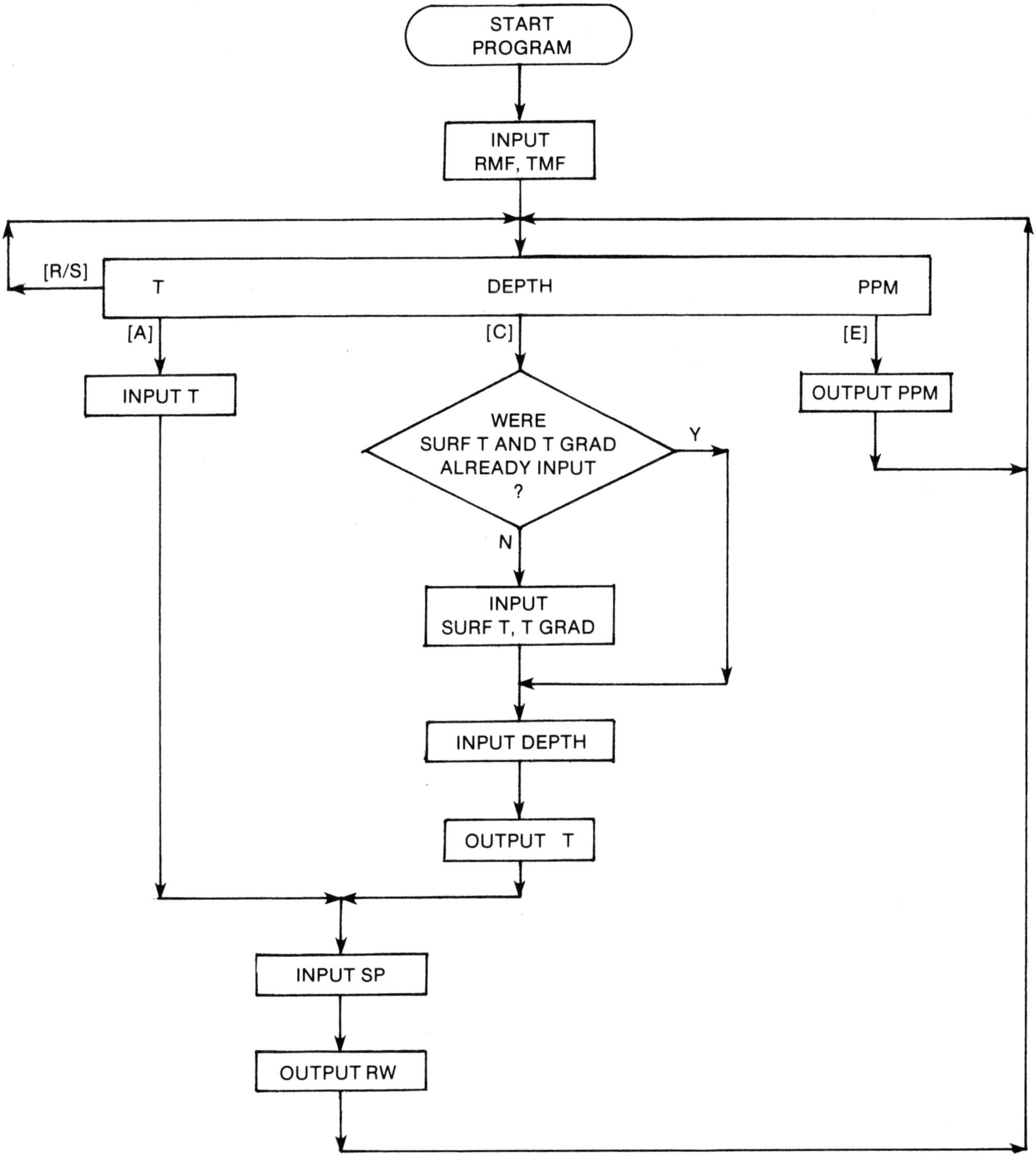

Program Listing

```
01♦LBL "RW"
"RW FROM SP" 28
XROM "TITLE" FC?C 25
PROMPT SF 27 25
STO 00 "OHM*M" ASTO 01
CLA ASTO 02 "RMF"
XEQ 07 XEQ 06 "TMF"
XROM "INU" 7 + 82 /
RCL 26 * STO 03 ADV

27♦LCL 15
" T DEPTH PPM" PROMPT
GTO 15

31♦LBL A
XROM "T" GTO 16

34♦LBL E
3.562 RCL 16 7 +
RCL 09 * 82 / .0123
- LOG - .955 / 10↑X
"PPM" XROM "OUT" ADV
GTO 15

54♦LBL B
55♦LBL C
FC?C 08 GTO 00 5
STO 00 XEQ 06 "SURF T"
XROM "INU" "F/100 FT"
ASTO 01 ASHF ASTO 02
"T GRAD" XEQ 07

69♦LBL 00
7 STO 00 "FT" ASTO 01
"M" ASTO Y CLA
```

```
ASTO 02 ASTO Z "DEPTH"
XROM "INU" RCL 07 * 1
% RCL 06 + STO 16
XEQ 06 "T" XROM "OUTU"

91♦LBL 16
3 STO 00 "MV" ASTO 01
CLA ASTO 02 "SP"
XEQ 07 RCL 16 7.5 /
60 + / 10↑X .1
RCL 03 X>Y? GTO 02
146 * 5 - R↑ / 337
RCL 03 * 77 + GTO 03

123♦LBL 02
85 % R↑

127♦LBL 03
/ .12 X<=Y? GTO 04
RCL Y 77 * 5 + 146
R↑ 337 * - / GTO 05

144♦LBL 04
X<>Y 69 % .24 -
10↑X .58 -

153♦LBL 05
82 * RCL 16 7 + /
STO 09 "OHM*M" ASTO 01
CLA ASTO 02 "RW"
XEQ 08 ADV GTO 15

169♦LBL 06
"F" ASTO 01 "C"
ASTO Y CLA ASTO 02
```

```
ASTO Z RTN

178♦LBL 07
AOFF ASTO 05 CF 22
ISG 00 CLD RCL IND 00
"⊢=?" CF 21 AVIEW CLA
FS? 55 SF 21 STOP
AOFF RCL IND 00 FC? 22
GTO 09 X<>Y STO IND 00
CF 21 FC? 55 RTN CLA
ARCL 05 "⊢=" ARCL X
GTO 10

206♦LBL 08
AOFF STO 00 "⊢="
ARCL X FS? 55 GTO 10
CF 21 AVIEW CLA
ARCL 01 ARCL 02 FS? 55
SF 21 STOP AOFF
RCL 00

223♦LBL 09
CF 21 FS? 55 SF 21
RTN

228♦LBL 10
"⊢ " ARCL 01 ARCL 02
SF 21 PRA END
```

29. XPLOT — Cross Plot Porosities

The formation porosity is used in a variety of reservoir engineering calculations and is normally determined from well-log analysis and perhaps core analysis. The three most common porosity tools — the sonic, density, and neutron logs — are all sensitive to formation lithology. Porosity determination by comparison of the responses of two porosity logging devices helps evaluate the effects of lithology and give a better estimate for the porosity. XPLOT allows the user to calculate cross plot porosities from a density-neutron or sonic-neutron pair of logs, as well as calculate RWa and the M and N values for use in a matrix identification plot.

Determining formation lithology is readily accomplished by comparing the apparent lithology values RHOMAa and DTMAa. This technique was originally described by Clavier and Rust. A matrix identification plot is shown in Figure 29–1. The value for M can be calculated by the program. RWa is also a very useful index. In 100% water-saturated formations, it is equal to RW. High values of RWa relative to RW indicate the possible presence of hydrocarbons.

Table 29–1 presents common values for bulk density for different substances. Table 29–2 presents typical sonic velocities and transit times for use with acoustic logs.

Table 29–1

Compound	*Bulk Density (G/CM3)*
Quartz	2.648
Calcite	2.71
Dolomite	2.876
Anhydrite	2.977
Anthracite coal	1.355-1.796
Bituminous coal	1.173-1.514
Fresh water	1.00
Salt water	1.135
Typical oil	0.85
Methane	1.335 RHO C1-0.188
Typical gas	1.325 RHO GAS-0.188

where RHO C1 and RHO GAS are the densities of methane and the typical gas, respectively.

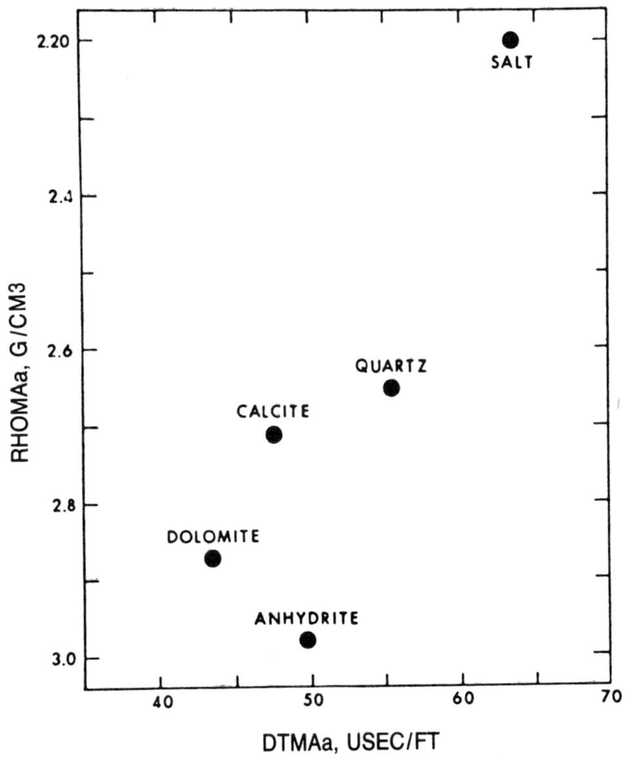

Figure 29–1
(after Bateman and Konen)

Table 29–2

Formation	VMA (FT/S)	DTMA (US/FT)
Sandstones		
Unconsolidated	17,000−	58.8+
Semiconsolidated	18,000	55.6
Consolidated	19,000	52.6
Limestone	21,000	47.6
Dolomite	23,000	43.5
Shale	6,000 to 16,000+	167 to 62.5−
Calcite	22,000	45.5
Anhydrite	20,000	50.0
Granite	20,000	50.0
Gypsum	19,000	52.6
Quartz	18,000	55.6
Salt	15,000	66.7
Fluids		
Water with 20% NaCl	5,300	189
Water with 10% NaCl	4,800	208
Fresh water	4,600	218
Typical oil	4,200	238
Methane	1,600	626
Air	1,100	910

Equations

Density-neutron equations:

$$\text{POR-X} = \frac{\text{POR-Da POR-N} - \text{POR-D POR-Na}}{\text{POR-Da} - \text{POR-Na}}$$

If POR-N ≥ POR-D,

POR-Da (apparent density porosity of pseudomineral)

$$= \frac{1.29}{\text{RHO MF} - 2.71}$$

POR-Na (apparent neutron porosity of pseudomineral)

$$= 0.7 - 10^{-(5\ \text{POR-N} + 0.16)}$$

If POR-N < POR-D

POR-Da = 1.0

$$\text{POR-Na} = -(2.06\ \text{POR-N} + 1.17) + 10^{-(16\ \text{POR-N} + 0.4)}$$

$$\text{POR-D} = \frac{2.71 - \text{RHO b}}{2.71 - \text{RHO MF}}$$

$$\text{RHOMAa} = \frac{\text{RHO b} - |\text{POR-X}|\ \text{RHO MF}}{1 - |\text{POR-X}|}$$

Sonic-neutron equations:

$$\text{POR-X} = \frac{\text{POR-Sa POR-N} - \text{POR-S POR-Na}}{\text{POR-Sa} - \text{POR-Na}}$$

If POR-N ≥ POR-S

POR-Sa (apparent sonic porosity of pseudomineral)

$$= -0.146$$

$$\text{POR-Na} = 0.5 - 10^{-(5\ \text{POR-N} + 0.3)}$$

If POR-N < POR-S,

POR-Sa = 0.50

$$\text{POR-Na} = -(0.62\ \text{POR-N} + 0.36) + 10^{-(18\ \text{POR-N} + 0.92)}$$

$$\text{POR-S} = \frac{\text{DT} - 47.6}{141.4}$$

$$\text{DTMAa} = \frac{\text{DT} - 189\ |\text{POR-X}|}{1 - |\text{POR-X}|}$$

$$\text{RWa} = \text{POR-X}^2\ \text{RT}$$

$$\text{M} = \frac{189 - \text{DT}}{100\ (\text{RHO b} - \text{RHO MF})}$$

$$\text{N} = \frac{1 - \text{POR-N}}{\text{RHO b} - \text{RHO MF}}$$

$$\text{POR-D} = \frac{\%\text{POR-D}}{100}$$

$$\text{POR-N} = \frac{\%\text{POR-N}}{100}$$

$$\text{POR-S} = \frac{\%\text{POR-S}}{100}$$

$$\text{POR-X} = \frac{\%\text{POR-X}}{100}$$

Nomenclature

Symbol	Variable Name	Input or Output	English Units	SI Units
DT	Transit time	I,O	US/FT*	US/FT*
DTMAa	Apparent matrix transit time	O	US/FT*	US/FT*
M	Lithology indicator	O	–	–

Symbol	Variable Name	Input or Output	English Units	SI Units
N	Lithology indicator	O	-	-
RHO-MAa†	Apparent matrix density	O	G/CM3	G/CM3
RHO MF†	Mud filtrate density	I	G/CM3	G/CM3
RHO b†	Bulk density	I,O	G/CM3	G/CM3
RT	True formation resistivity	O	OHM*M*	OHM*M*
RWa	Apparent water resistivity	O	OHM*M*	OHM*M*
%POR-D	Percent density limestone porosity	I,O	-	-
%POR-N	Percent neutron limestone porosity	I,O	-	-
%POR-S	Percent sonic limestone porosity	I,O	-	-
%POR-X	Percent cross plot porosity	I,O	-	-

*These units are not allowed by the Unit Management System in the Pac. Special input and output subroutines were included in this program to provide these units for annotation purposes only.
†The units for these variables are saved by the program.

Yes/No Questions

None

Example

XPLOT allows a variety of solution methods that are demonstrated in this example. Table 29-3 presents the sonic, density, neutron, and resistivity readings at two different points. Note that one point has sonic and density (limestone) porosities, while the other has transit times and measured densities. Calculate the cross plot porosities and RWA using the sonic-neutron cross plot porosity and M and N at the deeper point to use for entering a matrix identification plot.

Table 29-3

Depth (FT)	*5,102*	*5,126*
%POR-N	12.5	16.3
%POR-D	-	14.5
%POR-S	-	13.8
DT (US/FT)	59.2	-
RHO b (G/CM3)	2.59	-
RHO MF (G/CM3)	1.1	1.1
RT (OHM*M)	4.5	6.2

Keystrokes (SIZE> =022)	**Display**	**Comments**
[XEQ] [ALPHA] XPLOT [ALPHA]	*D.N S.N RT M N*	
[A]	*RHO MF=?*	Density-neutron option
1.1 [R/S]	*RHO b=?*	
2.59 [R/S]	*%POR-N=?*	%POR-D printed
12.5 [R/S]	*D.N S.N RT M N*	%POR-X and RHOMAa printed
[B]	*%POR-N=?*	Sonic-neutron option
[R/S]	*DT=?*	
59.2 [R/S]	*D.N S.N RT M N*	%POR-S, %POR-X, and DTMAa printed
[C]	*%POR-X=?*	True resistivity option
[R/S]	*RT=?*	
4.5 [R/S]	*D.N S.N RT M N*	RWa printed
[A]	*RHO MF=?*	
[R/S]	*RHO b=?*	
[R/S]	*%POR-D=?*	
14.5 [R/S]	*%POR-N=?*	
16.3 [R/S]	*D.N S.N RT M N*	%POR-X and DTMAa printed
[B]	*%POR-N=?*	
[R/S]	*DT=?*	
[R/S]	*%POR-S=?*	
13.8 [R/S]	*D.N S.N RT M N*	DT, %POR-X, and DTMAa printed
[C]	*%POR-X=?*	
[R/S]	*RT=?*	
6.2 [R/S]	*D.N S.N RT M N*	RWa printed
[D]	*RHO MF=?*	M option
[R/S]	*RHO b=?*	
[R/S]	*%POR-D=?*	
[R/S]	*DT=?*	
[R/S]	*%POR-S=?*	
[R/S]	*D.N S.N RT M N*	M printed
[E]	*RHO MF=?*	N option
[R/S]	*RHO b=?*	
[R/S]	*%POR-D=?*	
[R/S]	*%POR-N=?*	
[R/S]	*D.N S.N RT M N*	N printed

```
CROSSPLOT

RHO MF=1.1000 G/CM3
RHO b=2.5900 G/CM3
%POR-D=7.4534
%POR-N=12.5000
%POR-X=10.4773
RHOMAa=2.7644 G/CM3

DT=59.2000 US/FT
%POR-S=8.2037
%POR-X=9.3936

DTMAa=45.7430 US/FT

RT=4.5000 OHM*M
RWa=0.0397 OHM*M

%POR-D=14.5000
%POR-N=16.3000
%POR-X=15.5336
RHOMAa=2.7297 G/CM3

%POR-S=13.8000

DT=67.1132 US/FT
%POR-X=14.4412
DTMAa=46.5403 US/FT

RT=6.2000 OHM*M
RWa=0.1293 OHM*M

M=0.8855

N=0.6080
```

User Instructions

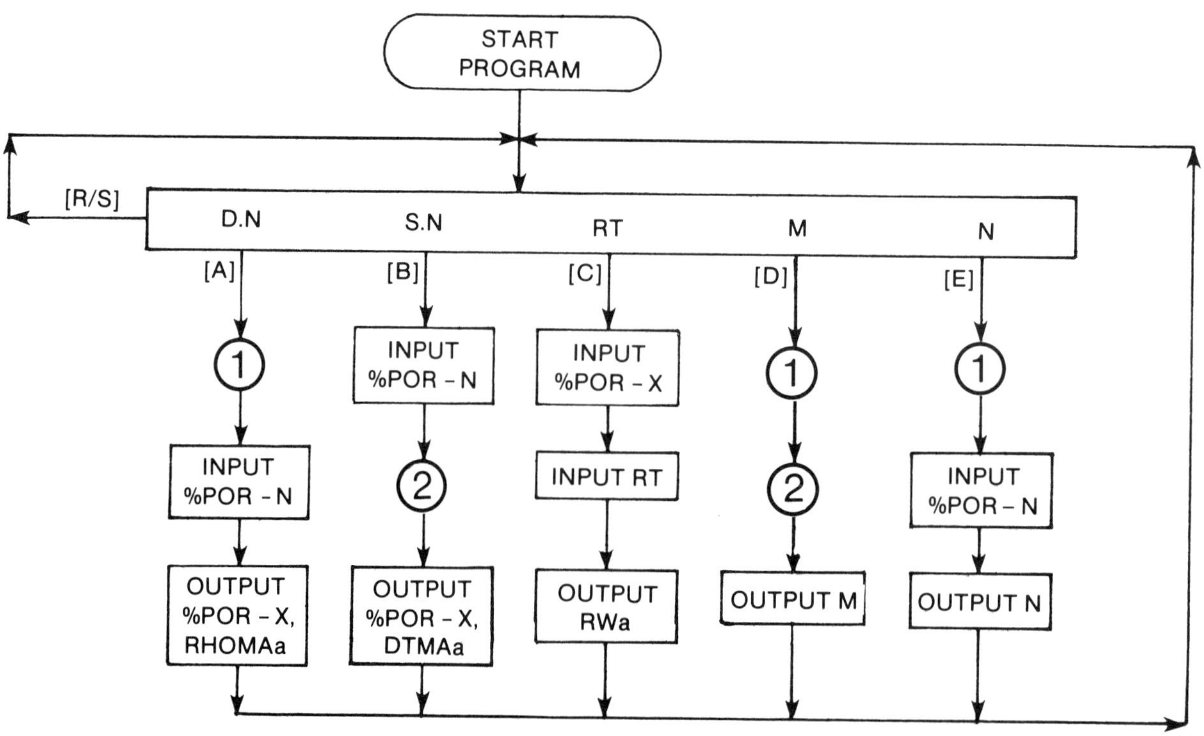

User Instructions

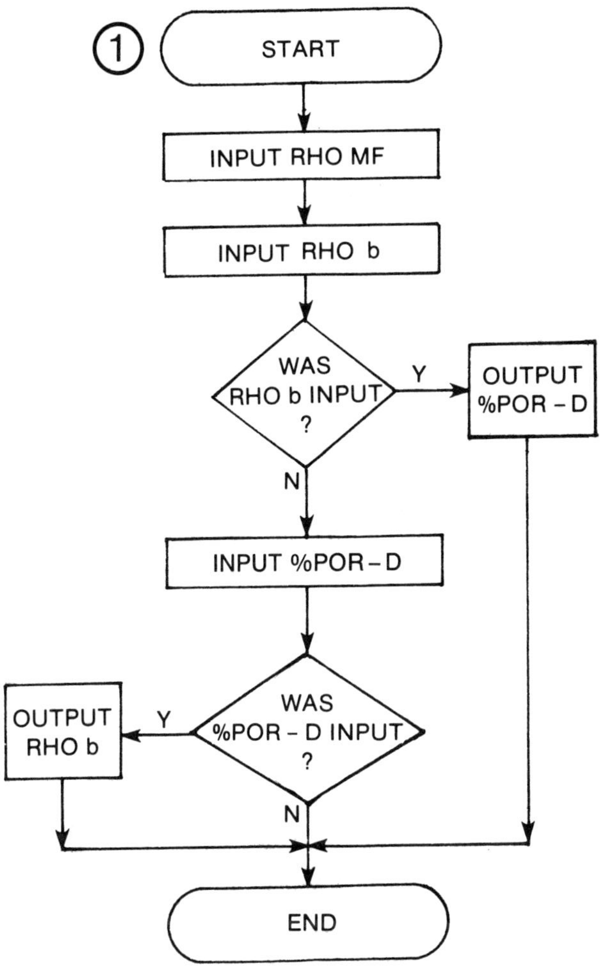

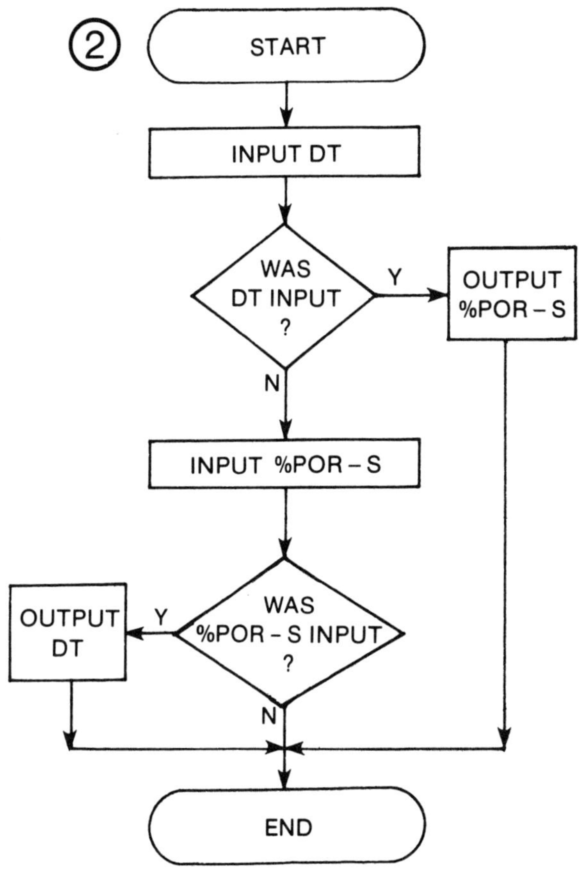

General Information

Memory Requirements

Program length:	715 bytes (4 cards)
Minimum size:	022
Minimum hardware:	41C + 1 memory module

Hidden Options

None

Pac Subroutines Called

TITLE, OUTK, OUT, IN, INK

Registers

03	Density units
04	Density units
09	RWa (OHM*M)
12	%POR-N
13	RHO MF (G/CM3)
14	RHO b (G/CM3)
15	%POR-D
18	%POR-X
19	RT (OHM*M)
20	DT (US/FT)
21	%POR-S

Registers 06–08, 10, 11, 16, 17 unused

Flags

None

Program Listing

```
 01♦LBL "XPLOT"
"CROSSPLOT" 22
XROM "TITLE" FC?C 25
PROMPT SF 27 "G/CM3"
ASTO 03 CLA ASTO 04

 12♦LBL 15
"D.N S.N RT M N" PROMPT
GTO 15

 16♦LBL A
XEQ 05 XEQ 09 RCL 15
RCL 12 X≠Y? X>Y?
GTO 00 16 % .4 +
CHS 10↑X .0206 RCL 12
* 1.17 + - 1 GTO 01

 38♦LBL 00
5 % .16 + CHS 10↑X
.7 X<>Y - 1.29
RCL 13 2.71 - /

 53♦LBL 01
15 XEQ 04 RCL 13 *
RCL 14 - X<>Y /
XEQ 08 "RHOMAa"
XROM "OUTK" XEQ 07 ADV
GTO 15

 68♦LBL B
XEQ 09 XEQ 10 RCL 21
RCL 12 X≠Y? X>Y?
GTO 02 18 % .92 +
CHS 10↑X RCL 12
62 E-4 * .36 + - .5
GTO 03

 90♦LBL 02
5 % .3 + CHS 10↑X
CHS .5 + -.146

101♦LBL 03
21 XEQ 04 189 *
RCL 20 - X<>Y /
"US/FT" ASTO 01 CLA
ASTO 02 "DTMAa" XEQ 13
ADV GTO 15

118♦LBL 04
RDN RCL IND T RCL Z *
X<>Y RCL 12 X<>Y *
LASTX ST- T RDN -
X<>Y / STO 18
"%POR-X" XROM "OUT" 1
% ABS STO Y 1 -
X<>Y RTN

144♦LBL C
17 STO 00 "%POR-X"
XROM "IN" "OHM*M"
ASTO 01 CLA ASTO 02
"RT" XEQ 12 RCL 18
100 / X↑2 * STO 09
"RWa" XEQ 13 ADV
GTO 15

165♦LBL D
XEQ 05 XEQ 10 189
RCL 20 - RCL 14
RCL 13 - / 1 % "M"
XROM "OUT" ADV GTO 15

181♦LBL E
XEQ 05 XEQ 09 RCL 12
1 % 1 - RCL 13
RCL 14 - / "N"
XROM "OUT" ADV GTO 15

197♦LBL 05
12 XEQ 08 "RHO MF"
XROM "INK" XEQ 07
"RHO b" XROM "INK"
XEQ 07 FS? 22 GTO 06
"%POR-D" XROM "IN"
FC? 22 RTN 1 % 2.71
RCL 13 - * 2.71 -
CHS STO 14 XEQ 08
"RHO b" XEQ 07 RTN

226♦LBL 06
2.71 - RCL 13 2.71 -
/ 100 * STO 15
"%POR-D" XROM "OUT"
RTN

239♦LBL 07
RDN STO 03 X<>Y
STO 04 X<>Y R↑ CF 08
RTN

248♦LBL 08
STO 00 "G/CM3" ASTO 01
CLA ASTO 02 RCL 04
RCL 03 RCL Z RTN

258♦LBL 09
11 STO 00 "%POR-N"
XROM "IN" RTN

264♦LBL 10
19 STO 00 "US/FT"
ASTO 01 CLA ASTO 02
"DT" XEQ 12 FS? 22
GTO 11 "%POR-S"
XROM "IN" FC? 22 RTN
1.414 * 47.6 +
STO 20 "DT" XEQ 13
RTN

287♦LBL 11
47.6 - 1.414 /
STO 21 "%POR-S"
XROM "OUT" RTN

296♦LBL 12
AOFF ASTO 05 CF 22
ISG 00 CLD RCL IND 00
"⊦=?" CF 21 AVIEW CLA
FS? 55 SF 21 STOP
AOFF RCL IND 00 FC? 22
GTO 14 X<>Y STO IND 00
CF 21 FC? 55 RTN CLA
ARCL 05 "⊦=" ARCL X
GTO 00

324♦LBL 13
AOFF STO 00 "⊦="
ARCL X FS? 55 GTO 00
CF 21 AVIEW CLA
ARCL 01 ARCL 02 FS? 55
SF 21 STOP AOFF
RCL 00

341♦LBL 14
CF 21 FS? 55 SF 21
RTN

346♦LBL 00
"⊦ " ARCL 01 ARCL 02
SF 21 PRA END
```

30. H2OSAT — Water Saturation Calculations

Calculation of oil or gas in place by volumetric methods requires an estimate of the water saturation and porosity. H2OSAT is quite useful in estimating water saturation to account for the influence of shale on the calculation of SW. For nonshaly formations, the Archie equation is used. The authors prefer the Indonesia equation to calculate water saturations for the effective shale (Poupon and Leveaux).

The program uses default values for *a* and *M* of 0.62 and 2.15, respectively, and a default value for *N* of 2. Other values may be input, but these variables will be initialized to these defaults each time the program is run. If a value for *M* equal to 0 is input, a variable cementation exponent is calculated. (See Equations)

Table 30—1 illustrates typical values of *M* for sandstones and limestones. For limestones and dolomites, the value for SW is generally calculated assuming a = 1 and M = 2. The measured resistivity factor versus porosity can be plotted (as wide a range as possible) to estimate *a* and *M* from the best straight line. The Humble equation sets a = 0.62 and M = 2.15 to represent typical sandstones, the program defaults.

Table 30–1

Lithology	*M*
Sandstones	
Loose, unconsolidated	1.3
Slightly consolidated	1.3–1.7
Moderately cemented	1.7–1.9
Well cemented	1.9–2.2
Limestones	
Moderately porous	2+
Oolitic limestones	2.5–2.8

H2OSAT allows the user to input a value for %VSH or calculate that value a number of ways. These are as follows:

1. Density-Neutron

The density-neutron option is useful due to the response differences of these two logs to gas and shale. When %VSH = 0, the ratio of %POR-D to %POR-N reaches its highest value. The maximum ratio of these two porosities (D/NMAX) can easily be read from logs or cross plots of density-neutron porosities. It is usually the case that %VSH is too high from this calculation. The value of D/NMAX should be selected for a single lithology and fluid content, since both of these factors affect the log. Although the %VSH from this option is usually an upper limit, adverse hole conditions may result in lower than actual %VSH. It is preferable not to use this option when hole conditions (e.g., washouts) affect the logs.

The sonic-density combination can be used the same way as the density-neutron but only when hole conditions are satisfactory. The neutron-sonic log responses to shale are so similar that this combination does not usually provide much help.

2. Spontaneous Potential

The SP curve can be used as an approximate measure of shaliness but tends to be too high when hydrocarbons are present. It is also too high for dispersed clay (as opposed to laminated clay). For shaly mixtures with no effective porosity, there is no SP deflection.

3. Resistivity

The resistivity of a clay (shale) mixture with a nonconductive material (such as quartz) depends on the resistivity of the clay and water. This method usually uses the induction or Laterolog devices, with the highest resistivity reading giving the more reliable shaliness indication.

4. Gamma Ray

5. Gamma Ray-Density

The gamma ray log is probably the best source for evaluating %VSH with nonradioactive formations or waters. The last option allows the user to further refine this technique by density-weighting the measured GR values. It is recommended that this option be used when possible. The GR MIN value is that found opposite shale beds (the shale baseline) with GR MAX equal to the GR reading in clean intervals, usually the highest value. If other radioactive minerals are present, this technique results in estimates that are too high. When shale without potassium-bearing clays is present, %VSH estimates will be too low.

When %VSH is calculated as ≤ 0 or > 100, %VSH is set equal to 0 or 100, respectively. Similarly, if %SW is calculated as > 100, it is set equal to 100.

Equations

Nonshaly formations:

$$SW = \left(\frac{a}{POR^M} \; \frac{RW}{RT} \right)^{1/N}$$

Shaly formations:

$$SW = \left[\frac{VSH^{(1 - VSH/2)}}{(RSH/RT)^{0.5}} + \left| \frac{POR^M}{a} \frac{RT}{RW} \right|^{0.5} \right]^{-2/N}$$

If M = 0 is input, M is calculated as $M = 1.87 + \frac{0.019}{POR}$

VSH equations:

Density-neutron:

$$VSH = \frac{POR\text{-}N\ D/NMAX - POR\text{-}D}{P\text{-}NSH\ D/NMAX - P\text{-}DSH}$$

Spontaneous potential:

$$VSH = 1 - \frac{PSP}{SSP}$$

Resistivity:

$VSH = (RSH/RT)^A$

If $RSH/RT \geq 0.5$, $A = 1$

If $RSH/RT < 0.5$, $A = 2 - (2\ RSH/RT)^{1/4}$

Gamma ray:

$$VSH = \frac{GR - GR\ MIN}{GR\ MAX - GR\ MIN}$$

Gamma ray-density:

$$VSH = \frac{GR\ (RHO\ b) - GR\ MIN\ (RHOMIN)}{GR\ MAX(RHOMAX) - GR\ MIN(RHOMIN)}$$

$$SW = \frac{\%SW}{100}$$

$$VSH = \frac{\%VSH}{100}$$

$$POR\text{-}N = \frac{\%POR\text{-}N}{100}$$

$$POR\text{-}D = \frac{\%POR\text{-}D}{100}$$

$$P\text{-}NSH = \frac{\%P\text{-}NSH}{100}$$

$$P\text{-}DSH = \frac{\%P\text{-}DSH}{100}$$

Nomenclature

Symbol	*Variable Name*	*Input or Output*	*English Units*	*SI Units*
a	Constant in Archie equation for nonshaly formations (default = 0.62)	I	–	–
D/N MAX	Maximum ratio of %POR-D and %POR-N	I	–	–
GR	Gamma ray reading	I	–	–
GR MAX	Gamma ray reading corresponding to clean sands	I	–	–
GR MIN	Gamma ray reading corresponding to shale	I	–	–
M	Cementation exponent (default = 2.15)	I	–	–
N	Saturation exponent (default = 2)	I	–	–
PSP	Spontaneous potential at the shale baseline	I	MV*	MV*
RHO MAX†	Maximum bulk density	I	G/CM3	G/CM3
RHO MIN†	Minimum bulk density	I	G/CM3	G/CM3
RHO b†	Bulk density	I	G/CM3	G/CM3
RSH	Shale resistivity	I	OHM*M*	OHM*M*
RT	True formation resistivity	I	OHM*M*	OHM*M*
RW	Water resistivity	I	OHM*M*	OHM*M*
SSP	Static spontaneous potential	I	OHM*M*	OHM*M*
%POR-D	Percent density limestone porosity	I	–	–
%POR-N	Percent neutron limestone porosity	I	–	–
%P-DSH	Percent density shale porosity	I	–	–
%P-NSH	Percent neutron shale porosity	I	–	–
%VSH	Volume percent shale	I,O	–	–

*These units are not allowed by the Unit Management System in the Pac. Special input and output subroutines were included in this program to provide these units for annotation purposes only.
†The units for these variables are saved by the program.

Yes/No Questions

SHALY? Yes: Shaly formation.
No: Nonshaly formation.

Example 1

Calculate %SW assuming default values for a (0.62), M (2.15), and N (2) and a clean formation with RW = 0.039 OHM*M. Use the values of %POR and RT shown below.

%POR	*RT (OHM*M)*
18.2	5.5
12.2	8.5
9.2	7.6
9.1	4.2
4.5	7.0

Keystrokes (SIZE> = 029)	Display	Comments
[XEQ] [ALPHA] H2OSAT [ALPHA]	*a=?*	
[R/S]	*M=?*	
[R/S]	*N=?*	
[R/S]	*RW=?*	
.039 [R/S]	*SHALY? Y/N:*	Last character is Y or N
N [R/S]	*%POR=?*	
18.2 [R/S]	*RT=?*	
5.5 [R/S]	*%POR=?*	%SW printed
12.2 [R/S]	*RT=?*	
8.5 [R/S]	*%POR=?*	%SW printed
9.2 [R/S]	*RT=?*	
7.6 [R/S]	*%POR=?*	%SW printed
9.1 [R/S]	*RT=?*	
4.2 [R/S]	*%POR=?*	%SW printed
4.5 [R/S]	*RT=?*	
7 [R/S]	*%POR=?*	%SW values >100 are set to 100

```
WATER SAT

RW=0.0390 OHM*M
SHALY: NO

%POR=18.2000
RT=5.5000 OHM*M
%SW=41.3971

%POR=12.2000
RT=8.5000 OHM*M
%SW=51.1896

%POR=9.2000
RT=7.6000 OHM*M
%SW=73.3246

%POR=9.1000
RT=4.2000 OHM*M
%SW=99.8008

%POR=4.5000
RT=7.0000 OHM*M
%SW=100.0000
```

Example 2

Use a = 0.75, N = 2, and calculate M from porosity (input M = 0). Assume a clean formation with RW = 0.029 OHM*M and the porosities and true resistivities shown below.

%POR	*RT (OHM*M)*
15.4	6.7
12.5	4.35

Keystrokes	Display	Comments
[XEQ] [ALPHA] H2OSAT [ALPHA]	*a=?*	
.75 [R/S]	*M=?*	
0 [R/S]	*N=?*	
[R/S]	*RW=?*	
.029 [R/S]	*SHALY? Y/N:N*	
[R/S]	*%POR=?*	
15.4 [R/S]	*RT=?*	
6.7 [R/S]	*%POR=?*	%SW printed
12.5 [R/S]	*RT=?*	
4.35 [R/S]	*%POR=?*	%SW printed

```
WATER SAT

a=0.7500
M=0.0000
RW=0.0290 OHM*M

%POR=15.4000
RT=6.7000 OHM*M
%SW=36.7689

%POR=12.5000
RT=4.3500 OHM*M
%SW=56.1803
```

Example 3

Use the defaults for *a*, *M*, and *N*. Assume a shaly formation with the following values:

RW = 0.018 OHM*M
RSH = 1.1 OHM*M
Porosity = 15.4%
RT = 7.5 OHM*M

Calculate %SW by inputting %VSH and by calculating %VSH, using the values shown below for each option. The gamma ray and gamma ray-density options are recommended.

1. Input %VSH = 15
2. Density-neutron

   D/NMAX = 2.5
   %POR-N = 18.6
   %POR-D = 12.25
   %P-NSH = 27.5
   %P-DSH = 20.1
3. Spontaneous potential

   PSP = 20 MV
   SSP = 48 MV
4. Gamma ray

   GR MIN = 28
   GR MAX = 115
   GR = 78
5. Gamma ray-density

   GR MIN = 28
   GR MAX = 115
   GR = 78
   RHOMIN = 2.52 G/CM3
   RHOMAX = 2.64 G/CM3
   RHO b = 2.68 G/CM3

Keystrokes	Display	Comments
[XEQ] [ALPHA] H2OSAT [ALPHA]	*a=?*	
[R/S]	*M=?*	
[R/S]	*N=?*	
[R/S]	*RW=?*	
.018 [R/S]	*SHALY? Y/N:N*	
Y [R/S]	*RSH=?*	
1.1 [R/S]	*%POR=?*	
15.4 [R/S]	*RT=?*	
7.5 [R/S]	*D.N SP R G G.D*	
[R/S]	*%VSH=?*	[R/S] to input %VSH directly
15 [R/S]	*%POR=?*	%SW printed
[R/S]	*RT=?*	
[R/S]	*D.N SP R G G.D*	
[A]	*D/NMAX=?*	Density-neutron option
2.5 [R/S]	*%POR-N=?*	
18.6 [R/S]	*%POR-D=?*	
12.25 [R/S]	*%P-NSH=?*	
27.5 [R/S]	*%P-DSH=?*	
20.1 [R/S]	*%POR=?*	%VSH and %SW printed
[R/S]	*RT=?*	
[R/S]	*D.N SP R G G.D*	
[B]	*PSP=?*	SP option
20 [R/S]	*SSP=?*	
48 [R/S]	*%POR=?*	%VSH and %SW printed
[R/S]	*RT=?*	
[R/S]	*D.N SP R G G.D*	
[C]	*%POR=?*	Resistivity option; %VSH and %SW printed
[R/S]	*RT=?*	
[R/S]	*D.N SP R G G.D*	
[D]	*GR MIN=?*	Gamma-ray option
28 [R/S]	*GR MAX=?*	
115 [R/S]	*GR=?*	
78 [R/S]	*%POR=?*	%VSH and %SW printed
[R/S]	*RT=?*	
[R/S]	*D.N SP R G G.D*	
[E]	*GR MIN=?*	Gamma ray-density option
[R/S]	*GR MAX=?*	
[R/S]	*GR=?*	
[R/S]	*RHOMIN=?*	
2.52 [R/S]	*RHOMAX=?*	
2.64 [R/S]	*RHO b=?*	
2.68 [R/S]	*%POR=?*	%VSH and %SW printed

```
WATER SAT

RW=0.0180 OHM*M
SHALY: YES
RSH=1.1000 OHM*M

%POR=15.4000
RT=7.5000 OHM*M

%VSH=15.0000
%SW=25.5024

D/NMAX=2.5000
%POR-N=18.6000
%POR-D=12.2500
%P-NSH=27.5000
%P-DSH=20.1000
```

```
%VSH=70.4008
%SW=18.0192

PSP=20.0000 MV
SSP=48.0000 MV
```

```
%VSH=58.3333
%SW=19.0400

%VSH=21.9021
%SW=24.1254
```

```
GR MIN=28.0000
GR MAX=115.0000
GR=78.0000
%VSH=57.4713
%SW=19.1234
```

```
RHOMIN=2.5200 G/CM3
RHOMAX=2.6400 G/CM3
RHO b=2.6800 G/CM3
%VSH=59.4233
%SW=18.9366
```

User Instructions

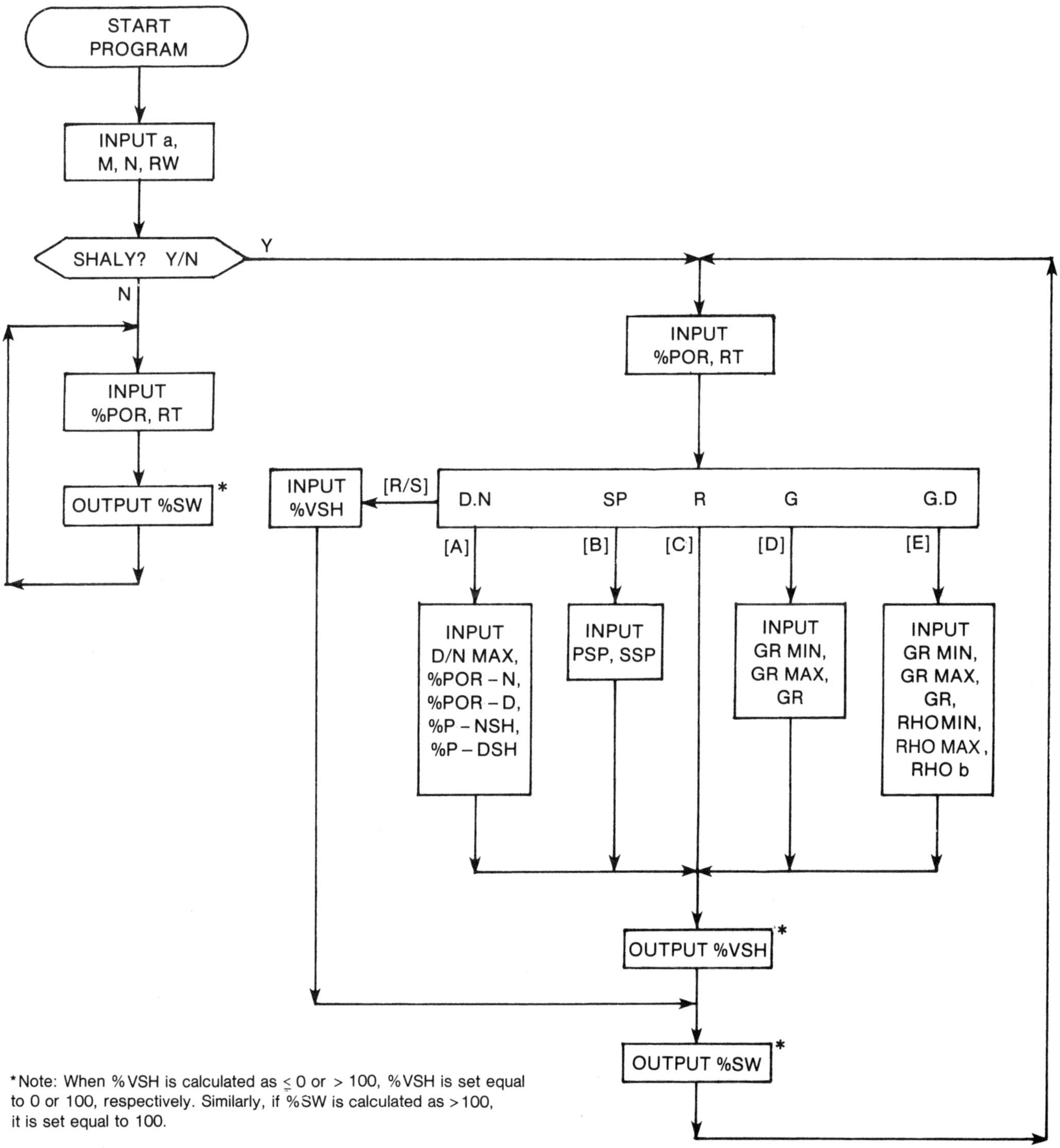

*Note: When %VSH is calculated as ≤ 0 or > 100, %VSH is set equal to 0 or 100, respectively. Similarly, if %SW is calculated as > 100, it is set equal to 100.

General Information

Memory Requirements

Program length: 610 bytes (3 cards)
Minimum size: 028
Minimum hardware: 41C + 1 memory module

Hidden Options

None

Pac Subroutines Called

TITLE, IN, Y/N?, %POR, INK, OUT

Registers

03 Density units
04 Density units
06 D/NMAX, PSP (MV), GR MIN
07 %P-NSH, SSP (MV), GR MAX
08 %P-DSH, GR
09 RW (OHM*M)
12 %POR-N
14 RHO b (G/CM3)
15 %POR-D
19 RT (OHM*M)
20 %VSH
22 RHOMIN (G/CM3)
23 RHOMAX (G/CM3)
25 a
26 M
27 N
28 RSH (OHM*M)

Registers 10, 11, 16, 17, and 24 unused

Flags

04 Set: Shaly formation.
Clear: Nonshaly formation.

Program Listing

```
 01♦LBL "H2OSAT"
"WATER SAT"  29
XROM "TITLE"  FC?C 25
PROMPT  SF 27  "G/CM3"
ASTO 03  CLA  ASTO 04
24  STO 00  .62  STO 25
2.15  STO 26  2  STO 27
"a"  XROM "IN"  "M"
XROM "IN"  "N"
XROM "IN"  8  STO 00
"OHM*M"  ASTO 01  CLA
ASTO 02  "RW"  XEQ 03
"SHALY"  4  XROM "Y/N?"
27  STO 00  "RSH"
FS? 04  XEQ 03

 42♦LBL 15
ADV  XROM "%POR"
"OHM*M"  ASTO 01  CLA
ASTO 02  "RT"  XEQ 03
1.9  RCL 18  /  1.87  +
RCL 26  X<=0?  X<>Y
STO 26  FS? 04  GTO 00
RCL 25  RCL 09  *
RCL 18  100  /  RCL 26
Y↑X  RCL 19  *  /  1
GTO 18

 75♦LBL 00
5  STO 00
 D.N SP R G G.D"  PROMPT
19  STO 00  "%VSH"
XROM "IN"  GTO 17

 85♦LBL D
XEQ 01  RCL 06  -
RCL 07  RCL 06  -  /
GTO 16

 94♦LBL E
XEQ 01  21  STO 00
"G/CM3"  ASTO 01  CLA
ASTO 02  RCL 04  RCL 03
RCL Z  "RHOMIN"  XEQ 02
CF 08  "RHOMAX"  XEQ 02
13  STO 00  RDN  "RHO b"
XEQ 02  RCL 08  *
RCL 22  RCL 06  *  -
LASTX  RCL 07  RCL 23  *
X<>Y  -  /  GTO 16

129♦LBL 01
"GR MIN"  XROM "IN"
"GR MAX"  XROM "IN"
"GR"  XROM "IN"  RTN

137♦LBL 02
XROM "INK"  RDN  STO 03
X<>Y  STO 04  X<>Y  R↑
RTN

146♦LBL C
RCL 28  RCL 19  /  ST+ X
ENTER↑  SQRT  SQRT  2
X<>Y  -  1/X  X<>Y  1
X<Y?  STO Z  RDN  2  /
X<>Y  Y↑X  GTO 16

168♦LBL B
"MV"  ASTO 01  CLA
ASTO 02  "PSP"  XEQ 03
"SSP"  XEQ 03  1  RCL 06
RCL 07  /  -  GTO 16

183♦LBL A
"D/NMAX"  XROM "IN"  11
STO 00  "%POR-N"
XROM "IN"  14  STO 00
"%POR-D"  XROM "IN"  6
STO 00  "%P-NSH"
XROM "IN"  "%P-DSH"
XROM "IN"  RCL 12
RCL 06  *  RCL 15  -
RCL 07  RCL 06  *
RCL 08  -  /

211♦LBL 16
1  X>Y?  *  X<0?  0
ENTER↑  100  *  STO 28
"%VSH"  XROM "OUT"

223♦LBL 17
RCL 20  1  %  1  RCL Y
2  /  -  Y↑X  RCL 28
SQRT  /  RCL 18  100  /
RCL 26  Y↑X  RCL 25  /
RCL 09  /  SQRT  +
RCL 19  SQRT  *  -2

251♦LBL 18
RCL 27  /  Y↑X  100  *
LASTX  X>Y?  X<>Y
STO 21  "%SW"
XROM "OUT"  GTO 15

264♦LBL 03
AOFF  ASTO 05  CF 22
ISG 00  CLD  RCL IND 00
"⊢=?"  CF 21  AVIEW  CLA
FS? 55  SF 21  STOP
AOFF  RCL IND 00  FC? 22
GTO 04  X<>Y  STO IND 00
CF 21  FC? 55  RTN  CLA
ARCL 05  "⊢="  ARCL X
"⊢ "  ARCL 01  ARCL 02
SF 21  PRA  RTN

297♦LBL 04
CF 21  FS? 55  SF 21
END
```

31. SLANT — True Stratigraphic and Vertical Thicknesses

Many offshore and an increasing number of onshore wells involve deliberately (or otherwise) deviated wells with angles as high as 70° or more from normal. The measured thickness of a formation from a well log made in a deviated hole does not represent its true stratigraphic thickness or its true vertical thickness. Measured thickness can be greater or even less than true vertical thickness, complicating reserve estimations, pressure transient analysis, and geological studies. SLANT provides a simple method to solve these calculations to help both the geologist and engineer.

A good discussion of this problem is presented by Bateman and Konen. Figures 31-1 through 31-3 briefly illustrate the concepts of a deviated wellbore, dipping beds, and the combined effects of deviating wellbores and dipping beds.

Equations

MEAS H = BOTTOM — TOP

TVDDIF = MEAS H cos(WELDEV)

TST = cos(WELDEV) cos(BEDDIP) — MEAS H × cos(HOLEAZ — DIP AZ) sin(WELDEV) × sin(BEDDIP)

TVT = TST/cos(BEDDIP)

Nomenclature

Symbol	*Variable Name*	*Input or Output*	*English Units*	*SI Units*
BEDDIP	Bed dip	I	–	–
BOT-TOM	Measured bottom of the formation	I	FT	M
DIP AZ	Dip azimuth	I	–	–
HOLEAZ	Hole azimuth	I	–	–
MEAS H	Measured thickness	O	FT	M
TOP	Measured top of the formation	I	FT	M
TST	True stratigraphic thickness	O	FT	M
TVDDIF	Difference betweeen measured depth and true vertical depth	O	FT	M
TVT	True vertical thickness	O	FT	M
WELDEV	Well deviation from vertical	I	–	–

Yes/No Questions

None

Example

Use SLANT to solve the example problem presented by Bateman and Konen. Data are as follows:

Well deviation = 30°
Hole azimuth = 128°
Bed dip = 25°
Dip azimuth = 45°
Measured top of formation = 5,642 FT
Measured bottom of formation = 5,878 FT

Keystrokes (SIZE> =010)	**Display**	**Comments**
[XEQ] [ALPHA] SLANT [ALPHA]	*TOP=?*	
5642 [R/S]	*BOTTOM=?*	
5878 [R/S]	*WELDEV=?*	
30 [R/S]	*HOLEAZ=?*	
128 [R/S]	*BEDDIP=?*	
25 [R/S]	*DIP AZ=?*	
45 [R/S]	*TOP=?*	MEAS H, TVDDIF, TST, and TVT printed

```
TST, TVT

TOP=5,642.0000 FT
BOTTOM=5,878.0000 FT
WELDEV=30.0000
HOLEAZ=128.0000
BEDDIP=25.0000
DIP AZ=45.0000

MEAS H=236.0000 FT
TVDDIF=204.3820 FT
TST=179.1555 FT
TVT=197.6762 FT
```

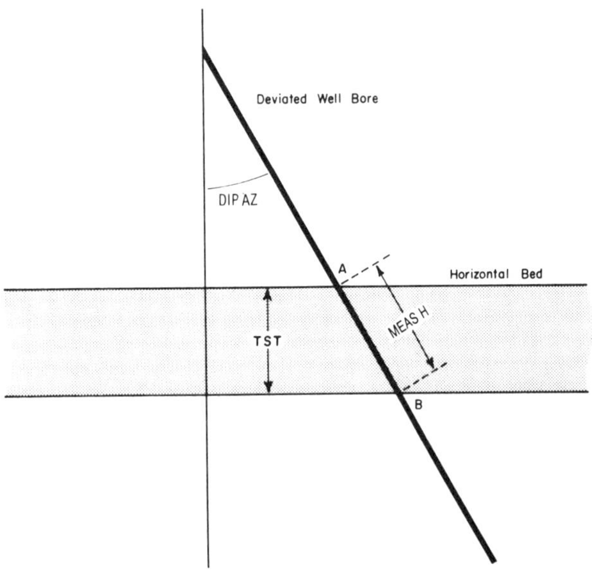

Figure 31–1 *(after Bateman and Konen)*

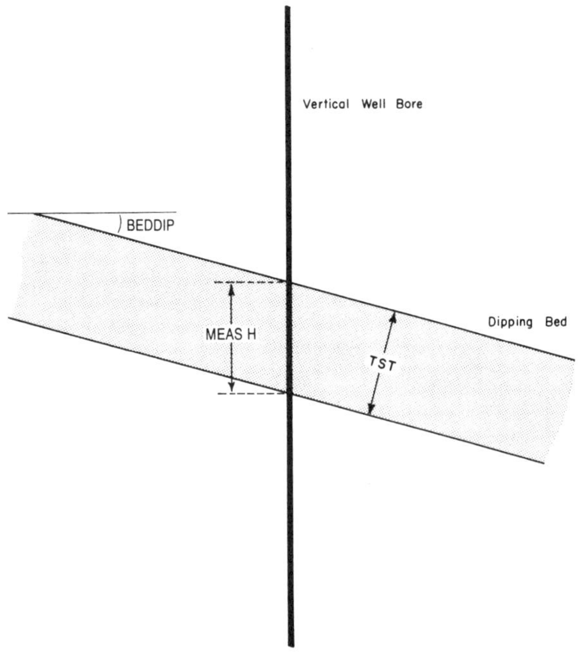

Figure 31–2 *(after Bateman and Konen)*

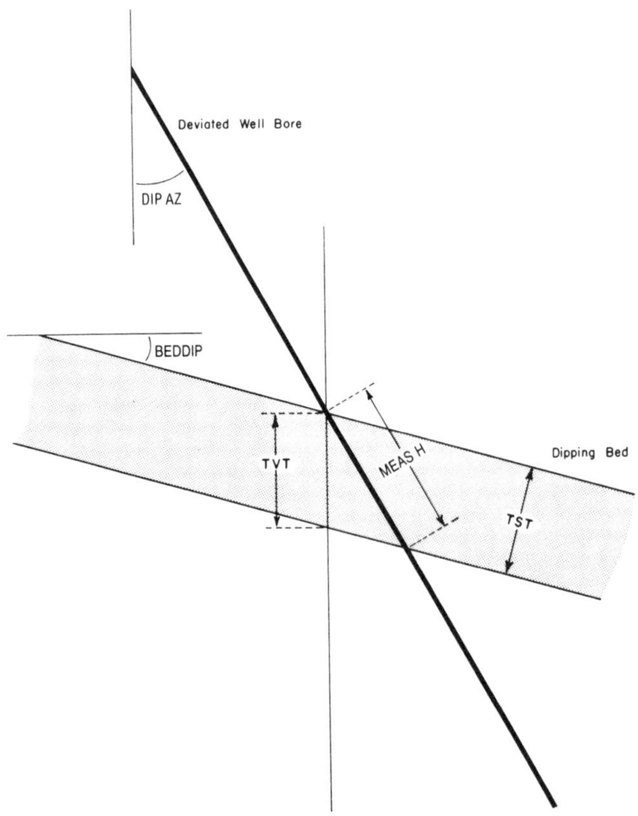

Figure 31–3 *(after Bateman and Konen)*

User Instructions

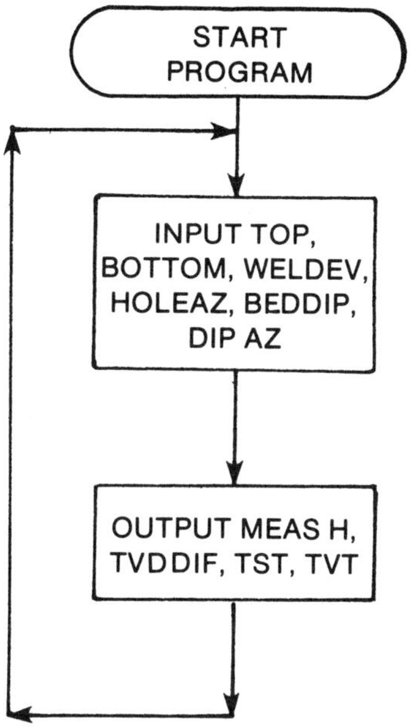

General Information

Memory Requirements

Program length: 182 bytes (1 card)
Minimum size: 010
Minimum hardware: 41C

Hidden Options

None

Pac Subroutines Called

TITLE, INU, IN, OUTU

Registers

03 TOP (FT)
04 BOTTOM (FT)
06 WELDEV
07 HOLEAZ
08 BEDDIP
09 DIP AZ

Flags

None

Note: The program sets the trig mode to DEG.

Program Listing

```
 01♦LBL "SLANT"
10  "TST,TVT"
XROM "TITLE"  FC?C 25
PROMPT  DEG

 08♦LBL 15
2  STO 00  XEQ 00  "TOP"
XROM "INU"  XEQ 00
"BOTTOM"  XROM "INU"  5
STO 00  "WELDEV"
XROM "IN"  "HOLEAZ"
XROM "IN"  "BEDDIP"
XROM "IN"  "DIP AZ"
XROM "IN"  RCL 04
RCL 03  -  XEQ 00  ADV
"MEAS H"  XROM "OUTU"
RCL 06  COS  *  XEQ 00
"TVDDIF"  XROM "OUTU"
RCL 06  COS  RCL 08  COS
*  RCL 06  SIN  RCL 08
SIN  *  RCL 07  RCL 09
-  COS  *  -  RCL 04
RCL 03  -  *  XEQ 00
"TST"  XROM "OUTU"
RCL 08  COS  /  XEQ 00
"TVT"  XROM "OUTU"  ADV
GTO 15

 71♦LBL 00
"FT"  ASTO 01  CLA
ASTO 02  ASTO Z  "M"
ASTO Y  END
```

Section 9
Economics

32. DISC — Discounted Cash Flow Analysis

Calculates a variety of economic indices from annual net cash flows, including payout, net present value, and discounted cash flow rate of return.

32. DISC — Discounted Cash Flow Analysis

As a group, reservoir engineers usually do not perform all of the calculations we have discussed just for fun, although they are often quite enjoyable. We are all, presumably, quite interested in maximizing our total wealth. The decisions made by reservoir engineers (e.g., waterfloods, hydraulic fracture treatments, enhanced-recovery schemes, infill drilling) affect the amounts of reserves ultimately recovered and the rates of recovery and investments required.

These investments must earn more than their total associated costs to be worthwhile endeavors. The larger the difference between cash returned and cash expended, the more profitable the venture and the greater the justification for putting capital at risk. Several indices for measuring the financial merit of various projects are calculated by DISC. Each of the economic parameters described in this section is a quantitative measure of the desirability of undertaking a venture. A number of sophisticated commercial software packages are available for comprehensive economic analyses. DISC calculates three of the most common economic indices: PAYOUT, NPV, and DCFROR. An excellent discussion of economics and risk analysis for the practicing engineer is presented by Newendorp.

Characteristics of Economic Parameters

No single economic parameter used to measure the attractiveness of investment opportunities considers all of the factors or dimensions of investment projects that are pertinent to the decisionmaker. Therefore, it is important to select the economic parameters that most nearly represent the opportunity's financial characteristics. Listed below are some of the characteristiics that a realistic economic index should have:

1. It must be suitable for comparing and ranking the attractiveness of investment opportunities.
2. It should reflect the firm's time value of capital. That is, it should represent the firm's future investment opportunities.
3. It should provide a means of determining whether profitability exceeds some minimum level, such as cost of capital or the firm's average earning rates.
4. It should include quantitative statements of risk (probability numbers).
5. It would be desirable to have the parameter reflect other factors, such as corporate goals, decisionmakers' risk preferences, and the firm's asset position.

The following order of presentation is meant to describe the indices calculated by DISC, but does not rank one over the other. In fact, each index yields information of particular importance on a project's earning power. The terms used in the program are also defined.

Net Cash Flow

The term NET CASH FLOW (NCF) refers to the sum of:

cash flow from operations
(revenues — cash expenses — taxes)
— fixed & working capital investments
± terminal value

Typically, one would multiply oil and gas prices times net production volumes (after royalty) and subtract operating expenses; state, local, and windfall profit taxes; etc. The calculation of federal income taxes requires the calculation of depreciation, investment tax credits, etc. The taxes and investments (and perhaps salvage cost) are subtracted to yield annual NCFs. These are required to calculate any of the parameters in the program and are considered inputs.

Payout

Defined in general terms, PAYOUT is the length of time required to reach an undiscounted cumulative NET CASH FLOW of zero for an investment opportunity. Stated differently, PAYOUT is the length of time it takes to get back the capital investment (fixed and working capital).

PAYOUT is an easy number to calculate and is used by decisionmakers as a simple indicator of the riskiness of an investment opportunity. It serves this purpose because PAYOUT is an approximate measure of the rate at which cash flows are generated in the early years of the business or project.

PAYOUT has two major weaknesses as a measure of attractiveness. First, it ignores cash flows beyond the time of payout; therefore, decisionmakers have no information about the total economic wealth created by the investment or the rate of cash flow generation after the payout time. Second, it is nearly impossible to rank investment opportunities of different classifications.

PROJECT PAYOUT occasionally is used to refer to payout of a group of investments. For example, a development drilling program might be initiated in a certain field. Individual wells will pay out at various times, but PROJECT PAYOUT would be the time at which the cumulative NET CASH FLOW for all wells would equal zero. Some individual wells (and all dry holes) need not have a PAYOUT for a PROJECT PAYOUT to exist.

Time Value of Money Considerations

PAYOUT is an economic parameter that evaluates only the magnitude of cash flows. However, timing and reinvestment rate are also relevant aspects of cash flow streams because cash has a cost — the cost of capital. Other than PAYOUT, virtually all economic parameters include some recognition of timing and reinvestment rate of cash inflows and outflows.

Cost of Capital

The prices that investors are willing to pay for a firm's securities (stocks, bonds, etc.) and the return that they demand from those securities determine a market cost for capital raised through debt and equity. The cost of debt is easy to see because it is simply the interest that must be paid. There is a corresponding cost for shareholder's equity since investors in the firm want to earn a satisfactory rate of return on their investment. This cost applies both to new investments made in the firm through purchases of stock and to cash that is retained in the company rather than paid out through dividends. These costs, combined with the capital structure of the firm, result in a weighted average cost of capital. (Normally, the analyst need not calculate the cost of capital for each project. Additional information on the weighted average cost of capital is given in Brigham and Weston.) Since this is the average rate demanded by the capital markets for the investment funds, the firm should consider only those investments whose cash flows will yield at least that rate.

As a final comment regarding the cost of capital, generally it is inappropriate to mix the decisions of whether an investment should be made and how the investment will be financed. That is, with rare exceptions, the cost of capital should not be the marginal cost of debt or equity just because the next available dollar will come from a loan or stock issue, respectively. Capital structure decisions are long-term policy decisions. In essence, if not in fact, most corporations will continually be borrowing funds, reusing internally generated funds, and issuing stock to finance its businesses.

Present Value and Discounting

With PAYOUT, the focus of economic evaluation is the accumulated value of cash flows out to some point in time. This is a very natural way to think about evaluating cash flows, since this so closely parallels everyday savings account computations. But it does have drawbacks in business applications. One of these is that, in the comparison of alternatives, we are dealing in "future" dollars. Considering the typical pattern of project lives for investment opportunities, that future could be very distant. To many decisionmakers, it would seem rather unnatural to think in terms of dollars 20 years from now, particularly when the decision is being made with "now" dollars. If the perspective is changed from future dollars to now dollars, this problem is eliminated.

The process of making now dollars out of future dollars is known as *discounting*. Generally, discounting is a technique whereby the value of any time period's cash flow is determined for any other time period. Normally, an analyst is concerned with the present value — now dollars — of some future period's cash flow. The *present value* of a future cash flow is the amount that makes us indifferent between receiving the present value now or waiting for the future payment. Or, in other words, if X dollars are to be received in the future, what amount of cash is of equivalent value today at a Y% rate of return?

Please note that any one period's cash flow is a tangible item; the corporate treasurer can count it and deposit it in a bank. Today, the present value of that same cash flow is a conceptual quantity; only in that time period does it become tangible.

The computation of the present value of some period's cash flow is described mathematically by:

$$PV_T = \text{period's net cash flow}_{T+t} * \text{discount factor}_t$$

where T = base period

t = number of time periods from period T

Discount Rate

When cash flows are discounted to their present values, there is a need to do so at some discount rate. The question in the previous section would use Y% as the discount rate. The discount rate is used to calculate the discount factor that makes now dollars out of future dollars.

Cash that becomes available does not lie idle; rather, it is recommitted to other activities throughout the corporation and these give rise to future cash flows. Strictly speaking, the discount rate that is sought is the rate that must be earned on a dollar so that management is indifferent between receiving the dollar now or receiving the dollar plus its earnings in the future.

The calculation of certain economic indices, such as NET PRESENT VALUE, requires an explicit assumption as to the corporate discount rate. Over the years, numerous authors have written thousands of pages on the subject of the proper discount rate to use in cash flow analysis. Analysts have been asked to use the cost of capital, a corporate cutoff rate (hurdle rate), or whatever the boss says. We suggest using the cost of capital as the appropriate corporate discount rate.

Various methods of discounting exist, and their effect on present value depends upon how the discount factor is calculated. The method used by DISC is annual end-of-period discounting, in which cash flows are modeled to occur at the end of each time period. This is the most conservative of the discounting methods used for discounting positive NET CASH FLOWS.

$$\text{Discount factor} = 1/(1 + r)^t$$

where r = corporate discount rate
t = number of time periods from time zero

Discounted Cash Flow Rate of Return (DCFROR)

Conceptually, the discounted cash flow rate of return (DCFROR) is the discount rate that will make the present value of annual NET CASH FLOWS sum to zero. DCFROR is given different names including the internal yield, internal rate of return (IRR), profitability index, marginal efficiency of capital, and the investor's method.

The DCFROR calculation is made after the series of anticipated NET CASH FLOWS has been defined. The calculation is a trial-and-error process that begins by selecting a discount rate and discounting all the NET CASH FLOWS back to time zero, i.e., finding the present value of all cash flows associated with an investment. If the sum of the present values of the NET CASH FLOWS is greater than zero, the discount rate selected was too low. If the sum of present values is exactly equal to zero, the discounting rate selected is by definition the rate of return.

Specific characteristics of DCFROR include the following:

1. Computation of the rate of return requires a series of trial-and-error computations because the mathematical equation for the rate of return cannot be solved explicitly. This is one reason it is generally only calculated by computer or calculator solutions.

2. The DCFROR concept accounts for timing differences in cash inflows and outflows. To be able to compute rate of return, the analyst must predict a cash flow time rate schedule over the entire life of the investment opportunity.

3. DCFROR is an indicator that is independent of the absolute magnitude of the cash flows.

4. There are certain types of cash flow schedules in which there is more than one discount rate that satisfies the definition of DCFROR. Examples of cash flow schedules that sometimes lead to multiple rates of return include some rate acceleration projects and projects requiring a major expenditure at a later point in the life of the project. In all cases when there is not a unique solution to the rate of return algorithm, it is because the arithmetic sign of the cumulative cash flows has changed more than once. In other words, the cumulative NET CASH FLOWS are negative early in the project, turn positive at some future date, and then turn negative again or vice versa. More than one arithmetic sign change of the cumulative cash flows is a necessary, but not a sufficient condition for multiple solutions to the rate of return algorithm.

5. The DCFROR cannot be calculated for the following situations:

   a. Cash flows are all negative; for example, a dry hole.
   b. Cash flows are all positive; for example, an investment is paid out of future revenues.
   c. Total undiscounted NET CASH FLOWS are less than the investment; for example, a marginal producing well or field depleted before reaching payout.

   For these situations, the rate of return is mathematically undefined and cannot be computed. Negative rates of return have no meaning. A zero rate of return corresponds to a cash flow schedule for which the undiscounted net cash flows exactly equal the undiscounted investment.

6. Cash flows received early in the project are weighted more heavily than later cash flows. This becomes particularly pronounced as the discount rate increases. Cash flows received or disbursed late in the life of the project (after 20 years or so) have very little effect on the computed rate of return. The higher the rate of return, the less the effect of late-in-the-life cash flows.

The computed rate of return is relatively sensitive to errors in estimating initial cash outflows (investments) and early cash inflows (revenue-related streams). When uncertainty is present, say about drilling cost, it is recommended that DCFROR be computed for several possible variations in the initial investment. A small variation on a percentage basis in the initial investment can sometimes cause a much larger percentage variation in the resultant rate of return.

DCFROR is a convenient measure of financial attractiveness to compare with minimum criteria such as the cost of capital or corporate objectives for annual growth. Management can easily relate a rate of return to interest on loans, etc. This is one of the reasons for its wide popularity as an economic index.

However, DCFROR has certain weaknesses when viewed as being similar to an interest rate on a loan. In this context, DCFROR includes the implicit assumption that all cash inflows will be reinvested at the computed rate of return when received. If they are not reinvested at that rate, the initial expenditure will not have the earning power of the rate of return as calculated. This is an extremely important characteristic of DCFROR, which is often misunderstood or ignored by those who assume the criterion to be a realistic measure of financial attractiveness in the same sense as interest rates.

For example, if future reinvestment opportunities are in the range of 12 to 15%, then higher rate of return projects, say 30 to 40%, will actually yield something less than that. Lower rate-of-return projects, say 8%, will actually yield a higher figure.

DCFROR is not a completely realistic parameter to rank the desirability of competing investments because risk or probability numbers cannot be incorporated mathematically into the rate of return calculation. However, there are methods that will allow risk weighting of NET CASH FLOW to calculate DCFROR.

In summary, DCFROR is certainly a more realistic measure of attractiveness than PAYOUT, primarily because it includes the *time value of money* concept. It is a useful measure of the relative attractiveness of investments having approximately the same total life and cash flow patterns. DCFROR's weaknesses as a measure of financial attractiveness are the frequent problems of satisfying the underlying assumption of reinvestment at the computed rate of return and the failure of the index to be capable of considering explicitly the dimensions of risk and uncertainty.

Net Present Value (NPV)

If one accepts the primary responsibility of corporate management is the creation of economic wealth, then the NET PRESENT VALUE concept is the most important economic criterion that senior management has for evaluating the attractiveness of an investment opportunity. NET PRESENT VALUE, sometimes called PRESENT WORTH, is the amount of wealth that an investment opportunity will create after all costs have been paid, including the cost of capital.

The NET PRESENT VALUE at a given discount rate (usually that discount rate will be the cost of capital) is determined by discounting all the NET CASH FLOWS related to an investment opportunity to their present value at time zero. The word "net" appears in the name of the concept because the present value of cash outflows are subtracted from the present value of cash inflows. These cash outflows are generally associated with investments or capitalized items. However, negative cash outflows can be associated with expense items, such as feedstocks or fixed operating expenses, and still lend themselves to being evaluated on a NET PRESENT VALUE basis.

The concept of NET PRESENT VALUE indicates to management how much additional cash (in present-denominated dollars) will be in the corporate treasury by the end of the project life after all costs, including the cost of debt and equity capital, have been paid. It is the only economic index that will indicate the total amount of wealth being created. Other economic indices address the efficiency at which cash is generated. Such numbers tend to be ratios, but these economic indices are not sufficient to project the total wealth that will be created by an investment decision.

Beyond its ability to indicate wealth creation, NET PRESENT VALUE has other attractive features as an economic indicator. Unlike the DCFROR calculation, the NET PRESENT VALUE calculation is not a trial-and-error solution. For a given discount rate, multiple values are impossible. Furthermore, NET PRESENT VALUE is suitable for use with probability numbers to consider risk in a quantitative and explicit manner. It also can be used to evaluate purchase versus leasing alternatives — a case in which all cash flows are negative. NET PRESENT VALUE's greatest disadvantage as an economic index is that it does not indicate the rate of cash generation, an important consideration under conditions of capital limitation.

In determining the NET PRESENT VALUE, the appropriate parameter for discounting is cash and, even more precisely, NET CASH FLOW. It is inappropriate to calculate NET PRESENT VALUE on profit or on cash flow from operations because profit is not the equivalent of cash, and cash from operations ignores the relevant cash flow for investments in fixed or working capital.

Net Present Value and DCFROR

For all investment opportunities in which the rate of return is calculated to be higher than the assumed cost of capital, and assuming the company is not operating on a capital allocation basis, both DCFROR and NET PRESENT VALUE will yield the same accept or reject decisions. The priority assigned to projects, however, could vary, depending upon the actual figures for NET PRESENT VALUE and DCFROR.

If one accepts the statement that management's responsibility is to increase the wealth of the shareholders (and this is one of the reasons that financial markets allocate capital to a corporation), then the criteria of accepting the NET PRESENT VALUE rather than the DCFROR index is the most suitable action for management to take. Under the assumption that management is interested in increasing economic wealth, fixation upon a rate of return can lead to poor decisions as can be seen from the following example (Figure 32–1).

If the cost of capital (corporate discount rate) lies to the left of the intersection of the NET PRESENT VALUE curves of the two mutually exclusive projects, then accepting Project B (the higher DCFROR project) will result in a lower NET PRESENT VALUE project being accepted. If the cost of capital lies to the right of the intersection — in other words, if the cost of capital is higher than the discount rate at which the two mutually exclusive projects have an equal NET PRESENT VALUE — then accepting the higher rate of return project (Project B) is consistent with the objective of wealth maximization because its NET PRESENT VALUE is higher than Project A. Under the assumption that management's task is to increase stockholder's wealth, giving priority to DCFROR as an economic indicator has the potential for leaving money on the table (*a* minus *b* at the corporate discount rate), but using NET PRESENT VALUE does not. Other factors such as the relative riskiness of the projects or the timing of the cash flows or other characteristics pertinent to the decision may, of course, favor one project over the other, and these factors should be brought to the attention of senior management so that prudent decisions can be made. Nonetheless, the NET PRESENT VALUE index is preferred to DCFROR.

Present Value Ratio

To sidestep the weakness of NPV being dependent on the absolute size of the cash flows, the PRESENT VALUE RATIO index can be used. This is the ratio of discounted cash inflows to discounted cash outlays, both at the corporate discount rate. A similar criterion is the DISCOUNTED PROFIT TO INVESTMENT RATIO, also known as investment efficiency. The DISCOUNTED PROFIT TO INVESTMENT RATIO is simply the PRESENT VALUE RATIO minus one. A PRESENT VALUE RATIO equal to one is equivalent to a project with a NET PRESENT VALUE of zero and the rate of return equal to the discount rate.

PRESENT VALUE RATIO has many of the advantages of NET PRESENT VALUE (such as realistic

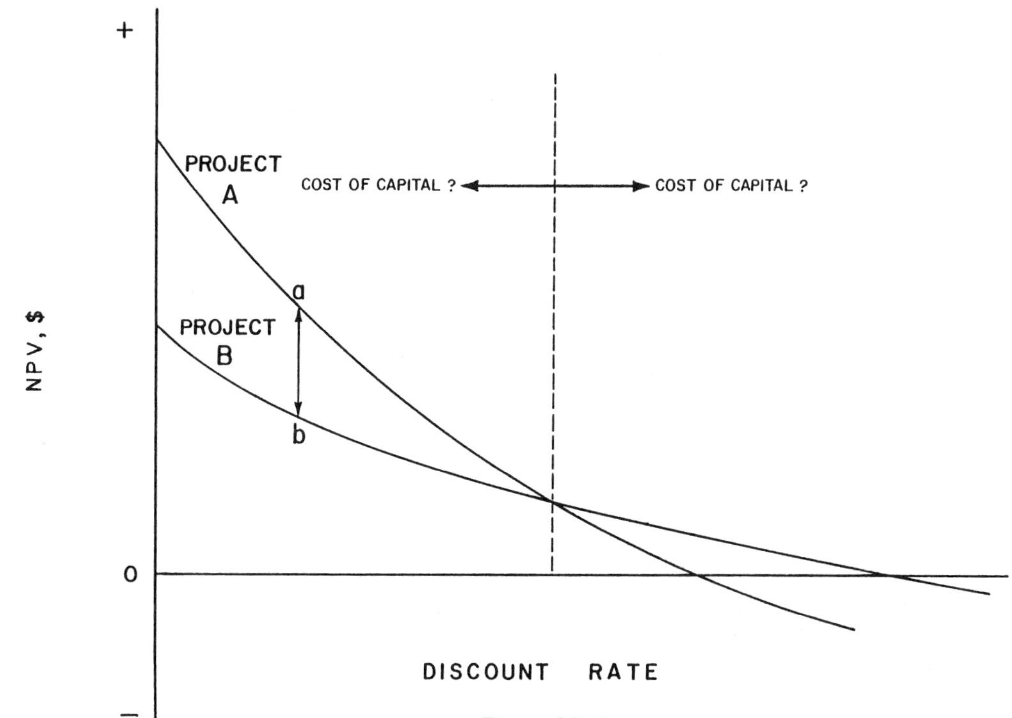

Figure 32–1

reinvestment rate, no multiple rates, not a trial and error solution, etc.) and also provides a measure of profitability per dollar investment. This is a particularly important consideration for selecting projects from a list that contains more opportunities than the available funds can cover. The PRESENT VALUE RATIO is one of the most important economic indices in cases where capital is limited and the efficiency of investments must be high so cash can be generated short term to supply funds for subsequent investments. If there are investments at other than time zero, DISC can calculate the NPV of investments in a separate run. This allows the user to calculate the PRESENT VALUE RATIO at any discount rate (see Example 2).

Equations

$$\text{CUM CF} = \text{CF 0} + \sum_{i=1}^{n} \text{NCF}_i$$

$$\text{PAYOUT} = i - \frac{\text{CUM CF}_i}{\text{NCF}_{i+1}}$$

where i = the last period number in which CUM CF < 0
CUM CF_i = CUM CF at period i
NCF_{i+1} = NCF at period i + 1

$$\text{NPV} = \text{CF 0} + \sum_{i=1}^{n} \frac{\text{NCF}_i}{\left(1 + \frac{\%R}{100}\right)^i}$$

DCFROR is calculated iteratively using Newton's method as follows:

$$\text{DCFROR}_{j+1} = \text{DCFROR}_j - \frac{100\ \text{NPV}_j}{\sum_{i=1}^{n} \frac{\text{NCF}_i}{i\left(1 + \frac{\text{DCFROR}_j}{100}\right)^{-i-1}}}$$

NPV_j = NPV at %R = $DCFROR_j$
n = number of NCF values input by the user

Nomenclature

Symbol	Variable Name	Input or Output	English Units	SI Units
CF 0	Cash flow at time zero (initial investment)	I	–	–
CUM CF	Cumulative net cash flows	O	–	–
DCFROR	Discounted cash flow rate of return	O	–	–
EST %R	Estimated DCFROR to begin iterative solution (set to 25% if no value input)	I	–	–
NCF_i	Net cash flow at $time_i$	I	–	–
NPV	Net present value	O	–	–
NPV10	Net present value at %R = 10	O	–	–
PAY-OUT	Time at which cumulative net cash flows equal zero	O	YR	YR
%R	Rate at which NPV is to be evaluated	I	–	–

Yes/No Questions

DO ROR? Yes: Calculate DCFROR.
No: Don't calculate DCFROR.

EDIT? Yes: Allow editing of NCF values.
No: No editing necessary.

Example 1

These examples illustrate the procedures used to calculate economic indices from NET CASH FLOW. The forecasts of production, prices, expenses, taxes, etc., must all have been completed previously. Consider the following problem. An expenditure of $1,000 is expected to result in the following net cash inflows. Calculate PAYOUT, DCFROR, and NET PRESENT VALUE at 10, 15, and 20%.

Time (YR)	Cash Outflow	Cash Inflow	Net Cash Flow
0	1,000	0	–1,000
1	–	450	450
2	–	300	300
3	–	250	250
4	–	200	200
5	–	175	175
6	–	150	150
	1,000	1,525	525

Keystrokes (SIZE > =039)	Display	Comments
[XEQ] [ALPHA] DISC [ALPHA]	*CF 0=?*	
1000 [CHS] [R/S]	*NCF 1=?*	
450 [R/S]	*NCF 2=?*	
300 [R/S]	*NCF 3=?*	
250 [R/S]	*NCF 4=?*	
200 [R/S]	*NCF 5=?*	
175 [R/S]	*NCF 6=?*	
150 [R/S]	*NCF 7=?*	
[R/S]	*EDIT? Y/N:N*	
[R/S]	*DO ROR? Y/N:*	CUM CF, PAYOUT, and NPV10 printed; last character is Y or N
Y [R/S]	*EST %R=?*	Guess higher than 10% since NPV10 is positive
15 [R/S]	*%R=?*	DCFROR printed
15 [R/S]	*%R=?*	NPV at 15% printed
20 [R/S]	*%R=?*	NPV at 20% printed

```
DISCASH

CF 0=-1,000.00
NCF 1=450.00
NCF 2=300.00
NCF 3=250.00
NCF 4=200.00
NCF 5=175.00
NCF 6=150.00

CUM CF=525.00
PAYOUT=3.00 YR
NPV10=174.79

DO ROR: YES
EST %R=15.00
DCFROR=17.23

%R=15.00
NPV=48.73

%R=20.00
NPV=-54.98
```

Example 2

In Example 1, the PRESENT VALUE RATIO (PV RATIO) is equal to the NPV plus the investment divided by the investment, or (1,000 + 174.79)/1,000 = 1.17. This works easily if the only cash outflows are at time zero. However, if year two's cash flow had been an investment of $100, what would the PV10 and PV15 ratios have been? The NPV of the investments must be calculated.

Keystrokes	Display	Comments
[R/S]	*CF 0=?*	
[R/S]	*NCF 1=?*	
0 [R/S]	*NCF 2=?*	
100 [CHS] [R/S]	*NCF 3=?*	
[R/S]	*EDIT? Y/N:N*	
[R/S]	*%R=?*	CUM CF and NPV10 printed; PAYOUT not output and DO ROR? Y/N prompt does not appear since CUM CF<0
15 [R/S]	*%R=?*	NPV at 15% printed

The PV ratios can be calculated as follows:

$$\text{PV10 ratio} = \frac{1{,}082.64 + 174.79}{1{,}082.64} = 1.16$$

$$\text{PV15 ratio} = \frac{1{,}075.61 + 48.73}{1{,}075.61} = 1.05$$

```
NCF 1=0.00
NCF 2=-100.00

CUM CF=-1,100.00
NPV10=-1,082.64

%R=15.00
NPV=-1,075.61
```

User Instructions

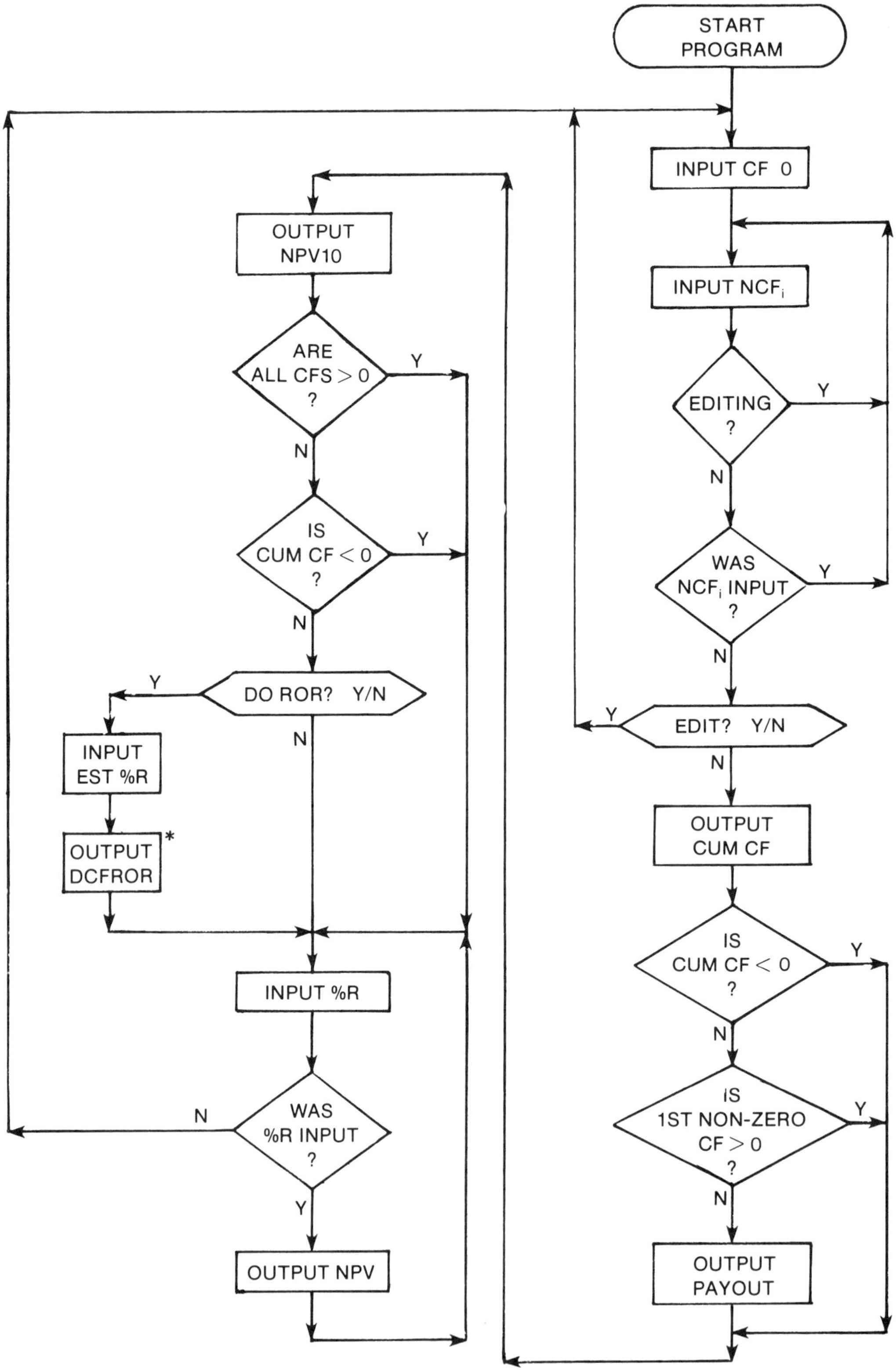

Note: for NCF_i, i = 1, 2, 3,..., n, where n is the number of NCF values input by the user.

* Tones will sound while DCFROR is calculated iteratively.

General Information

Memory Requirements

Program length: 378 bytes (2 cards)
Minimum size: 039*
Minimum hardware: 41C + 1 memory module

*This size will allow up to 25 NCF values. To accommodate v NCF values, use size 15 + v.

Hidden Options

None

Pac Subroutines Called

TITLE, IN, Y/N?, OUT, OUTU

Registers

06	Scratch
07	Scratch
08	EST %R, %R
09	Pointer
10	Scratch
11	Scratch
12	CUM CF
13	CF 0
14	NCF 1
15	NCF 2
16	NCF 3
	etc.

Registers 03 and 04 unused

Flags

00	Set:	All cash flows are > 0.
	Clear:	Not all cash flows are > 0.
03	Set:	Allow editing of cash flows.
	Clear:	No editing necessary.
04	Set:	Calculate DCFROR.
	Clear:	Do not calculate DCFROR.
07	Set:	Flag 29 set.
	Clear:	Flag 29 clear.

Note: The program sets the display mode to FIX 2.

Program Listing

```
 01♦LBL "DISC"
FIX 2  "DISCASH"  39
XROM "TITLE"  FC?C 25
PROMPT  CF 07

 09♦LBL 15
CF 03  1.3  STO 09

 13♦LBL 14
12  STO 00  "CF 0"
XROM "IN"

 18♦LBL 13
"NCF "  FS?C 29  SF 07
RCL 09  FIX 0  ARCL X
FIX 2  FS?C 07  SF 29
XROM "IN"  FC? 03
FS? 22  GTO 00  GTO 01

 33♦LBL 00
ISG 09  GTO 13

 36♦LBL 01
RCL 09  INT  1  -  1 E3
+  LASTX  /  STO 09  ADV
"EDIT"  3  XROM "Y/N?"
FS? 03  GTO 14  12.013
ST+ 09  SF 00  RCL 09  0

 57♦LBL 12
RCL IND Y  X<0?  CF 00
+  X≠0?  X>0?  GTO 02
STO 11  X<>Y  STO 10
X<>Y

 69♦LBL 02
ISG Y  GTO 12  STO 12
"CUM CF"  XROM "OUT"
X<0?  GTO 04  RCL 09

 78♦LBL 11
RCL IND X  X≠0?  GTO 03
RDN  ISG X  GTO 11

 85♦LBL 03
X>0?  GTO 04  RCL 10
INT  1  +  RCL 11
RCL IND Y  /  -  14  -
"YR"  ASTO 01  ASTO Y
CLA  ASTO 02  ASTO Z
"PAYOUT"  XROM "OUTU"

106♦LBL 04
10  STO 08  XEQ 16
"NPV10"  XROM "OUT"
RCL 12  FC? 00  X<0?
GTO 08  ADV  "DO ROR"  4
XROM "Y/N?"  FC? 04
GTO 08  7  STO 00
"EST %R"  XROM "IN"
FC? 22  25  STO 08

129♦LBL 10
TONE 5  XEQ 16  STO 06
RCL 09  1  STO 11  +
STO 10  CLX

139♦LBL 09
RCL 07  RCL 11  1  +
CHS  Y↑X  RCL 11  *
RCL IND 10  *  +  ISG 11
CLD  ISG 10  GTO 09
RCL 06  X<>Y  /  100  *
ENTER↑  X<> 08  ST+ 08
/  ABS  1 E-4  X<=Y?
GTO 10  RCL 08  TONE 9
"DCFROR"  XROM "OUT"

172♦LBL 08
ADV  7  STO 00  "%R"
XROM "IN"  FC? 22
GTO 15  XEQ 16  "NPV"
XROM "OUT"  GTO 08

184♦LBL 16
RCL 08  100  /  1  +
STO 07  RCL 09  1
STO 11  +  STO 10
RCL 13

197♦LBL 07
RCL IND 10  RCL 07
RCL 11  Y↑X  /  +
ISG 11  CLD  ISG 10
GTO 07  END
```

Section 10
Bar Code

Note: Without protection, bar code will wear with use. To protect your bar code, place one of the transparent sheets you received with your Wand over the bar code, glossy side down, before you begin scanning. If you want to use other types of protective coverings for your bar code, choose those that will not present a glossy surface to the Wand tip. A glossy surface may cause reflections that can reduce the Wand's ability to read the bar code you are scanning.

PROGRAM 1: OILPVT
SIZE: 028
PROGRAM REGISTERS NEEDED: 51

D. NATHAN MEEHAN
ERIC L. VOGEL

PAGE 1 OF 2

ROW 1 (1 : 2)

ROW 2 (2 : 7)

ROW 3 (7 : 11)

ROW 4 (12 : 19)

ROW 5 (19 : 23)

ROW 6 (24 : 29)

ROW 7 (30 : 35)

ROW 8 (36 : 41)

ROW 9 (42 : 51)

ROW 10 (51 : 59)

ROW 11 (59 : 66)

ROW 12 (66 : 73)

ROW 13 (73 : 83)

ROW 14 (83 : 89)

ROW 15 (89 : 97)

ROW 16 (98 : 102)

ROW 17 (102 : 107)

ROW 18 (108 : 112)

PROGRAM 1: OILPVT
SIZE: 028

D. NATHAN MEEHAN
ERIC L. VOGEL

PAGE 2
OF 2

ROW 19 (113 : 118)

ROW 20 (119 : 124)

ROW 21 (125 : 134)

ROW 22 (135 : 144)

ROW 23 (145 : 151)

ROW 24 (152 : 156)

ROW 25 (157 : 161)

ROW 26 (162 : 167)

ROW 27 (167 : 173)

PROGRAM 2: GASPVT
SIZE: 045
PROGRAM REGISTERS NEEDED: 56

2
D. NATHAN MEEHAN
ERIC L. VOGEL

PAGE 1
OF 2

ROW 1 (1 : 2)

ROW 2 (2 : 7)

ROW 3 (7 : 13)

ROW 4 (13 : 18)

ROW 5 (19 : 19)

ROW 6 (20 : 25)

ROW 7 (25 : 29)

ROW 8 (30 : 37)

ROW 9 (37 : 41)

ROW 10 (41 : 45)

ROW 11 (46 : 52)

ROW 12 (52 : 59)

ROW 13 (59 : 66)

ROW 14 (66 : 73)

ROW 15 (74 : 80)

ROW 16 (81 : 86)

ROW 17 (87 : 93)

ROW 18 (94 : 101)

PROGRAM 2: GASPVT
SIZE: 045

2
D. NATHAN MEEHAN
ERIC L. VOGEL

PAGE 2
OF 2

ROW 19 (102 : 110)

ROW 20 (111 : 120)

ROW 21 (120 : 125)

ROW 22 (126 : 130)

ROW 23 (131 : 136)

ROW 24 (136 : 140)

ROW 25 (140 : 145)

ROW 26 (146 : 153)

ROW 27 (153 : 157)

ROW 28 (158 : 168)

ROW 29 (168 : 177)

ROW 30 (177 : 182)

PROGRAM 3: DEW
SIZE: 045
PROGRAM REGISTERS NEEDED: 67

3
D. NATHAN MEEHAN
ERIC L. VOGEL

PAGE 1 OF 2

ROW 1 (1 : 3)

ROW 2 (3 : 7)

ROW 3 (8 : 11)

ROW 4 (11 : 18)

ROW 5 (18 : 23)

ROW 6 (23 : 28)

ROW 7 (29 : 32)

ROW 8 (33 : 36)

ROW 9 (36 : 43)

ROW 10 (44 : 50)

ROW 11 (51 : 56)

ROW 12 (57 : 60)

ROW 13 (61 : 64)

ROW 14 (64 : 70)

ROW 15 (71 : 76)

ROW 16 (76 : 82)

ROW 17 (82 : 84)

ROW 18 (84 : 92)

PROGRAM 3: DEW
SIZE: 045

3
D. NATHAN MEEHAN
ERIC L. VOGEL

PAGE 2
OF 2

ROW 19 (92 : 95)

ROW 20 (96 : 98)

ROW 21 (98 : 104)

ROW 22 (105 : 107)

ROW 23 (107 : 112)

ROW 24 (112 : 118)

ROW 25 (118 : 123)

ROW 26 (123 : 127)

ROW 27 (127 : 132)

ROW 28 (132 : 138)

ROW 29 (138 : 141)

ROW 30 (141 : 145)

ROW 31 (146 : 153)

ROW 32 (154 : 162)

ROW 33 (162 : 169)

ROW 34 (169 : 175)

ROW 35 (175 : 180)

ROW 36 (180 : 184)

PROGRAM 4: GOR
SIZE: 029
PROGRAM REGISTERS NEEDED: 96

4
D. NATHAN MEEHAN
ERIC L. VOGEL

PAGE
OF

ROW 1 (1 : 2)

ROW 2 (2 : 5)

ROW 3 (6 : 10)

ROW 4 (10 : 16)

ROW 5 (16 : 20)

ROW 6 (21 : 25)

ROW 7 (25 : 29)

ROW 8 (29 : 33)

ROW 9 (34 : 40)

ROW 10 (41 : 44)

ROW 11 (45 : 50)

ROW 12 (51 : 59)

ROW 13 (60 : 64)

ROW 14 (65 : 70)

ROW 15 (70 : 77)

ROW 16 (78 : 80)

ROW 17 (80 : 85)

ROW 18 (86 : 90)

PROGRAM 4: GOR
SIZE: 029

4
D. NATHAN MEEHAN
ERIC L. VOGEL

PAGE 2
OF 3

ROW 19 (90 : 95)

ROW 20 (96 : 100)

ROW 21 (101 : 108)

ROW 22 (108 : 112)

ROW 23 (112 : 118)

ROW 24 (118 : 123)

ROW 25 (123 : 127)

ROW 26 (127 : 134)

ROW 27 (135 : 140)

ROW 28 (140 : 146)

ROW 29 (147 : 153)

ROW 30 (153 : 157)

ROW 31 (158 : 161)

ROW 32 (161 : 166)

ROW 33 (166 : 171)

ROW 34 (171 : 175)

ROW 35 (175 : 180)

ROW 36 (180 : 185)

PROGRAM 4: GOR
SIZE: 029

4
D. NATHAN MEEHAN
ERIC L. VOGEL

ROW 37 (185 : 190)

ROW 38 (190 : 200)

ROW 39 (201 : 210)

ROW 40 (211 : 219)

ROW 41 (220 : 226)

ROW 42 (227 : 237)

ROW 43 (238 : 248)

ROW 44 (249 : 258)

ROW 45 (259 : 268)

ROW 46 (269 : 277)

ROW 47 (278 : 285)

ROW 48 (286 : 295)

ROW 49 (295 : 304)

ROW 50 (304 : 314)

ROW 51 (314 : 318)

ROW 52 (319 : 323)

PROGRAM 5: H2OPVT
SIZE: 020
PROGRAM REGISTERS NEEDED: 41

D. NATHAN MEEHAN
ERIC L. VOGEL

PAGE 1
OF 2

ROW 1 (1 : 2)

ROW 2 (2 : 7)

ROW 3 (7 : 11)

ROW 4 (12 : 16)

ROW 5 (16 : 22)

ROW 6 (22 : 30)

ROW 7 (30 : 37)

ROW 8 (38 : 44)

ROW 9 (45 : 51)

ROW 10 (52 : 59)

ROW 11 (59 : 68)

ROW 12 (69 : 73)

ROW 13 (74 : 77)

ROW 14 (78 : 83)

ROW 15 (83 : 88)

ROW 16 (89 : 92)

ROW 17 (93 : 97)

ROW 18 (97 : 101)

PROGRAM 5: H2OPVT
SIZE: 020

D. NATHAN MEEHAN
ERIC L. VOGEL

ROW 19 (101 : 106)

ROW 20 (107 : 117)

ROW 21 (118 : 127)

ROW 22 (127 : 132)

PROGRAM 6: GASPROD
SIZE: 027
PROGRAM REGISTERS NEEDED: 55

D. NATHAN MEEHAN
ERIC L. VOGEL

PAGE 1
OF 2

ROW 1 (1 : 2)

ROW 2 (2 : 5)

ROW 3 (6 : 11)

ROW 4 (11 : 16)

ROW 5 (16 : 20)

ROW 6 (20 : 26)

ROW 7 (27 : 30)

ROW 8 (31 : 38)

ROW 9 (38 : 43)

ROW 10 (43 : 46)

ROW 11 (46 : 50)

ROW 12 (50 : 54)

ROW 13 (54 : 60)

ROW 14 (61 : 70)

ROW 15 (70 : 79)

ROW 16 (79 : 87)

ROW 17 (88 : 97)

ROW 18 (98 : 107)

PROGRAM 6: GASPROD
SIZE: 027

D. NATHAN MEEHAN
ERIC L. VOGEL

PAGE 2
OF 2

ROW 19 (108 : 113)

ROW 20 (114 : 122)

ROW 21 (122 : 124)

ROW 22 (125 : 131)

ROW 23 (132 : 137)

ROW 24 (138 : 142)

ROW 25 (142 : 146)

ROW 26 (147 : 152)

ROW 27 (152 : 159)

ROW 28 (160 : 164)

ROW 29 (164 : 167)

ROW 30 (168 : 171)

PROGRAM 7: DECLINE
SIZE: 049
PROGRAM REGISTERS NEEDED: 268

D. NATHAN MEEHAN
ERIC L. VOGEL

PAGE 1
OF 8

ROW 1 (1 : 2)

ROW 2 (2 : 6)

ROW 3 (7 : 12)

ROW 4 (12 : 17)

ROW 5 (18 : 22)

ROW 6 (23 : 29)

ROW 7 (29 : 30)

ROW 8 (31 : 36)

ROW 9 (36 : 38)

ROW 10 (38 : 43)

ROW 11 (43 : 47)

ROW 12 (47 : 50)

ROW 13 (50 : 58)

ROW 14 (58 : 62)

ROW 15 (62 : 68)

ROW 16 (69 : 75)

ROW 17 (76 : 81)

ROW 18 (81 : 84)

PROGRAM 7: DECLINE
SIZE: 049

D. NATHAN MEEHAN
ERIC L. VOGEL

ROW 19 (84 : 88)

ROW 20 (88 : 93)

ROW 21 (94 : 99)

ROW 22 (99 : 102)

ROW 23 (102 : 106)

ROW 24 (106 : 111)

ROW 25 (111 : 115)

ROW 26 (115 : 120)

ROW 27 (120 : 124)

ROW 28 (125 : 129)

ROW 29 (129 : 134)

ROW 30 (134 : 138)

ROW 31 (138 : 140)

ROW 32 (141 : 147)

ROW 33 (148 : 152)

ROW 34 (152 : 157)

ROW 35 (157 : 159)

ROW 36 (160 : 164)

PROGRAM 7: DECLINE
SIZE: 049

D. NATHAN MEEHAN
ERIC L. VOGEL

PAGE 3
OF 8

ROW 37 (165 : 169)

ROW 38 (170 : 174)

ROW 39 (174 : 178)

ROW 40 (179 : 183)

ROW 41 (184 : 187)

ROW 42 (188 : 190)

ROW 43 (190 : 195)

ROW 44 (195 : 201)

ROW 45 (201 : 205)

ROW 46 (205 : 210)

ROW 47 (211 : 217)

ROW 48 (217 : 222)

ROW 49 (223 : 233)

ROW 50 (233 : 243)

ROW 51 (244 : 250)

ROW 52 (250 : 258)

ROW 53 (258 : 267)

ROW 54 (268 : 274)

PROGRAM 7: DECLINE
SIZE: 049

D. NATHAN MEEHAN
ERIC L. VOGEL

ROW 55 (275 : 283)

ROW 56 (284 : 290)

ROW 57 (291 : 297)

ROW 58 (298 : 309)

ROW 59 (310 : 318)

ROW 60 (319 : 328)

ROW 61 (329 : 338)

ROW 62 (339 : 348)

ROW 63 (348 : 357)

ROW 64 (358 : 367)

ROW 65 (368 : 376)

ROW 66 (377 : 383)

ROW 67 (384 : 385)

ROW 68 (386 : 393)

ROW 69 (394 : 398)

ROW 70 (398 : 402)

ROW 71 (402 : 406)

ROW 72 (407 : 412)

PROGRAM 7: DECLINE
SIZE: 049

D. NATHAN MEEHAN
ERIC L. VOGEL

ROW 73 (412 : 416)

ROW 74 (416 : 424)

ROW 75 (424 : 431)

ROW 76 (431 : 440)

ROW 77 (441 : 446)

ROW 78 (446 : 453)

ROW 79 (453 : 462)

ROW 80 (463 : 469)

ROW 81 (469 : 478)

ROW 82 (478 : 488)

ROW 83 (489 : 499)

ROW 84 (499 : 508)

ROW 85 (508 : 517)

ROW 86 (518 : 526)

ROW 87 (527 : 538)

ROW 88 (539 : 542)

ROW 89 (543 : 549)

ROW 90 (550 : 559)

PROGRAM 7: DECLINE
SIZE: 049

D. NATHAN MEEHAN
ERIC L. VOGEL

ROW 91 (560 : 569)

ROW 92 (570 : 579)

ROW 93 (580 : 588)

ROW 94 (589 : 597)

ROW 95 (598 : 606)

ROW 96 (607 : 618)

ROW 97 (619 : 629)

ROW 98 (630 : 636)

ROW 99 (637 : 647)

ROW 100 (648 : 657)

ROW 101 (658 : 665)

ROW 102 (666 : 673)

ROW 103 (674 : 684)

ROW 104 (684 : 693)

ROW 105 (693 : 699)

ROW 106 (699 : 705)

ROW 107 (706 : 710)

ROW 108 (711 : 721)

PROGRAM 7: DECLINE
SIZE: 049

D. NATHAN MEEHAN
ERIC L. VOGEL

PAGE 7
OF 8

ROW 109 (721 : 726)

ROW 110 (726 : 730)

ROW 111 (731 : 739)

ROW 112 (740 : 745)

ROW 113 (746 : 754)

ROW 114 (755 : 761)

ROW 115 (762 : 770)

ROW 116 (770 : 776)

ROW 117 (776 : 781)

ROW 118 (782 : 787)

ROW 119 (788 : 797)

ROW 120 (798 : 802)

ROW 121 (802 : 810)

ROW 122 (811 : 818)

ROW 123 (819 : 825)

ROW 124 (825 : 829)

ROW 125 (829 : 837)

ROW 126 (838 : 845)

ROW 127 (845 : 847)

ROW 128 (848 : 854)

ROW 129 (855 : 863)

ROW 130 (863 : 872)

ROW 131 (873 : 881)

ROW 132 (882 : 888)

ROW 133 (889 : 897)

ROW 134 (897 : 905)

ROW 135 (906 : 908)

ROW 136 (908 : 915)

ROW 137 (915 : 920)

ROW 138 (921 : 928)

ROW 139 (928 : 932)

ROW 140 (933 : 941)

ROW 141 (942 : 947)

ROW 142 (947 : 951)

ROW 143 (951 : 958)

ROW 144 (958 : 962)

PROGRAM 8: OIP
SIZE: 041
PROGRAM REGISTERS NEEDED: 87

D. NATHAN MEEHAN
ERIC L. VOGEL

ROW 1 (1 : 2)

ROW 2 (2 : 5)

ROW 3 (6 : 11)

ROW 4 (11 : 16)

ROW 5 (17 : 23)

ROW 6 (24 : 30)

ROW 7 (31 : 36)

ROW 8 (37 : 41)

ROW 9 (42 : 47)

ROW 10 (47 : 53)

ROW 11 (53 : 59)

ROW 12 (60 : 67)

ROW 13 (68 : 75)

ROW 14 (75 : 83)

ROW 15 (84 : 88)

ROW 16 (88 : 92)

ROW 17 (92 : 98)

ROW 18 (98 : 103)

PROGRAM 8: OIP
SIZE: 041

D. NATHAN MEEHAN
ERIC L. VOGEL

ROW 19 (103 : 108)

ROW 20 (108 : 115)

ROW 21 (115 : 119)

ROW 22 (119 : 125)

ROW 23 (126 : 133)

ROW 24 (133 : 140)

ROW 25 (141 : 146)

ROW 26 (147 : 153)

ROW 27 (153 : 159)

ROW 28 (159 : 166)

ROW 29 (167 : 171)

ROW 30 (172 : 176)

ROW 31 (177 : 183)

ROW 32 (184 : 190)

ROW 33 (191 : 196)

ROW 34 (197 : 203)

ROW 35 (204 : 209)

ROW 36 (209 : 215)

PROGRAM 8: OIP
SIZE: 041

D. NATHAN MEEHAN
ERIC L. VOGEL

PAGE 3
OF 3

ROW 37 (215 : 221)

ROW 38 (221 : 227)

ROW 39 (228 : 233)

ROW 40 (234 : 239)

ROW 41 (239 : 246)

ROW 42 (246 : 251)

ROW 43 (252 : 259)

ROW 44 (259 : 264)

ROW 45 (265 : 272)

ROW 46 (272 : 279)

ROW 47 (280 : 286)

PROGRAM 9: GIP
SIZE: 040
PROGRAM REGISTERS NEEDED: 63

D. NATHAN MEEHAN
ERIC L. VOGEL

PAGE 1
OF 2

ROW 1 (1 : 2)

ROW 2 (2 : 5)

ROW 3 (6 : 11)

ROW 4 (11 : 16)

ROW 5 (16 : 23)

ROW 6 (24 : 30)

ROW 7 (30 : 36)

ROW 8 (36 : 40)

ROW 9 (41 : 47)

ROW 10 (47 : 53)

ROW 11 (53 : 58)

ROW 12 (59 : 67)

ROW 13 (67 : 73)

ROW 14 (74 : 82)

ROW 15 (83 : 91)

ROW 16 (91 : 97)

ROW 17 (97 : 102)

ROW 18 (102 : 107)

PROGRAM 9: GIP
SIZE: 040

D. NATHAN MEEHAN
ERIC L. VOGEL

PAGE 2
OF 2

ROW 19 (107 : 113)

ROW 20 (113 : 121)

ROW 21 (121 : 130)

ROW 22 (130 : 135)

ROW 23 (135 : 139)

ROW 24 (139 : 144)

ROW 25 (144 : 148)

ROW 26 (149 : 158)

ROW 27 (159 : 165)

ROW 28 (165 : 169)

ROW 29 (170 : 175)

ROW 30 (176 : 181)

ROW 31 (182 : 188)

ROW 32 (189 : 196)

ROW 33 (197 : 205)

ROW 34 (206 : 212)

PROGRAM 10: OILMBE
SIZE: 044
PROGRAM REGISTERS NEEDED: 128

D. NATHAN MEEHAN
ERIC L. VOGEL

ROW 1 (1 : 2)

ROW 2 (2 : 4)

ROW 3 (4 : 11)

ROW 4 (11 : 15)

ROW 5 (15 : 21)

ROW 6 (21 : 26)

ROW 7 (26 : 33)

ROW 8 (33 : 39)

ROW 9 (40 : 43)

ROW 10 (44 : 48)

ROW 11 (49 : 53)

ROW 12 (53 : 59)

ROW 13 (59 : 64)

ROW 14 (64 : 69)

ROW 15 (69 : 73)

ROW 16 (73 : 77)

ROW 17 (78 : 83)

ROW 18 (84 : 90)

PROGRAM 10: OILMBE
SIZE: 044

D. NATHAN MEEHAN
ERIC L. VOGEL

PAGE 2
OF 4

ROW 19 (91 : 98)

ROW 20 (98 : 103)

ROW 21 (103 : 110)

ROW 22 (110 : 117)

ROW 23 (117 : 123)

ROW 24 (123 : 129)

ROW 25 (130 : 136)

ROW 26 (136 : 142)

ROW 27 (142 : 149)

ROW 28 (149 : 155)

ROW 29 (156 : 161)

ROW 30 (161 : 166)

ROW 31 (166 : 172)

ROW 32 (172 : 177)

ROW 33 (178 : 181)

ROW 34 (182 : 186)

ROW 35 (187 : 190)

ROW 36 (190 : 194)

PROGRAM 10: OILMBE
SIZE: 044

D. NATHAN MEEHAN
ERIC L. VOGEL

ROW 37 (194 : 200)

ROW 38 (201 : 206)

ROW 39 (206 : 208)

ROW 40 (209 : 214)

ROW 41 (214 : 221)

ROW 42 (221 : 226)

ROW 43 (226 : 233)

ROW 44 (233 : 239)

ROW 45 (239 : 246)

ROW 46 (246 : 254)

ROW 47 (255 : 262)

ROW 48 (262 : 269)

ROW 49 (269 : 277)

ROW 50 (278 : 286)

ROW 51 (287 : 294)

ROW 52 (295 : 304)

ROW 53 (304 : 312)

ROW 54 (313 : 324)

PROGRAM 10: OILMBE
SIZE: 044

D. NATHAN MEEHAN
ERIC L. VOGEL

PAGE 4
OF 4

ROW 55 (324 : 330)

ROW 56 (331 : 338)

ROW 57 (338 : 342)

ROW 58 (342 : 348)

ROW 59 (349 : 355)

ROW 60 (355 : 360)

ROW 61 (360 : 368)

ROW 62 (369 : 375)

ROW 63 (375 : 382)

ROW 64 (383 : 389)

ROW 65 (390 : 396)

ROW 66 (397 : 404)

ROW 67 (404 : 414)

ROW 68 (415 : 422)

ROW 69 (422 : 429)

PROGRAM 11: KG/KO
SIZE: 041
PROGRAM REGISTERS NEEDED: 95

D. NATHAN MEEHAN
ERIC L. VOGEL

PAGE 1 OF 3

ROW 1 (1 : 2)

ROW 2 (2 : 8)

ROW 3 (9 : 13)

ROW 4 (13 : 15)

ROW 5 (15 : 20)

ROW 6 (20 : 25)

ROW 7 (25 : 32)

ROW 8 (33 : 41)

ROW 9 (41 : 46)

ROW 10 (47 : 53)

ROW 11 (54 : 63)

ROW 12 (63 : 71)

ROW 13 (71 : 76)

ROW 14 (77 : 80)

ROW 15 (81 : 85)

ROW 16 (85 : 92)

ROW 17 (92 : 97)

ROW 18 (98 : 101)

PROGRAM 11: KG/KO
SIZE: 041

D. NATHAN MEEHAN
ERIC L. VOGEL

PAGE 2
OF 3

ROW 19 (102 : 108)

ROW 20 (109 : 112)

ROW 21 (113 : 117)

ROW 22 (117 : 123)

ROW 23 (124 : 130)

ROW 24 (131 : 135)

ROW 25 (135 : 140)

ROW 26 (140 : 146)

ROW 27 (147 : 155)

ROW 28 (156 : 161)

ROW 29 (162 : 169)

ROW 30 (169 : 174)

ROW 31 (174 : 181)

ROW 32 (181 : 190)

ROW 33 (190 : 196)

ROW 34 (196 : 203)

ROW 35 (203 : 208)

ROW 36 (209 : 214)

PROGRAM 11: KG/KO
SIZE: 041

D. NATHAN MEEHAN
ERIC L. VOGEL

PAGE 3
OF 3

ROW 37 (214 : 221)

ROW 38 (222 : 230)

ROW 39 (231 : 234)

ROW 40 (234 : 240)

ROW 41 (240 : 245)

ROW 42 (246 : 255)

ROW 43 (255 : 264)

ROW 44 (265 : 270)

ROW 45 (271 : 275)

ROW 46 (276 : 282)

ROW 47 (283 : 293)

ROW 48 (293 : 302)

ROW 49 (303 : 311)

ROW 50 (311 : 320)

ROW 51 (321 : 327)

PROGRAM 12: OILPRED
SIZE: 063
PROGRAM REGISTERS NEEDED: 216

D. NATHAN MEEHAN
ERIC L. VOGEL

PAGE 1
OF 7

ROW 1 (1 : 2)

ROW 2 (2 : 3)

ROW 3 (4 : 10)

ROW 4 (11 : 15)

ROW 5 (16 : 19)

ROW 6 (20 : 25)

ROW 7 (25 : 32)

ROW 8 (32 : 39)

ROW 9 (40 : 45)

ROW 10 (46 : 50)

ROW 11 (51 : 56)

ROW 12 (57 : 64)

ROW 13 (64 : 70)

ROW 14 (71 : 75)

ROW 15 (76 : 81)

ROW 16 (81 : 87)

ROW 17 (88 : 92)

ROW 18 (92 : 96)

PROGRAM 12: OILPRED
SIZE: 063

D. NATHAN MEEHAN
ERIC L. VOGEL

ROW 19 (97 : 100)

ROW 20 (100 : 105)

ROW 21 (105 : 113)

ROW 22 (113 : 119)

ROW 23 (120 : 124)

ROW 24 (125 : 128)

ROW 25 (129 : 134)

ROW 26 (134 : 139)

ROW 27 (139 : 144)

ROW 28 (144 : 152)

ROW 29 (153 : 159)

ROW 30 (159 : 167)

ROW 31 (168 : 177)

ROW 32 (177 : 182)

ROW 33 (183 : 188)

ROW 34 (189 : 194)

ROW 35 (195 : 202)

ROW 36 (203 : 209)

PROGRAM 12: OILPRED
SIZE: 063

D. NATHAN MEEHAN
ERIC L. VOGEL

PAGE 3
OF 7

ROW 37 (209 : 214)

ROW 38 (214 : 219)

ROW 39 (219 : 224)

ROW 40 (224 : 229)

ROW 41 (229 : 234)

ROW 42 (235 : 240)

ROW 43 (241 : 249)

ROW 44 (250 : 258)

ROW 45 (259 : 268)

ROW 46 (269 : 275)

ROW 47 (276 : 280)

ROW 48 (280 : 285)

ROW 49 (285 : 291)

ROW 50 (292 : 298)

ROW 51 (298 : 304)

ROW 52 (305 : 311)

ROW 53 (312 : 318)

ROW 54 (319 : 324)

PROGRAM 12: OILPRED
SIZE: 063

D. NATHAN MEEHAN
ERIC L. VOGEL

ROW 55 (325 : 329)

ROW 56 (329 : 336)

ROW 57 (337 : 342)

ROW 58 (343 : 347)

ROW 59 (347 : 353)

ROW 60 (354 : 360)

ROW 61 (360 : 367)

ROW 62 (367 : 374)

ROW 63 (375 : 383)

ROW 64 (384 : 392)

ROW 65 (392 : 400)

ROW 66 (401 : 408)

ROW 67 (409 : 416)

ROW 68 (416 : 421)

ROW 69 (421 : 428)

ROW 70 (429 : 436)

ROW 71 (437 : 445)

ROW 72 (446 : 453)

PROGRAM 12: OILPRED
SIZE: 063

D. NATHAN MEEHAN
ERIC L. VOGEL

PAGE 5
OF 7

ROW 73 (454 : 462)

ROW 74 (463 : 471)

ROW 75 (472 : 481)

ROW 76 (481 : 487)

ROW 77 (488 : 495)

ROW 78 (496 : 503)

ROW 79 (504 : 512)

ROW 80 (513 : 519)

ROW 81 (519 : 526)

ROW 82 (526 : 530)

ROW 83 (530 : 535)

ROW 84 (536 : 543)

ROW 85 (543 : 551)

ROW 86 (551 : 554)

ROW 87 (555 : 562)

ROW 88 (562 : 569)

ROW 89 (570 : 577)

ROW 90 (577 : 584)

PROGRAM 12: OILPRED
SIZE: 063

D. NATHAN MEEHAN
ERIC L. VOGEL

PAGE 6
OF 7

ROW 91 (584 : 591)

ROW 92 (592 : 602)

ROW 93 (603 : 610)

ROW 94 (611 : 619)

ROW 95 (620 : 628)

ROW 96 (629 : 635)

ROW 97 (636 : 643)

ROW 98 (643 : 650)

ROW 99 (650 : 658)

ROW 100 (658 : 666)

ROW 101 (667 : 675)

ROW 102 (676 : 683)

ROW 103 (683 : 693)

ROW 104 (694 : 701)

ROW 105 (702 : 711)

ROW 106 (711 : 719)

ROW 107 (720 : 729)

ROW 108 (730 : 736)

PROGRAM 12: OILPRED
SIZE: 063

D. NATHAN MEEHAN
ERIC L. VOGEL

PAGE 7
OF 7

ROW 109 (737 : 743)

ROW 110 (743 : 751)

ROW 111 (752 : 760)

ROW 112 (760 : 767)

ROW 113 (768 : 776)

ROW 114 (777 : 783)

ROW 115 (783 : 788)

ROW 116 (788 : 795)

PROGRAM 13: QOVST
SIZE: 057
PROGRAM REGISTERS NEEDED: 128

D. NATHAN MEEHAN
ERIC L. VOGEL

PAGE 1 OF 4

ROW 1 (1 : 2)

ROW 2 (2 : 5)

ROW 3 (6 : 11)

ROW 4 (11 : 15)

ROW 5 (15 : 22)

ROW 6 (23 : 28)

ROW 7 (28 : 33)

ROW 8 (34 : 41)

ROW 9 (42 : 47)

ROW 10 (47 : 52)

ROW 11 (53 : 57)

ROW 12 (58 : 63)

ROW 13 (64 : 69)

ROW 14 (69 : 75)

ROW 15 (76 : 81)

ROW 16 (82 : 89)

ROW 17 (89 : 96)

ROW 18 (96 : 100)

PROGRAM 13: QOVST
SIZE: 057

D. NATHAN MEEHAN
ERIC L. VOGEL

PAGE 2
OF 4

ROW 19 (101 : 106)

ROW 20 (107 : 112)

ROW 21 (113 : 115)

ROW 22 (115 : 120)

ROW 23 (120 : 125)

ROW 24 (126 : 134)

ROW 25 (135 : 140)

ROW 26 (141 : 147)

ROW 27 (148 : 154)

ROW 28 (155 : 161)

ROW 29 (161 : 168)

ROW 30 (168 : 173)

ROW 31 (173 : 181)

ROW 32 (181 : 189)

ROW 33 (190 : 196)

ROW 34 (196 : 202)

ROW 35 (203 : 207)

ROW 36 (208 : 212)

PROGRAM 13: QOVST
SIZE: 057

D. NATHAN MEEHAN
ERIC L. VOGEL

ROW 37 (212 : 218)

ROW 38 (218 : 223)

ROW 39 (224 : 229)

ROW 40 (230 : 236)

ROW 41 (236 : 240)

ROW 42 (241 : 248)

ROW 43 (248 : 252)

ROW 44 (252 : 257)

ROW 45 (258 : 264)

ROW 46 (264 : 269)

ROW 47 (269 : 277)

ROW 48 (277 : 283)

ROW 49 (284 : 293)

ROW 50 (294 : 302)

ROW 51 (303 : 311)

ROW 52 (312 : 316)

ROW 53 (317 : 325)

ROW 54 (326 : 334)

PROGRAM 13: QOVST
SIZE: 057

D. NATHAN MEEHAN
ERIC L. VOGEL

PAGE 4
OF 4

ROW 55 (335 : 343)

ROW 56 (344 : 350)

ROW 57 (351 : 358)

ROW 58 (358 : 364)

ROW 59 (365 : 373)

ROW 60 (373 : 381)

ROW 61 (382 : 389)

ROW 62 (390 : 397)

ROW 63 (398 : 405)

ROW 64 (406 : 414)

ROW 65 (415 : 423)

ROW 66 (424 : 430)

ROW 67 (430 : 437)

ROW 68 (437 : 443)

ROW 69 (444 : 448)

PRGMS 14&15: INFCOEF & INFLUX
SIZES: 130 & 130
PROGRAM REGISTERS NEEDED: 147

D. NATHAN MEEHAN
ERIC L. VOGEL

PAGE 1
OF 5

ROW 1 (1 : 2)

ROW 2 (2 : 3)

ROW 3 (3 : 10)

ROW 4 (10 : 13)

ROW 5 (13 : 16)

ROW 6 (17 : 22)

ROW 7 (22 : 29)

ROW 8 (29 : 36)

ROW 9 (36 : 44)

ROW 10 (45 : 51)

ROW 11 (52 : 58)

ROW 12 (58 : 60)

ROW 13 (60 : 62)

ROW 14 (62 : 68)

ROW 15 (69 : 76)

ROW 16 (76 : 81)

ROW 17 (81 : 85)

ROW 18 (85 : 91)

PRGMS 14&15: INFCOEF & INFLUX
SIZES: 130 & 130

D. NATHAN MEEHAN
ERIC L. VOGEL

PAGE 2
OF 5

ROW 19 (91 : 96)

ROW 20 (97 : 101)

ROW 21 (102 : 106)

ROW 22 (107 : 110)

ROW 23 (111 : 116)

ROW 24 (117 : 123)

ROW 25 (123 : 127)

ROW 26 (128 : 133)

ROW 27 (134 : 141)

ROW 28 (141 : 147)

ROW 29 (148 : 153)

ROW 30 (154 : 160)

ROW 31 (161 : 167)

ROW 32 (167 : 174)

ROW 33 (174 : 181)

ROW 34 (182 : 188)

ROW 35 (188 : 194)

ROW 36 (195 : 200)

PRGMS 14&15: INFCOEF & INFLUX
SIZES: 130 & 130

D. NATHAN MEEHAN
ERIC L. VOGEL

PAGE 3
OF 5

ROW 37 (200 : 207)

ROW 38 (208 : 214)

ROW 39 (215 : 222)

ROW 40 (223 : 229)

ROW 41 (230 : 236)

ROW 42 (237 : 245)

ROW 43 (245 : 245)

ROW 44 (245 : 250)

ROW 45 (250 : 257)

ROW 46 (257 : 261)

ROW 47 (261 : 269)

ROW 48 (270 : 277)

ROW 49 (277 : 283)

ROW 50 (283 : 290)

ROW 51 (290 : 298)

ROW 52 (298 : 304)

ROW 53 (304 : 311)

ROW 54 (312 : 318)

PRGMS 14&15: INFCOEF & INFLUX
SIZES: 130 & 130

D. NATHAN MEEHAN
ERIC L. VOGEL

PAGE 4
OF 5

ROW 55 (319 : 323)

ROW 56 (323 : 329)

ROW 57 (329 : 339)

ROW 58 (340 : 345)

ROW 59 (345 : 352)

ROW 60 (353 : 361)

ROW 61 (362 : 370)

ROW 62 (371 : 380)

ROW 63 (380 : 385)

ROW 64 (386 : 393)

ROW 65 (393 : 398)

ROW 66 (398 : 404)

ROW 67 (405 : 412)

ROW 68 (413 : 418)

ROW 69 (419 : 426)

ROW 70 (427 : 430)

ROW 71 (431 : 438)

ROW 72 (439 : 448)

PRGMS 14&15: INFCOEF & INFLUX
SIZES: 130 & 130

D. NATHAN MEEHAN
ERIC L. VOGEL

ROW 73 (449 : 458)

ROW 74 (459 : 466)

ROW 75 (467 : 476)

ROW 76 (477 : 485)

ROW 77 (485 : 490)

ROW 78 (491 : 499)

ROW 79 (500 : 505)

PROGRAM 16: GASMBE
SIZE: 052
PROGRAM REGISTERS NEEDED: 177

D. NATHAN MEEHAN
ERIC L. VOGEL

ROW 1 (1 : 2)

ROW 2 (2 : 4)

ROW 3 (4 : 11)

ROW 4 (11 : 16)

ROW 5 (17 : 20)

ROW 6 (21 : 27)

ROW 7 (27 : 33)

ROW 8 (33 : 39)

ROW 9 (40 : 43)

ROW 10 (43 : 49)

ROW 11 (50 : 55)

ROW 12 (55 : 58)

ROW 13 (58 : 62)

ROW 14 (62 : 64)

ROW 15 (64 : 68)

ROW 16 (68 : 75)

ROW 17 (75 : 81)

ROW 18 (81 : 85)

PROGRAM 16: GASMBE
SIZE: 052

D. NATHAN MEEHAN
ERIC L. VOGEL

ROW 19 (85 : 90)

ROW 20 (90 : 96)

ROW 21 (97 : 104)

ROW 22 (104 : 111)

ROW 23 (111 : 115)

ROW 24 (115 : 121)

ROW 25 (121 : 126)

ROW 26 (127 : 130)

ROW 27 (131 : 135)

ROW 28 (135 : 142)

ROW 29 (142 : 145)

ROW 30 (145 : 149)

ROW 31 (150 : 156)

ROW 32 (156 : 162)

ROW 33 (163 : 168)

ROW 34 (168 : 173)

ROW 35 (173 : 181)

ROW 36 (181 : 186)

PROGRAM 16: GASMBE
SIZE: 052

D. NATHAN MEEHAN
ERIC L. VOGEL

ROW 37 (187 : 193)

ROW 38 (193 : 200)

ROW 39 (200 : 205)

ROW 40 (205 : 210)

ROW 41 (210 : 218)

ROW 42 (219 : 221)

ROW 43 (221 : 228)

ROW 44 (228 : 234)

ROW 45 (234 : 240)

ROW 46 (241 : 248)

ROW 47 (248 : 251)

ROW 48 (252 : 258)

ROW 49 (258 : 264)

ROW 50 (264 : 270)

ROW 51 (270 : 275)

ROW 52 (276 : 285)

ROW 53 (286 : 292)

ROW 54 (292 : 298)

PROGRAM 16: GASMBE
SIZE: 052

D. NATHAN MEEHAN
ERIC L. VOGEL

ROW 55 (298 : 306)

ROW 56 (307 : 313)

ROW 57 (314 : 321)

ROW 58 (322 : 330)

ROW 59 (330 : 339)

ROW 60 (339 : 344)

ROW 61 (344 : 350)

ROW 62 (351 : 360)

ROW 63 (361 : 366)

ROW 64 (366 : 368)

ROW 65 (368 : 373)

ROW 66 (373 : 381)

ROW 67 (382 : 387)

ROW 68 (388 : 395)

ROW 69 (396 : 401)

ROW 70 (401 : 409)

ROW 71 (410 : 416)

ROW 72 (417 : 422)

PROGRAM 16: GASMBE
SIZE: 052

D. NATHAN MEEHAN
ERIC L. VOGEL

PAGE 5
OF 6

ROW 73 (423 : 431)

ROW 74 (432 : 442)

ROW 75 (442 : 451)

ROW 76 (451 : 458)

ROW 77 (459 : 466)

ROW 78 (467 : 473)

ROW 79 (473 : 482)

ROW 80 (482 : 491)

ROW 81 (491 : 499)

ROW 82 (500 : 505)

ROW 83 (506 : 515)

ROW 84 (515 : 520)

ROW 85 (520 : 529)

ROW 86 (530 : 535)

ROW 87 (536 : 542)

ROW 88 (542 : 549)

ROW 89 (550 : 556)

ROW 90 (557 : 566)

PROGRAM 16: GASMBE
SIZE: 052

D. NATHAN MEEHAN
ERIC L. VOGEL

ROW 91 (566 : 572)

ROW 92 (573 : 580)

ROW 93 (581 : 584)

ROW 94 (585 : 589)

ROW 95 (589 : 596)

PRGMS 17&18: BHPWHP & GASDEL
SIZES: 044 & 058
PROGRAM REGISTERS NEEDED: 192

D. NATHAN MEEHAN
ERIC L. VOGEL

PAGE 1
OF 6

ROW 1 (1 : 2)

ROW 2 (2 : 7)

ROW 3 (7 : 12)

ROW 4 (13 : 16)

ROW 5 (17 : 22)

ROW 6 (22 : 28)

ROW 7 (29 : 36)

ROW 8 (37 : 44)

ROW 9 (44 : 50)

ROW 10 (51 : 57)

ROW 11 (58 : 63)

ROW 12 (64 : 71)

ROW 13 (71 : 75)

ROW 14 (75 : 80)

ROW 15 (81 : 86)

ROW 16 (86 : 90)

ROW 17 (91 : 97)

ROW 18 (97 : 104)

PRGMS 17&18: BHPWHP & GASDEL
SIZES: 044 & 058

D. NATHAN MEEHAN
ERIC L. VOGEL

PAGE 2
OF 6

ROW 19 (105 : 113)

ROW 20 (113 : 119)

ROW 21 (120 : 125)

ROW 22 (126 : 128)

ROW 23 (128 : 130)

ROW 24 (130 : 134)

ROW 25 (134 : 139)

ROW 26 (139 : 144)

ROW 27 (145 : 149)

ROW 28 (150 : 155)

ROW 29 (156 : 160)

ROW 30 (160 : 165)

ROW 31 (165 : 171)

ROW 32 (172 : 178)

ROW 33 (179 : 187)

ROW 34 (188 : 195)

ROW 35 (196 : 204)

ROW 36 (205 : 213)

PRGMS 17&18: BHPWHP & GASDEL
SIZES: 044 & 058

D. NATHAN MEEHAN
ERIC L. VOGEL

PAGE 3
OF 6

ROW 37 (214 : 223)

ROW 38 (223 : 227)

ROW 39 (227 : 235)

ROW 40 (236 : 244)

ROW 41 (244 : 251)

ROW 42 (252 : 258)

ROW 43 (259 : 266)

ROW 44 (267 : 275)

ROW 45 (276 : 285)

ROW 46 (286 : 296)

ROW 47 (297 : 306)

ROW 48 (307 : 315)

ROW 49 (315 : 318)

ROW 50 (319 : 327)

ROW 51 (328 : 332)

ROW 52 (332 : 333)

ROW 53 (333 : 339)

ROW 54 (340 : 345)

PRGMS 17&18: BHPWHP & GASDEL
SIZES: 044 & 058

D. NATHAN MEEHAN
ERIC L. VOGEL

PAGE 4
OF 6

ROW 55 (345 : 352)

ROW 56 (352 : 358)

ROW 57 (358 : 364)

ROW 58 (364 : 366)

ROW 59 (366 : 372)

ROW 60 (372 : 377)

ROW 61 (378 : 382)

ROW 62 (383 : 390)

ROW 63 (390 : 395)

ROW 64 (396 : 402)

ROW 65 (403 : 403)

ROW 66 (404 : 410)

ROW 67 (410 : 416)

ROW 68 (417 : 422)

ROW 69 (423 : 428)

ROW 70 (428 : 432)

ROW 71 (433 : 438)

ROW 72 (439 : 443)

PRGMS 17&18: BHPWHP & GASDEL
SIZES: 044 & 058

D. NATHAN MEEHAN
ERIC L. VOGEL

PAGE 5
OF 6

ROW 73 (443 : 448)

ROW 74 (448 : 449)

ROW 75 (449 : 457)

ROW 76 (458 : 464)

ROW 77 (464 : 468)

ROW 78 (468 : 474)

ROW 79 (474 : 479)

ROW 80 (480 : 486)

ROW 81 (486 : 490)

ROW 82 (491 : 495)

ROW 83 (496 : 500)

ROW 84 (500 : 505)

ROW 85 (505 : 514)

ROW 86 (515 : 522)

ROW 87 (523 : 529)

ROW 88 (530 : 538)

ROW 89 (539 : 548)

ROW 90 (548 : 555)

ROW 91 (556 : 562)

ROW 92 (563 : 569)

ROW 93 (570 : 578)

ROW 94 (579 : 586)

ROW 95 (586 : 594)

ROW 96 (595 : 602)

ROW 97 (603 : 612)

ROW 98 (612 : 621)

ROW 99 (621 : 628)

ROW 100 (629 : 635)

ROW 101 (635 : 644)

ROW 102 (645 : 651)

ROW 103 (652 : 660)

ROW 104 (660 : 660)

PROGRAM 19: STAB
SIZE: 049
PROGRAM REGISTERS NEEDED: 112

D. NATHAN MEEHAN
ERIC L. VOGEL

PAGE 1
OF 4

ROW 1 (1 : 2)

ROW 2 (2 : 5)

ROW 3 (6 : 10)

ROW 4 (10 : 15)

ROW 5 (15 : 21)

ROW 6 (21 : 26)

ROW 7 (27 : 33)

ROW 8 (33 : 38)

ROW 9 (39 : 43)

ROW 10 (44 : 49)

ROW 11 (50 : 55)

ROW 12 (55 : 61)

ROW 13 (61 : 67)

ROW 14 (68 : 74)

ROW 15 (75 : 79)

ROW 16 (79 : 83)

ROW 17 (83 : 88)

ROW 18 (88 : 91)

PROGRAM 19: STAB
SIZE: 049

D. NATHAN MEEHAN
ERIC L. VOGEL

ROW 19 (91 : 95)

ROW 20 (96 : 99)

ROW 21 (100 : 103)

ROW 22 (103 : 109)

ROW 23 (110 : 120)

ROW 24 (121 : 129)

ROW 25 (130 : 137)

ROW 26 (137 : 144)

ROW 27 (144 : 151)

ROW 28 (151 : 157)

ROW 29 (157 : 164)

ROW 30 (165 : 175)

ROW 31 (175 : 182)

ROW 32 (183 : 191)

ROW 33 (192 : 195)

ROW 34 (195 : 199)

ROW 35 (200 : 206)

ROW 36 (206 : 210)

PROGRAM 19: STAB
SIZE: 049

D. NATHAN MEEHAN
ERIC L. VOGEL

ROW 37 (210 : 214)

ROW 38 (215 : 219)

ROW 39 (219 : 224)

ROW 40 (224 : 231)

ROW 41 (232 : 240)

ROW 42 (240 : 246)

ROW 43 (246 : 252)

ROW 44 (252 : 260)

ROW 45 (261 : 268)

ROW 46 (269 : 278)

ROW 47 (279 : 286)

ROW 48 (287 : 295)

ROW 49 (296 : 304)

ROW 50 (304 : 310)

ROW 51 (311 : 318)

ROW 52 (319 : 327)

ROW 53 (328 : 336)

ROW 54 (337 : 345)

PROGRAM 19: STAB
SIZE: 049

D. NATHAN MEEHAN
ERIC L. VOGEL

ROW 55 (346 : 354)

ROW 56 (355 : 364)

ROW 57 (364 : 370)

ROW 58 (371 : 378)

ROW 59 (379 : 385)

ROW 60 (385 : 392)

ROW 61 (392 : 393)

PROGRAM 20: BLDUTIL
SIZE: 063
PROGRAM REGISTERS NEEDED: 112

D. NATHAN MEEHAN
ERIC L. VOGEL

PAGE 1
OF 4

ROW 1 (1 : 2)

ROW 2 (2 : 3)

ROW 3 (4 : 10)

ROW 4 (10 : 15)

ROW 5 (15 : 20)

ROW 6 (20 : 25)

ROW 7 (26 : 29)

ROW 8 (29 : 33)

ROW 9 (33 : 38)

ROW 10 (39 : 44)

ROW 11 (44 : 50)

ROW 12 (50 : 55)

ROW 13 (55 : 61)

ROW 14 (61 : 67)

ROW 15 (68 : 72)

ROW 16 (73 : 79)

ROW 17 (80 : 87)

ROW 18 (87 : 92)

PROGRAM 20: BLDUTIL
SIZE: 063

D. NATHAN MEEHAN
ERIC L. VOGEL

ROW 19 (92 : 100)

ROW 20 (100 : 106)

ROW 21 (106 : 112)

ROW 22 (112 : 119)

ROW 23 (120 : 128)

ROW 24 (129 : 137)

ROW 25 (138 : 145)

ROW 26 (145 : 151)

ROW 27 (152 : 157)

ROW 28 (158 : 165)

ROW 29 (165 : 174)

ROW 30 (175 : 183)

ROW 31 (184 : 190)

ROW 32 (191 : 198)

ROW 33 (198 : 203)

ROW 34 (203 : 206)

ROW 35 (207 : 215)

ROW 36 (216 : 224)

PROGRAM 20: BLDUTIL
SIZE: 063

D. NATHAN MEEHAN
ERIC L. VOGEL

PAGE 3
OF 4

ROW 37 (224 : 228)

ROW 38 (228 : 236)

ROW 39 (237 : 245)

ROW 40 (246 : 253)

ROW 41 (253 : 263)

ROW 42 (264 : 273)

ROW 43 (274 : 283)

ROW 44 (283 : 291)

ROW 45 (292 : 301)

ROW 46 (302 : 307)

ROW 47 (307 : 314)

ROW 48 (314 : 318)

ROW 49 (318 : 322)

ROW 50 (322 : 327)

ROW 51 (327 : 334)

ROW 52 (334 : 340)

ROW 53 (340 : 348)

ROW 54 (349 : 357)

PROGRAM 20: BLDUTIL
SIZE: 063

D. NATHAN MEEHAN
ERIC L. VOGEL

PAGE 4
OF 4

ROW 55 (358 : 365)

ROW 56 (365 : 372)

ROW 57 (372 : 380)

ROW 58 (381 : 389)

ROW 59 (390 : 398)

ROW 60 (398 : 405)

PROGRAM 21: BUILDUP
SIZE: 054
PROGRAM REGISTERS NEEDED: 198

D. NATHAN MEEHAN
ERIC L. VOGEL

PAGE 1
OF 6

ROW 1 (1 : 2)

ROW 2 (2 : 6)

ROW 3 (7 : 12)

ROW 4 (12 : 17)

ROW 5 (18 : 21)

ROW 6 (21 : 26)

ROW 7 (27 : 32)

ROW 8 (32 : 37)

ROW 9 (38 : 44)

ROW 10 (44 : 51)

ROW 11 (52 : 57)

ROW 12 (58 : 63)

ROW 13 (64 : 70)

ROW 14 (70 : 76)

ROW 15 (77 : 82)

ROW 16 (82 : 86)

ROW 17 (87 : 91)

ROW 18 (91 : 97)

PROGRAM 21: BUILDUP
SIZE: 054

D. NATHAN MEEHAN
ERIC L. VOGEL

ROW 19 (97 : 103)

ROW 20 (103 : 109)

ROW 21 (110 : 116)

ROW 22 (116 : 123)

ROW 23 (124 : 129)

ROW 24 (130 : 135)

ROW 25 (135 : 140)

ROW 26 (141 : 149)

ROW 27 (150 : 154)

ROW 28 (155 : 160)

ROW 29 (161 : 166)

ROW 30 (167 : 171)

ROW 31 (171 : 175)

ROW 32 (175 : 180)

ROW 33 (180 : 185)

ROW 34 (185 : 191)

ROW 35 (192 : 195)

ROW 36 (195 : 199)

PROGRAM 21: BUILDUP
SIZE: 054

D. NATHAN MEEHAN
ERIC L. VOGEL

ROW 37 (200 : 206)

ROW 38 (207 : 210)

ROW 39 (211 : 215)

ROW 40 (215 : 220)

ROW 41 (221 : 225)

ROW 42 (226 : 232)

ROW 43 (232 : 238)

ROW 44 (239 : 245)

ROW 45 (246 : 253)

ROW 46 (253 : 257)

ROW 47 (257 : 265)

ROW 48 (266 : 272)

ROW 49 (272 : 276)

ROW 50 (276 : 281)

ROW 51 (282 : 288)

ROW 52 (289 : 295)

ROW 53 (295 : 300)

ROW 54 (301 : 302)

PROGRAM 21: BUILDUP
SIZE: 054

D. NATHAN MEEHAN
ERIC L. VOGEL

ROW 55 (302 : 308)

ROW 56 (309 : 315)

ROW 57 (315 : 318)

ROW 58 (318 : 322)

ROW 59 (322 : 329)

ROW 60 (329 : 331)

ROW 61 (331 : 336)

ROW 62 (337 : 342)

ROW 63 (343 : 349)

ROW 64 (349 : 355)

ROW 65 (355 : 360)

ROW 66 (360 : 364)

ROW 67 (364 : 366)

ROW 68 (367 : 375)

ROW 69 (376 : 383)

ROW 70 (383 : 389)

ROW 71 (390 : 397)

ROW 72 (397 : 402)

PROGRAM 21: BUILDUP
SIZE: 054

D. NATHAN MEEHAN
ERIC L. VOGEL

ROW 73 (403 : 406)

ROW 74 (406 : 413)

ROW 75 (414 : 420)

ROW 76 (420 : 430)

ROW 77 (431 : 438)

ROW 78 (439 : 446)

ROW 79 (446 : 452)

ROW 80 (453 : 461)

ROW 81 (462 : 468)

ROW 82 (469 : 478)

ROW 83 (479 : 483)

ROW 84 (483 : 488)

ROW 85 (488 : 493)

ROW 86 (494 : 499)

ROW 87 (499 : 506)

ROW 88 (506 : 512)

ROW 89 (513 : 516)

ROW 90 (517 : 524)

PROGRAM 21: BUILDUP
SIZE: 054

D. NATHAN MEEHAN
ERIC L. VOGEL

ROW 91 (524 : 531)

ROW 92 (531 : 535)

ROW 93 (535 : 539)

ROW 94 (540 : 548)

ROW 95 (548 : 556)

ROW 96 (557 : 565)

ROW 97 (566 : 572)

ROW 98 (573 : 580)

ROW 99 (580 : 586)

ROW 100 (586 : 594)

ROW 101 (595 : 598)

ROW 102 (598 : 601)

ROW 103 (602 : 610)

ROW 104 (611 : 619)

ROW 105 (619 : 623)

ROW 106 (623 : 627)

ROW 107 (628 : 629)

PROGRAM 22: DRAW
SIZE: 089
PROGRAM REGISTERS NEEDED: 189

D. NATHAN MEEHAN
ERIC L. VOGEL

PAGE 1
OF 6

ROW 1 (1 : 2)

ROW 2 (2 : 7)

ROW 3 (8 : 13)

ROW 4 (14 : 18)

ROW 5 (19 : 22)

ROW 6 (23 : 29)

ROW 7 (29 : 30)

ROW 8 (30 : 36)

ROW 9 (37 : 42)

ROW 10 (43 : 50)

ROW 11 (50 : 53)

ROW 12 (53 : 59)

ROW 13 (59 : 64)

ROW 14 (64 : 70)

ROW 15 (71 : 77)

ROW 16 (77 : 83)

ROW 17 (84 : 89)

ROW 18 (90 : 96)

PROGRAM 22: DRAW
SIZE: 089

D. NATHAN MEEHAN
ERIC L. VOGEL

PAGE 2
OF 6

ROW 19 (96 : 103)

ROW 20 (103 : 110)

ROW 21 (110 : 114)

ROW 22 (115 : 119)

ROW 23 (120 : 123)

ROW 24 (124 : 130)

ROW 25 (131 : 135)

ROW 26 (135 : 142)

ROW 27 (142 : 149)

ROW 28 (149 : 156)

ROW 29 (156 : 162)

ROW 30 (162 : 169)

ROW 31 (169 : 176)

ROW 32 (176 : 181)

ROW 33 (182 : 187)

ROW 34 (187 : 191)

ROW 35 (191 : 196)

ROW 36 (196 : 201)

PROGRAM 22: DRAW
SIZE: 089

D. NATHAN MEEHAN
ERIC L. VOGEL

PAGE 3
OF 6

ROW 37 (201 : 206)

ROW 38 (207 : 211)

ROW 39 (212 : 218)

ROW 40 (219 : 225)

ROW 41 (226 : 233)

ROW 42 (233 : 240)

ROW 43 (241 : 247)

ROW 44 (248 : 255)

ROW 45 (256 : 262)

ROW 46 (262 : 263)

ROW 47 (264 : 270)

ROW 48 (271 : 280)

ROW 49 (281 : 289)

ROW 50 (290 : 297)

ROW 51 (298 : 304)

ROW 52 (305 : 311)

ROW 53 (312 : 320)

ROW 54 (320 : 330)

PROGRAM 22: DRAW
SIZE: 089

D. NATHAN MEEHAN
ERIC L. VOGEL

PAGE 4
OF 6

ROW 55 (330 : 333)

ROW 56 (334 : 339)

ROW 57 (339 : 343)

ROW 58 (344 : 350)

ROW 59 (350 : 355)

ROW 60 (355 : 363)

ROW 61 (363 : 366)

ROW 62 (367 : 373)

ROW 63 (374 : 380)

ROW 64 (381 : 387)

ROW 65 (387 : 393)

ROW 66 (394 : 400)

ROW 67 (400 : 406)

ROW 68 (407 : 413)

ROW 69 (414 : 419)

ROW 70 (420 : 426)

ROW 71 (427 : 434)

ROW 72 (434 : 442)

PROGRAM 22: DRAW
SIZE: 089

D. NATHAN MEEHAN
ERIC L. VOGEL

PAGE 5
OF 6

ROW 73 (443 : 450)

ROW 74 (451 : 458)

ROW 75 (459 : 466)

ROW 76 (467 : 476)

ROW 77 (476 : 484)

ROW 78 (485 : 491)

ROW 79 (492 : 498)

ROW 80 (498 : 504)

ROW 81 (505 : 509)

ROW 82 (510 : 516)

ROW 83 (516 : 521)

ROW 84 (521 : 524)

ROW 85 (525 : 533)

ROW 86 (534 : 542)

ROW 87 (542 : 547)

ROW 88 (548 : 555)

ROW 89 (556 : 563)

ROW 90 (564 : 570)

PROGRAM 22: DRAW
SIZE: 089

D. NATHAN MEEHAN
ERIC L. VOGEL

ROW 91 (571 : 581)

ROW 92 (582 : 591)

ROW 93 (592 : 599)

ROW 94 (600 : 609)

ROW 95 (610 : 618)

ROW 96 (619 : 625)

ROW 97 (626 : 631)

ROW 98 (631 : 637)

ROW 99 (637 : 646)

ROW 100 (646 : 656)

ROW 101 (656 : 663)

ROW 102 (663 : 664)

PROGRAM 23: FWVSW
SIZE: 051
PROGRAM REGISTERS NEEDED: 128

D. NATHAN MEEHAN
ERIC L. VOGEL

PAGE 1
OF 4

ROW 1 (1 : 2)

ROW 2 (2 : 6)

ROW 3 (7 : 12)

ROW 4 (12 : 19)

ROW 5 (19 : 26)

ROW 6 (27 : 29)

ROW 7 (29 : 32)

ROW 8 (32 : 36)

ROW 9 (37 : 45)

ROW 10 (45 : 54)

ROW 11 (54 : 61)

ROW 12 (62 : 65)

ROW 13 (66 : 72)

ROW 14 (73 : 77)

ROW 15 (77 : 83)

ROW 16 (83 : 89)

ROW 17 (89 : 93)

ROW 18 (93 : 97)

PROGRAM 23: FWVSW
SIZE: 051

D. NATHAN MEEHAN
ERIC L. VOGEL

ROW 19 (98 : 104)

ROW 20 (105 : 109)

ROW 21 (109 : 117)

ROW 22 (117 : 123)

ROW 23 (124 : 132)

ROW 24 (132 : 140)

ROW 25 (141 : 143)

ROW 26 (143 : 148)

ROW 27 (148 : 152)

ROW 28 (153 : 162)

ROW 29 (162 : 167)

ROW 30 (168 : 172)

ROW 31 (172 : 179)

ROW 32 (179 : 184)

ROW 33 (184 : 188)

ROW 34 (189 : 197)

ROW 35 (198 : 204)

ROW 36 (205 : 213)

PROGRAM 23: FWVSW
SIZE: 051

D. NATHAN MEEHAN
ERIC L. VOGEL

PAGE 3
OF 4

ROW 37 (213 : 222)

ROW 38 (223 : 229)

ROW 39 (229 : 235)

ROW 40 (235 : 244)

ROW 41 (245 : 249)

ROW 42 (250 : 256)

ROW 43 (257 : 263)

ROW 44 (264 : 271)

ROW 45 (271 : 273)

ROW 46 (273 : 277)

ROW 47 (278 : 284)

ROW 48 (285 : 293)

ROW 49 (294 : 298)

ROW 50 (299 : 305)

ROW 51 (306 : 315)

ROW 52 (316 : 322)

ROW 53 (323 : 330)

ROW 54 (330 : 337)

PROGRAM 23: FWVSW
SIZE: 051

D. NATHAN MEEHAN
ERIC L. VOGEL

PAGE 4
OF 4

ROW 55 (337 : 342)

ROW 56 (342 : 346)

ROW 57 (347 : 351)

ROW 58 (351 : 358)

ROW 59 (359 : 363)

ROW 60 (363 : 372)

ROW 61 (373 : 381)

ROW 62 (382 : 388)

ROW 63 (389 : 397)

ROW 64 (398 : 405)

ROW 65 (406 : 414)

ROW 66 (414 : 422)

ROW 67 (422 : 429)

ROW 68 (430 : 440)

ROW 69 (441 : 446)

PROGRAM 24: 5SPOT
SIZE: 065
PROGRAM REGISTERS NEEDED: 253

D. NATHAN MEEHAN
ERIC L. VOGEL

PAGE 1
OF 8

ROW 1 (1 : 2)

ROW 2 (2 : 6)

ROW 3 (7 : 13)

ROW 4 (13 : 17)

ROW 5 (17 : 23)

ROW 6 (23 : 28)

ROW 7 (29 : 32)

ROW 8 (33 : 37)

ROW 9 (38 : 43)

ROW 10 (43 : 47)

ROW 11 (48 : 51)

ROW 12 (52 : 57)

ROW 13 (57 : 62)

ROW 14 (62 : 67)

ROW 15 (67 : 71)

ROW 16 (72 : 75)

ROW 17 (75 : 81)

ROW 18 (81 : 87)

PROGRAM 24: 5SPOT
SIZE: 065

D. NATHAN MEEHAN
ERIC L. VOGEL

ROW 19 (88 : 94)

ROW 20 (94 : 101)

ROW 21 (102 : 107)

ROW 22 (107 : 114)

ROW 23 (115 : 122)

ROW 24 (122 : 129)

ROW 25 (129 : 134)

ROW 26 (134 : 135)

ROW 27 (135 : 143)

ROW 28 (144 : 145)

ROW 29 (145 : 152)

ROW 30 (153 : 157)

ROW 31 (157 : 161)

ROW 32 (162 : 171)

ROW 33 (171 : 178)

ROW 34 (179 : 182)

ROW 35 (183 : 185)

ROW 36 (185 : 186)

PROGRAM 24: 5SPOT
SIZE: 065

D. NATHAN MEEHAN
ERIC L. VOGEL

PAGE 3
OF 8

ROW 37 (187 : 195)

ROW 38 (195 : 198)

ROW 39 (199 : 205)

ROW 40 (206 : 209)

ROW 41 (210 : 213)

ROW 42 (213 : 218)

ROW 43 (219 : 227)

ROW 44 (228 : 236)

ROW 45 (236 : 241)

ROW 46 (241 : 247)

ROW 47 (248 : 255)

ROW 48 (255 : 264)

ROW 49 (264 : 271)

ROW 50 (271 : 279)

ROW 51 (279 : 286)

ROW 52 (287 : 295)

ROW 53 (296 : 297)

ROW 54 (297 : 298)

PROGRAM 24: 5SPOT
SIZE: 065

D. NATHAN MEEHAN
ERIC L. VOGEL

ROW 55 (299 : 306)

ROW 56 (306 : 311)

ROW 57 (312 : 316)

ROW 58 (317 : 320)

ROW 59 (321 : 324)

ROW 60 (325 : 327)

ROW 61 (327 : 328)

ROW 62 (328 : 333)

ROW 63 (333 : 338)

ROW 64 (338 : 342)

ROW 65 (343 : 350)

ROW 66 (351 : 355)

ROW 67 (355 : 360)

ROW 68 (360 : 367)

ROW 69 (367 : 372)

ROW 70 (373 : 378)

ROW 71 (379 : 386)

ROW 72 (387 : 396)

PROGRAM 24: 5SPOT
SIZE: 065

D. NATHAN MEEHAN
ERIC L. VOGEL

PAGE 5
OF 8

ROW 73 (396 : 399)

ROW 74 (399 : 400)

ROW 75 (400 : 405)

ROW 76 (406 : 409)

ROW 77 (410 : 414)

ROW 78 (414 : 418)

ROW 79 (418 : 423)

ROW 80 (424 : 427)

ROW 81 (428 : 433)

ROW 82 (433 : 441)

ROW 83 (441 : 446)

ROW 84 (447 : 451)

ROW 85 (452 : 459)

ROW 86 (460 : 464)

ROW 87 (464 : 471)

ROW 88 (472 : 478)

ROW 89 (479 : 488)

ROW 90 (489 : 500)

PROGRAM 24: 5SPOT
SIZE: 065

D. NATHAN MEEHAN
ERIC L. VOGEL

ROW 91 (501 : 510)

ROW 92 (510 : 517)

ROW 93 (517 : 521)

ROW 94 (522 : 528)

ROW 95 (529 : 537)

ROW 96 (538 : 542)

ROW 97 (542 : 550)

ROW 98 (551 : 557)

ROW 99 (558 : 566)

ROW 100 (567 : 573)

ROW 101 (573 : 581)

ROW 102 (581 : 587)

ROW 103 (587 : 593)

ROW 104 (594 : 601)

ROW 105 (601 : 609)

ROW 106 (610 : 617)

ROW 107 (618 : 625)

ROW 108 (625 : 631)

PROGRAM 24: 5SPOT
SIZE: 065

D. NATHAN MEEHAN
ERIC L. VOGEL

ROW 109 (631 : 636)

ROW 110 (637 : 644)

ROW 111 (644 : 648)

ROW 112 (648 : 652)

ROW 113 (652 : 661)

ROW 114 (661 : 669)

ROW 115 (670 : 676)

ROW 116 (677 : 684)

ROW 117 (684 : 692)

ROW 118 (692 : 700)

ROW 119 (701 : 706)

ROW 120 (707 : 715)

ROW 121 (716 : 721)

ROW 122 (721 : 729)

ROW 123 (729 : 738)

ROW 124 (738 : 743)

ROW 125 (744 : 753)

ROW 126 (753 : 759)

PROGRAM 24: 5SPOT
SIZE: 065

D. NATHAN MEEHAN
ERIC L. VOGEL

ROW 127 (759 : 764)

ROW 128 (764 : 770)

ROW 129 (770 : 775)

ROW 130 (775 : 779)

ROW 131 (780 : 791)

ROW 132 (792 : 798)

ROW 133 (799 : 806)

ROW 134 (806 : 812)

ROW 135 (813 : 820)

ROW 136 (821 : 826)

PROGRAM 25: SUMWF
SIZE: 061
PROGRAM REGISTERS NEEDED: 63

D. NATHAN MEEHAN
ERIC L. VOGEL

ROW 1 (1 : 2)

ROW 2 (2 : 6)

ROW 3 (7 : 9)

ROW 4 (10 : 16)

ROW 5 (17 : 24)

ROW 6 (25 : 31)

ROW 7 (31 : 35)

ROW 8 (36 : 46)

ROW 9 (47 : 52)

ROW 10 (53 : 60)

ROW 11 (60 : 69)

ROW 12 (69 : 76)

ROW 13 (77 : 85)

ROW 14 (85 : 92)

ROW 15 (93 : 99)

ROW 16 (99 : 105)

ROW 17 (106 : 111)

ROW 18 (112 : 118)

PROGRAM 25: SUMWF
SIZE: 061

D. NATHAN MEEHAN
ERIC L. VOGEL

PAGE 2
OF 2

ROW 19 (119 : 125)

ROW 20 (126 : 135)

ROW 21 (135 : 142)

ROW 22 (143 : 150)

ROW 23 (150 : 157)

ROW 24 (158 : 163)

ROW 25 (163 : 168)

ROW 26 (168 : 173)

ROW 27 (174 : 181)

ROW 28 (182 : 186)

ROW 29 (186 : 194)

ROW 30 (195 : 204)

ROW 31 (205 : 214)

ROW 32 (215 : 223)

ROW 33 (224 : 233)

ROW 34 (234 : 239)

PROGRAM 26: INJ
SIZE: 040
PROGRAM REGISTERS NEEDED: 82

D. NATHAN MEEHAN
ERIC L. VOGEL

ROW 1 (1 : 2)

ROW 2 (2 : 7)

ROW 3 (8 : 10)

ROW 4 (11 : 17)

ROW 5 (18 : 23)

ROW 6 (23 : 29)

ROW 7 (29 : 36)

ROW 8 (36 : 42)

ROW 9 (42 : 47)

ROW 10 (47 : 52)

ROW 11 (53 : 57)

ROW 12 (57 : 61)

ROW 13 (61 : 62)

ROW 14 (63 : 67)

ROW 15 (68 : 71)

ROW 16 (71 : 73)

ROW 17 (74 : 76)

ROW 18 (76 : 79)

PROGRAM 26: INJ
SIZE: 040

D. NATHAN MEEHAN
ERIC L. VOGEL

PAGE 2
OF 3

ROW 19 (80 : 84)

ROW 20 (84 : 86)

ROW 21 (87 : 91)

ROW 22 (91 : 93)

ROW 23 (93 : 100)

ROW 24 (100 : 104)

ROW 25 (104 : 108)

ROW 26 (108 : 112)

ROW 27 (112 : 112)

ROW 28 (113 : 119)

ROW 29 (120 : 126)

ROW 30 (126 : 131)

ROW 31 (132 : 141)

ROW 32 (141 : 149)

ROW 33 (150 : 159)

ROW 34 (159 : 168)

ROW 35 (169 : 177)

ROW 36 (178 : 184)

PROGRAM 26: INJ
SIZE: 040

D. NATHAN MEEHAN
ERIC L. VOGEL

PAGE 3
OF 3

ROW 37 (184 : 192)

ROW 38 (192 : 201)

ROW 39 (201 : 210)

ROW 40 (210 : 214)

ROW 41 (215 : 223)

ROW 42 (223 : 232)

ROW 43 (233 : 241)

ROW 44 (242 : 248)

PROGRAM 27: CUTCUM
SIZE: 055
PROGRAM REGISTERS NEEDED: 151

D. NATHAN MEEHAN
ERIC L. VOGEL

PAGE 1
OF 5

ROW 1 (1 : 2)

ROW 2 (2 : 6)

ROW 3 (7 : 13)

ROW 4 (13 : 17)

ROW 5 (17 : 24)

ROW 6 (24 : 31)

ROW 7 (32 : 35)

ROW 8 (35 : 40)

ROW 9 (41 : 47)

ROW 10 (47 : 54)

ROW 11 (54 : 62)

ROW 12 (62 : 70)

ROW 13 (71 : 79)

ROW 14 (79 : 87)

ROW 15 (87 : 93)

ROW 16 (94 : 102)

ROW 17 (103 : 110)

ROW 18 (111 : 112)

PROGRAM 27: CUTCUM
SIZE: 055

D. NATHAN MEEHAN
ERIC L. VOGEL

PAGE 2
OF 5

ROW 19 (112 : 117)

ROW 20 (118 : 123)

ROW 21 (123 : 129)

ROW 22 (130 : 138)

ROW 23 (138 : 143)

ROW 24 (144 : 148)

ROW 25 (148 : 154)

ROW 26 (154 : 163)

ROW 27 (164 : 171)

ROW 28 (171 : 178)

ROW 29 (178 : 185)

ROW 30 (186 : 191)

ROW 31 (191 : 198)

ROW 32 (199 : 206)

ROW 33 (207 : 213)

ROW 34 (214 : 218)

ROW 35 (219 : 228)

ROW 36 (229 : 241)

PROGRAM 27: CUTCUM
SIZE: 055

D. NATHAN MEEHAN
ERIC L. VOGEL

ROW 37 (242 : 252)

ROW 38 (252 : 258)

ROW 39 (258 : 264)

ROW 40 (265 : 268)

ROW 41 (268 : 271)

ROW 42 (272 : 275)

ROW 43 (276 : 280)

ROW 44 (280 : 286)

ROW 45 (287 : 293)

ROW 46 (293 : 298)

ROW 47 (299 : 307)

ROW 48 (307 : 311)

ROW 49 (312 : 320)

ROW 50 (320 : 325)

ROW 51 (325 : 331)

ROW 52 (332 : 340)

ROW 53 (341 : 349)

ROW 54 (349 : 354)

PROGRAM 27: CUTCUM
SIZE: 055

D. NATHAN MEEHAN
ERIC L. VOGEL

ROW 55 (354 : 361)

ROW 56 (361 : 366)

ROW 57 (367 : 371)

ROW 58 (372 : 380)

ROW 59 (381 : 387)

ROW 60 (388 : 397)

ROW 61 (398 : 404)

ROW 62 (405 : 414)

ROW 63 (415 : 422)

ROW 64 (422 : 426)

ROW 65 (426 : 431)

ROW 66 (432 : 437)

ROW 67 (437 : 442)

ROW 68 (442 : 444)

ROW 69 (445 : 452)

ROW 70 (452 : 460)

ROW 71 (461 : 469)

ROW 72 (470 : 477)

PROGRAM 27: CUTCUM
SIZE: 055

D. NATHAN MEEHAN
ERIC L. VOGEL

ROW 73 (477 : 486)

ROW 74 (486 : 494)

ROW 75 (495 : 498)

ROW 76 (498 : 506)

ROW 77 (507 : 515)

ROW 78 (516 : 520)

ROW 79 (520 : 529)

ROW 80 (529 : 535)

ROW 81 (536 : 545)

ROW 82 (545 : 545)

PROGRAM 28: RW
SIZE: 028
PROGRAM REGISTERS NEEDED: 66

D. NATHAN MEEHAN
ERIC L. VOGEL

PAGE 1
OF 2

ROW 1 (1 : 2)

ROW 2 (2 : 7)

ROW 3 (8 : 13)

ROW 4 (13 : 17)

ROW 5 (17 : 25)

ROW 6 (26 : 28)

ROW 7 (28 : 33)

ROW 8 (33 : 39)

ROW 9 (40 : 47)

ROW 10 (47 : 53)

ROW 11 (53 : 60)

ROW 12 (60 : 63)

ROW 13 (63 : 67)

ROW 14 (67 : 72)

ROW 15 (72 : 78)

ROW 16 (79 : 85)

ROW 17 (86 : 92)

ROW 18 (93 : 99)

PROGRAM 28: RW
SIZE: 028

D. NATHAN MEEHAN
ERIC L. VOGEL

PAGE 2
OF 2

ROW 19 (99 : 106)

ROW 20 (107 : 115)

ROW 21 (116 : 124)

ROW 22 (124 : 132)

ROW 23 (133 : 140)

ROW 24 (141 : 149)

ROW 25 (150 : 158)

ROW 26 (159 : 164)

ROW 27 (165 : 170)

ROW 28 (171 : 178)

ROW 29 (179 : 185)

ROW 30 (185 : 193)

ROW 31 (194 : 201)

ROW 32 (202 : 209)

ROW 33 (209 : 216)

ROW 34 (216 : 224)

ROW 35 (225 : 231)

ROW 36 (232 : 234)

PROGRAM 29: XPLOT
SIZE: 022
PROGRAM REGISTERS NEEDED: 103

D. NATHAN MEEHAN
ERIC L. VOGEL

PAGE 1
OF 4

ROW 1 (1 : 2)

ROW 2 (2 : 6)

ROW 3 (7 : 11)

ROW 4 (12 : 13)

ROW 5 (13 : 17)

ROW 6 (18 : 26)

ROW 7 (26 : 33)

ROW 8 (33 : 41)

ROW 9 (42 : 50)

ROW 10 (50 : 57)

ROW 11 (58 : 63)

ROW 12 (63 : 69)

ROW 13 (69 : 76)

ROW 14 (76 : 83)

ROW 15 (83 : 91)

ROW 16 (92 : 100)

ROW 17 (100 : 106)

ROW 18 (106 : 112)

PROGRAM 29: XPLOT
SIZE: 022

D. NATHAN MEEHAN
ERIC L. VOGEL

ROW 19 (113 : 117)

ROW 20 (117 : 126)

ROW 21 (127 : 134)

ROW 22 (134 : 142)

ROW 23 (143 : 147)

ROW 24 (148 : 152)

ROW 25 (153 : 158)

ROW 26 (159 : 164)

ROW 27 (165 : 169)

ROW 28 (170 : 180)

ROW 29 (180 : 186)

ROW 30 (187 : 196)

ROW 31 (196 : 200)

ROW 32 (200 : 204)

ROW 33 (204 : 208)

ROW 34 (208 : 214)

ROW 35 (215 : 222)

ROW 36 (223 : 227)

PROGRAM 29: XPLOT
SIZE: 022

D. NATHAN MEEHAN
ERIC L. VOGEL

ROW 37 (227 : 233)

ROW 38 (234 : 239)

ROW 39 (240 : 250)

ROW 40 (250 : 256)

ROW 41 (257 : 262)

ROW 42 (262 : 268)

ROW 43 (268 : 274)

ROW 44 (274 : 278)

ROW 45 (279 : 283)

ROW 46 (284 : 289)

ROW 47 (290 : 293)

ROW 48 (293 : 300)

ROW 49 (301 : 307)

ROW 50 (308 : 315)

ROW 51 (316 : 322)

ROW 52 (323 : 330)

ROW 53 (330 : 337)

ROW 54 (338 : 347)

PROGRAM 29: XPLOT
SIZE: 022

D. NATHAN MEEHAN
ERIC L. VOGEL

ROW 55 (347 : 352)

PROGRAM 30: H2OSAT
SIZE: 029
PROGRAM REGISTERS NEEDED: 88

D. NATHAN MEEHAN
ERIC L. VOGEL

PAGE 1
OF 3

ROW 1 (1 : 2)

ROW 2 (2 : 5)

ROW 3 (6 : 11)

ROW 4 (11 : 16)

ROW 5 (17 : 23)

ROW 6 (24 : 29)

ROW 7 (29 : 34)

ROW 8 (34 : 39)

ROW 9 (40 : 45)

ROW 10 (45 : 50)

ROW 11 (50 : 56)

ROW 12 (56 : 64)

ROW 13 (65 : 72)

ROW 14 (73 : 78)

ROW 15 (78 : 81)

ROW 16 (82 : 86)

ROW 17 (86 : 94)

ROW 18 (95 : 99)

PROGRAM 30: H2OSAT
SIZE: 029

D. NATHAN MEEHAN
ERIC L. VOGEL

PAGE 2
OF 3

ROW 19 (99 : 105)

ROW 20 (105 : 108)

ROW 21 (108 : 113)

ROW 22 (113 : 122)

ROW 23 (123 : 130)

ROW 24 (130 : 132)

ROW 25 (133 : 140)

ROW 26 (141 : 150)

ROW 27 (150 : 161)

ROW 28 (162 : 169)

ROW 29 (170 : 175)

ROW 30 (175 : 182)

ROW 31 (182 : 186)

ROW 32 (186 : 190)

ROW 33 (191 : 196)

ROW 34 (196 : 198)

ROW 35 (198 : 208)

ROW 36 (209 : 218)

PROGRAM 30: H2OSAT
SIZE: 029

D. NATHAN MEEHAN
ERIC L. VOGEL

PAGE 3
OF 3

ROW 37 (219 : 224)

ROW 38 (224 : 234)

ROW 39 (235 : 242)

ROW 40 (243 : 252)

ROW 41 (252 : 261)

ROW 42 (261 : 267)

ROW 43 (267 : 273)

ROW 44 (274 : 281)

ROW 45 (282 : 289)

ROW 46 (289 : 295)

ROW 47 (295 : 301)

PROGRAM 31: SLANT
SIZE: 010
PROGRAM REGISTERS NEEDED: 26

D. NATHAN MEEHAN
ERIC L. VOGEL

PAGE 1
OF 1

ROW 1 (1 : 3)

ROW 2 (3 : 8)

ROW 3 (8 : 14)

ROW 4 (14 : 18)

ROW 5 (19 : 21)

ROW 6 (21 : 24)

ROW 7 (24 : 29)

ROW 8 (30 : 33)

ROW 9 (34 : 38)

ROW 10 (39 : 50)

ROW 11 (51 : 61)

ROW 12 (61 : 67)

ROW 13 (67 : 73)

ROW 14 (73 : 79)

PROGRAM 32: DISC
SIZE: 039
PROGRAM REGISTERS NEEDED: 54

D. NATHAN MEEHAN
ERIC L. VOGEL

PAGE 1
OF 2

ROW 1 (1 : 3)

ROW 2 (3 : 8)

ROW 3 (8 : 15)

ROW 4 (16 : 19)

ROW 5 (20 : 26)

ROW 6 (27 : 33)

ROW 7 (34 : 42)

ROW 8 (43 : 50)

ROW 9 (50 : 54)

ROW 10 (55 : 64)

ROW 11 (65 : 73)

ROW 12 (73 : 79)

ROW 13 (80 : 88)

ROW 14 (89 : 98)

ROW 15 (98 : 104)

ROW 16 (104 : 109)

ROW 17 (110 : 115)

ROW 18 (115 : 120)

PROGRAM 32: DISC
SIZE: 039

D. NATHAN MEEHAN
ERIC L. VOGEL

ROW 19 (120 : 125)

ROW 20 (125 : 132)

ROW 21 (133 : 145)

ROW 22 (146 : 154)

ROW 23 (155 : 163)

ROW 24 (164 : 170)

ROW 25 (170 : 176)

ROW 26 (176 : 181)

ROW 27 (181 : 187)

ROW 28 (188 : 199)

ROW 29 (200 : 208)

ROW 30 (208 : 208)

References

API. "A Statistical Study of Recovery Efficiency." *API Bulletin* D14 (October 1967).

Archie. "The Electrical Resistivity Log as an Aid in Determining Some Reservoir Characteristics." *JPT*, vol. 5 (1942).

Arps, J.J. *Trans. AIME*, vol. 160 (1945), p. 228.

Aziz, K. "Ways to Calculate Gas Flow and Static Head." *Petroleum Engineer* series (November 1960–September 1963).

Bateman, R.M., and C.E. Konen. "The Log Analyst and the Programmable Pocket Calculator." Amoco International Oil Co. (1979).

Brigham, E.F., and J.F. Weston. *Managerial Finance*. (Hinsdale, IL: Dryden Press, 1975).

Clavier, C., and D.H. Rust. "MDH Plot: A New Lithology Technique." *Log Analyst*. (November–December 1976).

Craig, F.F., Jr. *Reservoir Engineering Aspects of Waterflooding*. Spe Monograph Series, vol. 3 (1971).

Crawford, P.B., and G.H. Fancher. "Calculation of Flowing and Static Bottom-Hole Pressures of Natural Gas Wells from Surface Measurements." Bulletin 72. Texas Petroleum Research Committee. (1959).

Cullender, M.H., and R.V. Smith. "Practical Solution of Gas-Flow Equations for Wells and Pipelines with Large Temperature Gradients." *Trans. AIME*, vol. 207 (1956), pp. 281–287.

Deppe, J.C. "Injection Rate — The Effect of Mobility Ratio, Areal Sweep, and Pattern." *SPE Journal* (June 1951), pp. 81–91.

Earlougher, Robert C., Jr. *Advances in Well Test Analysis*. SPE Monograph Series, vol. 5 (1977).

Eilerts, C.K. "Phase Relations of Gas-Condensate Fluids." USBM Monograph 10 (1957).

Ershagi, I., and O. Omoregie. "A Method for Extrapolation of Cut vs. Recovery Curves." *JPT* (February 1978), pp. 203–204.

Fetkovich, M.J. "Decline Curve Analysis Using Type Curves." *JPT* (June 1980), pp. 1065–1077.

Frick, T.C. *Petroleum Production Handbook*. (New York: McGraw-Hill, 1962).

Gentry, R.W. "Decline Curve Analysis." *JPT* (January 1972), p. 38.

Horner, D.R. "Pressure Build-Up in Wells." Proceedings of the Third World Petroleum Conference (1951), Sec. II. pp. 503–523.

Ikoku, Chi U. *Natural Gas Engineering*. (Tulsa: PennWell Books, 1980).

Joiner, D.S., and G.B. Long. "Retrograde GOR Behavior of Condensate Reservoirs from Well Test Data." SPE Paper 7492, Fall Technical Conference, Houston (October 1978).

Katz, D.L., et al. *Handbook of Natural Gas Engineering*. (New York: McGraw-Hill, 1959).

Legatski, M.W., et al. "Displacement of Gas from Porous Media by Water." SPE, paper 899, 39th SPE meeting Houston (October 1964).

Matthews, C.S., and D.G. Russell. *Pressure Buildup and Flow Tests in Wells*. SPE Monograph Series, vol. 1 (1967).

Meehan, D. Nathan. "Calculator Program Simplifies Hyperbolic Decline Curve Analysis." *OGJ* (4 May 1981), pp. 247–250.

________. "Exponential, Harmonic Declines Programmed for Calculators." *OGJ* (26 May 1980), pp. 95–100.

________. "Forecast Made Easier for Developed Waterfloods." *OGJ* (7 July 1980), pp. 114–124.

________. "GOR Change Estimated for Condensate Reservoirs." *OGJ* (21 July 1980), pp. 72–73.

________. "Hyperbolic Oil Production Decline Analysis Programmed." *OGJ* (29 June 1980), pp. 52–56.

________. "Multiwell Production Forecast by Calculator." *OGJ* (23 June 1980), pp. 151–156.

________. "Program for HP-67/97 Calculator Used to find Hyperbolic Exponent." *OGJ* (11 May 1981), pp. 120–123.

________. "Stabilized Flow Coefficient Calculated for Gas Wells." *OGJ* (4 August 1980), pp. 53–54.

________. "Three New Programs Aid Decline Curve Analysis by Calculator." *OGJ* (20 April 1981), pp. 90–96.

Messer, P.H., R. Raghavan, and H.J. Ramey Jr. "Calculation of Bottom-Hole Pressures for Deep, Hot, Sour Gas Wells." *JPT* (January 1974), pp. 85–92.

Muskat, N. *Physical Principles of Oil Production*. (New York: McGraw-Hill, 1950), pp. 645–737.

Nemeth, L.K., and H.T. Kennedy. "A Correlation of Dew Point Pressure with Fluid Composition and Temperature." *JPT* (June 1967), p. 99.

Newendorp, Paul D. *Decision Analysis for Petroleum Exploration*. (Tulsa: PennWell Books, 1975).

Odeh, A.S., and F. Selig. "Pressure Build-up Analysis, Variable-Rate Case." *JPT* (July 1963), pp. 790–794.

Poettman, F.H., and D.L. Katz. "Calculation of Pressure Drop in the Flow of Natural Gas through Pipe." *Trans. AIME*, vol. 192 (1951), pp. 317–326.

Poettman, F.H., and R.E. Schilson. "Calculation of the Stabilized Performance Coefficient of Low Permeability Natural Gas Wells." *Trans. AIME*, vol. 216 (1959).

Poupon, A., and J. Leveaux. "Evaluation of Water Saturation in Shaly Formations." SPWLA 12th Annual Logging Symposium (May 1971).

Rawlins, E.L., and M.A. Schellhardt. "Back Pressure Data on Natural Gas Wells and Their Application to Production Practices." USBM Monograph 7.

Rzasa, M.J., and D.L. Katz. "Calculation of Static Pressure Gradients in Gas Wells." *Trans. AIME*, vol. 160 (1945), pp. 100–113.

Schlumberger. *The Essentials of Log Interpretation Practice* (1979).

Slider, H.C. *Practical Petroleum Reservoir Engineering Methods*. (Tulsa: PennWell Books, 1976).

SPE. "Oil and Gas Property Evaluation and Reserve Estimates." SPE Reprint Series 3 (1970), pp. 34–40.

Sukkar, U.K., and D. Comell. "Direct Calculation of Bottom-Hole Pressures in Natural Gas Wells." *Trans. AIME*, vol. 204 (1955), pp. 43–48.

Theis, Charles V. "The Relation Between the Lowering of the Piezometric Surface and the Rate and Duration of Discharge of a Well Using Groundwater Storage." *Trans. AGU* (1935), pp. 519–523.

Theory and Practice of the Testing of Gas Wells. Energy Resources Conservation Board, 3 ed (1975).

van Everdingen, A.F., and W. Hurst. "The Application of the Laplace Transformation to Flow Problems in Reservoirs." *Trans. AIME* (1949).

Vitter, A.L., Jr. "Back-Pressure Tests of Gas-Condensate Wells." *API Drilling and Production* (1942), p. 79.

Young, K.L. "Effect of Assumptions Used to Calculate Bottom-Hole Pressure in Gas Wells." *JPT* (April 1967), pp. 547-550.